W9-CRB-590

A User's Guide to Vacuum Technology

A User's Guide to Vacuum Technology

John F. O'Hanlon

Research Staff Member
IBM Thomas J. Watson Research Center

A Wiley-Interscience Publication

JOHN WILEY & SONS, New York • Chichester • Brisbane • Toronto

This book was set in press type on an Autologic APS-5
phototypesetter driven by an IBM VM/370 system. The
formatting was done with Termtext/Format, the Yorktown Formatting Language, and the Yorktown Mathematical Formula Processor.

Copyright © 1980 by John Wiley & Sons, Inc.

All rights reserved. Published simultaneously in Canada.

Reproduction or translation of any part of this work
beyond that permitted by Sections 107 or 108 of the
1976 United States Copyright Act without the permission
of the copyright owner is unlawful. Requests for
permission or further information should be addressed to
the Permissions Department, John Wiley & Sons, Inc.

Library of Congress Cataloging in Publication Data:

O'Hanlon, John F.   1937-
    A user's guide to vacuum technology.

    "A Wiley-Interscience publication."
    Includes index.
    1.   Vacuum technology—Handbooks, manuals, etc.

I.   Title.
TJ940.037   621.5'5   80-12589
ISBN 0-471-01624-1

Printed in the United States of America

10 9 8 7 6 5 4 3

TJ940
O37
PHYS

For Jean, Carol, Paul, and Amanda

Preface

This book is intended for the vacuum system user—the university student, technician, engineer, manager, or scientist—who wishes a fundamental understanding of modern vacuum technology and a user's perspective of current vacuum practice.

Vacuum technology is largely secondary in that it is a part of other technologies that are central to analysis, research, development, and manufacturing. It is used to provide a process environment. Many advances in vacuum technique have resulted from the demands of other technologies, although scientists and engineers have studied vacuum for its own sake. The average user is process-oriented and becomes immersed in vacuum technique only when problems develop with a process or new equipment purchases become necessary.

A User's Guide to Vacuum Technology focuses on the understanding, operation and selection of equipment for processes used in semiconductor, optics, and related technologies. It emphasizes subjects not adequately covered elsewhere while avoiding in-depth treatments of topics interesting only to the designer or curator. Residual gas analysis is an important topic whose treatment here goes beyond the usual explanation of mass filter theory. New components such as turbomolecular and helium gas refrigerator cryogenic pumps are widely used but not so well understood as diffusion pumps. New processes for film deposition and removal require the use of toxic, corrosive, or explosive gases. Special precautions need to be taken for safe use of these gases. The discussion of gauges, pumps, and materials is a prelude to the central discussion of the total system. Systems are grouped according to their common vacuum requirements of speed, working pressure, and gas throughput. The suitability of each pump is examined for several classes of systems, and basic operational procedures are given for each high vacuum pumping system. The economic analysis discusses the costs of purchasing, maintaining, and operating vacuum equipment and describes ways in which operating costs can be significantly reduced.

Thanks are due to Gordon Johnson of the IBM Data Systems Division, E. Fishkill, NY for many interesting discussions and for his reviews of several chapters. Larry Helweg and Harvey Yu of IBM DSD and Drs. Ned Chou, John Coburn, Jerry Cuomo, Richard Guarnieri, Takeshi Takamori, and Harold Winters of the IBM Research Division graciously reviewed chapters and pointed out many errors. I wish to thank all who provided illustrative material for their generosity. Much of the material in this book has been presented in lecture form to interested personnel from the IBM Research and Data Systems Divisions. Their participation has been most helpful in shaping the material.

The graphics department of the IBM T. J. Watson Research Center carefully and accurately prepared the artwork. Mrs. Georgianna K. Grant and Mrs. Alberta D. Meier typed the manuscript superbly, but it could not have been completed without the encouragement of Dr. Rick Dill. Ms. Meier formatted the text using macros written by Ms. Ann Gruhn. I am indebted to Ms. Beatrice Shube for her insight, advice, and excellent editorial supervision throughout its preparation.

J. F. O'Hanlon

Yorktown Heights, New York
April 1980

Contents

MATERIALS

SYSTEMS

APPENDIXES

Its Basis

An understanding of how vacuum components and systems function begins with an understanding of the behavior of gases at low pressures. Chapter 1 discusses the nature of vacuum technology. Chapter 2 reviews basic gas kinetics and the flow properties of gases at reduced pressures which form the foundation of vacuum technology.

CHAPTER 1

Vacuum Technology

Galileo was the first person to create a partial vacuum. He did so with a piston. This seventeenth-century discovery was followed in 1643 by the invention of the mercury barometer by Torricelli and in 1650 the first pump by von Guericke. Interest in the properties of gases at reduced pressures remained at a low level for more than 200 years, when a period of rapid discovery began with the invention of the compression gauge by McLeod. In 1905 Gaede, a prolific inventor, designed a rotary pump sealed with mercury. These developments, in addition to the thermal conductivity gauge, the diffusion pump, the ion gauge and pump, the liquefaction of helium, and the refinement of organic pumping fluids, formed the nucleus of a technology that has made possible everything from light bulbs to the simulation of outer space.

Vacuum technology is the systematic study of scientific ideas and the application of these principles to the production of practical, reduced-pressure environments. It has drawn on discoveries from many fields such as chemistry, physics, mathematics, ceramics, materials and surface science, and engineering. It has also made fundamental contributions in its own right.

A vacuum is a space from which air or other gas has been removed. In practice we know that all the gas can never be removed and we sometimes wish to remove only a particular fraction of that gas. Air is the most important gas to be pumped because it is in every system. It contains at least a dozen constituents, the concentrations of which are given in Table 1.1. It is useful to be aware of the content of air in

Table 1.1   Components of Dry Atmospheric Air

Constituent	Content (vol %)	Ppm	Pressure (Pa)
N_2	78.084±0.004		79,117
O_2	20.946±0.002		21,223
CO_2	0.033±0.001		33.437
Ar	0.934±0.001		946.357
Ne		18.18±0.04	1.842
He		5.24±0.004	0.51
Kr		1.14±0.01	0.116
Xe		0.087±0.001	0.009
H_2		0.5	0.051
CH_4		2.	0.203
N_2O		0.5±0.1	0.051

Source: Reprinted with permission from *The Handbook of Chemistry and Physics,* 59th ed., R. C. Weast, Ed., Copyright 1978, The Chemical Rubber Publishing Co., CRC Press, Inc., West Palm Beach, FL 33409.

order to predict the responses of pumps and gauges. The concentrations listed in Table 1.1 are those of dry atmospheric air at sea level [total pressure 101,323.2 Pa (760 torr)]. The partial pressure of water vapor is not given in this table because it is constantly changing. At 20°C a relative humidity of 50% is equivalent to a partial pressure of 1165 Pa (8.75 Torr) which makes it the third largest constituent of air. The total pressure changes rapidly with altitude, as shown in Fig. 1.1, its proportions, slowly but significantly. In outer space the atmosphere is thought to be mainly hydrogen with some helium [1].

For convenience it is customary to divide the pressure scale below atmospheric into several ranges and to relate phenomena and processes to them. Table 1.2 lists the ranges currently in use. Epitaxial growth of semiconductor films [2,3] takes place in the low vacuum range. Sputtering [4,5], plasma etching, plasma deposition [5], and low-pressure chemical vapor deposition [5–7] are examples of processes performed in the medium vacuum range. Pressures in the very high vacuum range are required for most thin-film preparation [5,8],

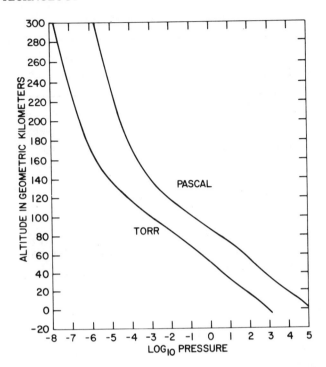

Fig. 1.1 Relation between the atmospheric pressure and the geometric altitude. Reprinted with permission from *The Handbook of Chemistry and Physics*, 59th ed., R. C. Weast, Ed. Copyright 1978, The Chemical Rubber Publishing Co., CRC Press, Inc., West Palm Beach, FL 33409.

electron microscopy [9], mass spectroscopy [10], crystal growth [11], x-ray and electron beam lithography [12,13], and the production of cathode ray and other vacuum tubes [14]. For ease of reading, we call the very high vacuum region *high vacuum* and the pumps, *high vacuum pumps*. Pressures in the ultrahigh vacuum range are necessary for surface and material studies [15].

SI units are used with few exceptions. Pumping speeds are given in L/s (high vacuum pumps and conductances) and in m³/h (mechanical pumps) instead of m³/s. Diffusion pumps, whose nomenclature is not well standardized, are referred to by the inch size of the top flange. This often bears little relation to their performance. Unless otherwise indicated, all formulas are given in the basic SI units.

Table 1.2   Vacuum Ranges

Degree of Vacuum	Pressure Range (Pa)
Low	$10^5 \quad > \text{P} \quad > 3.3 \times 10^3$
Medium	$3.3 \times 10^3 \geq \text{P} > \quad 10^{-1}$
High	$10^{-1} \quad \geq \text{P} > \quad 10^{-4}$
Veryhigh	$10^{-4} \quad \geq \text{P} > \quad 10^{-7}$
Ultrahigh	$10^{-7} \quad \geq \text{P} > \quad 10^{-10}$
Extreme Ultrahigh	$10^{-10} \quad > \text{P}$

Source: Reprinted with permission from *Dictionary for Vacuum Science and Technology*, M. Kaminsky and J. M. Lafferty, Eds., in preparation.

REFERENCES

1. D. J. Santeler et al., *Vacuum Technology and Space Simulation*, NASA SP-105, National Aeronautics and Space Administration, Washington, DC, 1966, p.34.

2. J. L. Deines and A. Spiro, *The Electrochemical Society Extended Abstracts*, May 1974, Abstract 62.

3. M. Ogirima et al., *J. Electrochem. Soc.*, **124**, 903 (1977).

4. H. F. Winters, *Advances in Chemistry Series No. 158, Radiation Effects on Solid Surfaces*, M. Kaminsky, Ed., American Chemical Society, 1976, p.1.

5. J. L. Vossen and W. Kerns, Eds., *Thin Film Processes*, Academic, New York, 1978.

6. L. F. Donaghey, P. Rai-Choudhuri, and R. N. Tauber, Eds., *Proc. 6th Int. Conf. on Vapor Deposition*, The Electrochemical Society, Princeton, NJ, 1977.

7. T. O. Sedgwick and H. Lydtin, Eds., *Proc. 7th Int. Conf. on Vapor Deposition*, The Electrochemical Society, Princeton, NJ, 1979.

8. R. Glang, in *Handbook of Thin Film Technology*, L. I. Maissel and R. Glang, Eds., McGraw-Hill, New York, 1970, Chapter 1.

9. A. M. Glauert, Ed., *Pract. Methods Electron Microsc.*, **2**, North Holland, Amsterdam, 1974.

10. F. A. White, *Mass Spectroscopy in Science and Technology*, Wiley, New York, 1968.

11. B. R. Pamplin, Ed., *Crystal Growth*, Pergamon, Oxford, 1975.

12. A. N. Broers, *Proc. 7th Symp. on Electron Beam Sci. Technol.*, The Electrochemical Society, Princeton, NJ, 1976, p. 587.

13. A. N. Broers, *Proc. 7th Int. Vacuum Congr.*, R. Dobrozemsky et al., Eds., Published by the Editors, 1977, p. 1521.

14. F. Rosebury, *Handbook of Electron Tube and Vacuum Techniques*, Addison-Wesley, Reading, MA, 1965.

15. P. A. Redhead, J. P. Hobson, and E. V. Kornelsen, *The Physical Basis of Ultrahigh Vacuum*, Chapman and Hall, London, 1968.

CHAPTER 2

Gas Properties

Phrases like *vacuum pump* and *vacuum system* are not particularly descriptive. In reality, a vacuum pump is a gas pump designed to operate at lower than atmospheric pressures. A vacuum system consists of pumps and a chamber connected by piping and ductwork. The low pressure in the chamber is maintained by the continued flow of gases from the chamber to the pumps, where they are entrained or expelled into the atmosphere. This chapter discusses static and dynamic gas properties and the flow of gases at reduced pressures.

2.1 THE KINETIC PICTURE OF A GAS

The kinetic picture of a gas is based on several assumptions. The volume of gas under consideration contains a large number of molecules. A cubic meter of gas at a pressure of 10^5 Pa and a temperature of 22°C contains 2.5×10^{25} molecules, whereas at a pressure of 10^{-7} Pa, a very high vacuum, it contains 2.5×10^{13} molecules. Indeed, any volume and pressure normally used in the laboratory will contain molecules in large numbers. Adjacent molecules are separated by distances that are large compared with their individual diameters. If we could stop all molecules instantaneously and place them on the coordinates of a grid, the average spacing between them would be about 3.4×10^{-9} m at atmospheric pressure (10^5 Pa). The diameter of most molecules is in the 2×10^{-10} to 6×10^{-10} m range and distances of about 6 to 15 times their diameter at atmospheric pressures separate them. For extremely low pressures, say 10^{-7} Pa, the separation distance is about 3×10^5 m. Molecules are in a constant state of motion. All directions of motion are equally likely and all speeds are possible, although not equally probable. Molecules exert no force on one another except when they collide. If this is true, then the molecules will be uniformly distributed throughout the volume and travel in straight lines until they collide with a wall or with one another.

Many interesting properties of ideal gases have been derived by using these assumptions. Some of the more elementary are reviewed here. As the individual molecules move about they collide with one another. These collisions are elastic; that is, they conserve energy, while the particles change velocity with each collision. Maxwell and Boltzmann calculated the average velocity of the particles as

$$v = \left(\frac{8kT}{\pi m} \right)^{\frac{1}{2}} \qquad (2.1)$$

where k is Boltzmann's constant, T is the absolute temperature, and m is the mass of the molecule. Here we observe the basic relationship between velocity and temperature. An increase in temperature causes molecules to collide with a wall or with one another with increased momentum and frequency. The average speeds of some gas and vapor molecules are given in Appendix B.2.

The fact that each molecule is randomly distributed and moving with a different velocity implies that each will travel a different straight-line distance, called a free path, before colliding with another. As illustrated in Fig. 2.1, not all free paths are the same length. The average, or mean, of the free paths λ, according to kinetic theory, is

$$\lambda = \frac{1}{2^{\frac{1}{2}} \pi d_o^2 n} \qquad (2.2)$$

where d_o is the molecular diameter and n, the gas density. The mean free path is clearly pressure-dependent. For air at room temperature the mean free path is

$$\lambda (\text{mm}) = \frac{6.6}{P} \qquad (2.3)$$

where λ has units in millimeters and P is the pressure in pascals. Kinetic theory also describes the distributions of free paths.

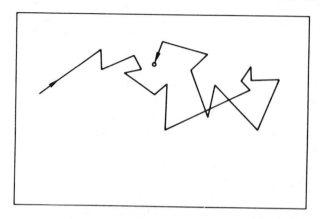

Fig. 2.1   Random motion of a molecule.

$$N = N'e^{-x/\lambda} \tag{2.4}$$

N' is the number of molecules in the volume and N is the number of molecules that traverse a distance x before suffering a collision. Equation 2.4 states that 63% of the collisions occur in a distance $0 \leq x \leq \lambda$, whereas about 37% of the collisions occur in range $\lambda \leq x \leq 5\lambda$. Only about 0.6% of the particles travel distances greater than 5λ without suffering a collision.

The concept of particle flux is helpful in understanding gas flow, pumping, and evaporation. According to kinetic theory the particle flux Γ of an ideal gas striking a unit surface or crossing an imaginary plane of unit area from one side is

$$\Gamma = \frac{nv}{4} \tag{2.5}$$

where n is the particle density and v, the average velocity. On substituting (2.1) we see that

$$\Gamma = n\left(\frac{kT}{2\pi m}\right)^{1/2} \tag{2.6}$$

The particle flux is directly proportional to the particle density and the square root of T/m.

The pressure on a surface is defined as the rate at which momentum is imparted to a unit surface. A molecule incident on a surface will impart a total impulse or pressure of $2mv \cos \Theta$. By integrating over all possible angles in the half-plane we find that the pressure is

$$P = \frac{1}{3}nmv^2 \qquad (2.7)$$

The total energy of a molecule, however, is proportional to its temperature

$$E = \frac{mv^2}{2} = \frac{3kT}{2} \qquad (2.8)$$

and (2.7) and (2.8) may be combined to yield the absolute pressure.

$$P = nkT \qquad (2.9)$$

If n is expressed in units of m^{-3}, k in joules per degree Kelvin, and T in degrees Kelvin, then P will be given in units of pascals (Pa). A pascal is a newton per square meter and the fundamental unit of pressure in System International. Simply divide the number of pascals by 133.32 to convert to units of torr, or by 100 to convert to units of millibars. A table for converting between different systems of units is included in Appendix A.3. Values of n, d', λ, and Γ for air at 22°C are tabulated in Table 2.1 for pressures that range from atmospheric to ultrahigh

Table 2.1   Low Pressure Properties of Air[a]

Pressure (Pa)	n (m^{-3})	d' (m)	λ (m)	Γ (m^{-2}-s^{-1})
1.01×10^5 (760 Torr)	2.48×10^{25}	3.43×10^{-9}	6.5×10^{-8}	2.86×10^{27}
100 (.75 Torr)	2.45×10^{22}	3.44×10^{-8}	6.6×10^{-5}	2.83×10^{24}
1 (7.5 mTorr)	2.45×10^{20}	1.6×10^{-7}	6.6×10^{-3}	2.83×10^{22}
10^{-3} (7.5×10^{-6} Torr)	2.45×10^{17}	1.6×10^{-6}	6.64	2.83×10^{19}
10^{-5} (7.5×10^{-8} Torr)	2.45×10^{15}	7.41×10^{-6}	664	2.83×10^{17}
10^{-7} (7.5×10^{-10} Torr)	2.45×10^{13}	3.44×10^{-5}	6.6×10^4	2.83×10^{15}

[a] Particle density, n; average molecular spacing, d'; mean free path, λ; and particle flux on a surface, Γ. $T = 22$°C.

vacuum. The pressure dependence of the mean free path is given for several gases in Appendix B.1.

Kinetic theory, as expressed in (2.9), summarizes all the earlier experimentally determined gas laws. In 1662 Robert Boyle demonstrated that when the temperature remained the same the volume occupied by a given quantity of gas varied inversely as its pressure:

$$P_1 V_1 = P_2 V_2 \quad (n, T \text{ constant}) \qquad (2.10)$$

This is easily derived from the general law by multiplying both sides by the volume V and noting that $N = nV$. The French physicist Charles found that gases expanded and contracted to the same extent under the same changes of temperature provided that no change in pressure occurred. Again by the same substitution in (2.9) we obtain

$$\frac{V_1}{T_1} = \frac{V_2}{T_2} \quad (N, P \text{ constant}) \qquad (2.11)$$

Avogadro observed that pressure and numbers of molecules were proportional for a given temperature and volume:

$$\frac{P_1}{N_1} = \frac{P_2}{N_2} \quad (T, V \text{ constant}) \qquad (2.12)$$

Dalton discovered that the total pressure of a mixture of gases was equal to the sum of the forces per unit area of each gas taken individually. By the same methods for a mixture of gases, we can develop the relation

$$P_t = n_t kT = n_1 kT + n_2 kT + n_3 kT + \dots n_i kT \qquad (2.13)$$

which reduces to

$$P_t = P_1 + P_2 + P_3 + \dots P_i \qquad (2.14)$$

where P_t, N_t are the total pressure and density and P_i, n_i are the partial pressures and densities. Equation 2.14 is called Dalton's law of partial pressures and is valid for pressures below atmospheric [1].

The primary assumption of a gas at rest in thermal equilibrium with its container is not always valid in practical situations; for example, a pressure gauge close to and facing a high vacuum cryogenic pumping surface will register a lower pressure than when it is close to and facing a warm surface in the same vessel [2].   Other nonequilibrium situations are discussed as the need arises.

2.2   ELEMENTARY GAS TRANSPORT PHENOMENA

In this section approximate views of viscosity, thermal conductivity, diffusion, and thermal transpiration are discussed. Results from kinetic theory are presented at the conclusion.

2.2.1   Viscosity

A viscous force is present in a gas when it is undergoing shearing motion.   Figure 2.2 illustrates two plane surfaces, one fixed and the other traveling in the x-direction with a uniform velocity.   The coefficient of viscosity η is defined by the equation

$$\frac{F_x}{A_{xz}} = \eta \frac{du}{dy} \qquad (2.15)$$

where F_x is the force in the x-direction, A_{xz} is the surface area in the x-z plane, and du/dy is the rate of change of the gas velocity at this position between the two surfaces. Because the gas stream velocity increases as the moving plate is approached, those molecules crossing the plane A_{xz} from below (1 in Fig. 2.2) will transport less momentum across the plane than will those crossing the same plane from above (2 in Fig. 2.2).   The result is that molecules crossing from below the plane will, on the average, reduce the momentum of the molecules above the plane, and in the same manner molecules crossing above the plane will increase the momentum of those molecules below the plane. To an observer this viscous force appears to be frictional; actually it is not.   It is merely the result of momentum transfer between the plates by successive molecular collisions.   Again, from kinetic theory the coefficient of viscosity is

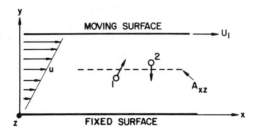

Fig. 2.2 Origin of the viscous force in a gas.

$$\eta = \frac{1}{3}nmv\lambda \qquad (2.16)$$

When the gas density is measured in units of m^{-3}, the molecular mass in kg, the velocity in m/s and the mean free path in m, η will have units of (N-s)/m^2, or Pa-s. One Pa-s is equal to 10 poise. A more rigorous treatment of viscosity [3] yields a result with a slightly different numerical coefficient:

$$\eta = 0.4999nmv\lambda \qquad (2.17)$$

Substituting (2.1) and (2.2) into this result yields

$$\eta = \frac{0.499(4mkT)^{1/2}}{\pi^{3/2}d_o^2} \qquad (2.18)$$

From (2.18) we see that kinetic theory predicts that viscosity should increase as $(mT)^{1/2}$ and decrease as the square of the molecular diameter. An interesting result of this simple theory is that viscosity is independent of gas density or pressure. This theory, however, is valid only in a limited pressure range. If there were a perfect vacuum between the two plates, there would be no viscous force because there would be no mechanism for transferring momentum from one plate to another. This leads to the conclusion that (2.18) is valid as long as the distance between the plates is of the order of the mean free path or greater.

For a rarified gas, in which the ratio of the mean free path to plate separation $\lambda/y \gg 1$ the viscous force can be expressed as

$$\frac{F}{A_{xz}} = \left(\frac{Pmv}{4kT} \right) \frac{U_1}{\beta} \tag{2.19}$$

where the term in parentheses is referred to as the *free-molecular viscosity*. The viscous force is directly proportional to the pressure or number of molecules available to transfer momentum between the plates and is valid in the region $\lambda \gg y$. The constant β in (2.19) is related to the slip of molecules on the surface of the plates. For most surfaces and gases involved in vacuum work $\beta \sim 1$.

Figure 2.3 illustrates the magnitude of the viscous force caused by air at 22°C between two plates moving with a relative velocity of 100 m/s for three plate separations. Equation 2.18 was used to calculate the asymptotic value of the viscous drag at high pressures and (2.19) was used to calculate the free molecular limit. A more complete treatment of the intermediate or viscous slip region is given elsewhere [4]. The viscous shear force is independent of the plate separation as long as the mean free path is larger than the largest dimension in question. This concept was used by Langmuir [5] to construct a viscosity gauge in which the damping was proportional to the pressure.

2.2.2 Thermal Conductivity

Heat conductivity is explained by kinetic theory in a manner analogous to that used to explain viscosity. The diagram in Fig. 2.2 could be relabeled to make the top plate stationary at temperature T_2, whereas the lower plate becomes a stationary plate at a temperature T_1 where $T_1 < T_2$. The phenomenon of heat conduction can be modeled by noting that the molecules moving across the plane toward the hotter surface carry less *energy* than those moving across the plane toward the cooler surface. The heat flow can be expressed as

$$H = AK \frac{dT}{dy} \tag{2.20}$$

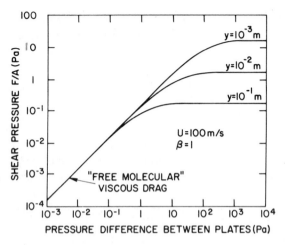

Fig. 2.3 Viscous shear force between two plates in air at 22°C.

where H is the heat flow and K, the heat conductivity. The simple theory predicts that the heat conductivity K is expressed by $K = \eta c_v$, where η is the viscosity and c_v, the specific heat at constant volume. This simple theory is correct only to an order of magnitude. A more detailed analysis, which accounts for the rotational and vibrational energy of the molecules, yields

$$K = \frac{1}{4}(9\gamma - 5)\eta c_v \qquad (2.21)$$

where γ is the ratio of specific heats at constant pressure and constant volume c_p/c_v. When η has the units of Pa-s and c_v has units (J/kg)/K, then K will have units of (W/m)/K. At room temperature the heat conductivity increases approximately as $(mT)^{k}/D^2$, as does the viscosity. Heat conductivity does not depend on pressure as long as the mean free path is smaller than the dimension of the chamber. This means that an object in a vacuum system can be thermally equilibrated with its surroundings by adding just enough gas to satisfy (2.3); any further increase in pressure will not hasten equilibration by collisional energy transfer. At very low pressures, at which the mean free path is much greater than the dimension of the system, molecules will bounce back and forth between surfaces, carrying heat from the hot surface to the cold surface without making any intermolecular collisions. The heat transfer in this case [6] has been calculated as

$$E_o = \alpha \Lambda P (T_2 - T_1)$$

where

$$\alpha = \frac{\alpha_1 \alpha_2}{\alpha_1 + \alpha_2 - \alpha_1 \alpha_2} \qquad \Lambda = \frac{1}{8} \frac{(\gamma + 1)\nu_1}{(\gamma - 1)T} \qquad (2.22)$$

This equation has the same general form as (2.19) for free-molecular viscosity. Λ is the free-molecular heat conductivity, and α_1, α_2, α are the accommodation coefficients of the cold surface, hot surface, and system, respectively. If the molecule is effective in thermally equilibrating with the surface, say by making many small collisions on a rough surface, α will have a value the approaches unity. If, however, the same surface is smooth and the molecule recoils without gaining or losing energy, α will approach zero. The kinetic picture of heat conductivity is rather like that viscosity except that the heat conductivity is determined by *energy* transfer, and the viscosity is determined by *momentum* transfer. Even so, Fig. 2.3 can be sketched for thermal conductivity where the vertical axis has dimensions of heat flow. Both thermocouple and Pirani gauges operate in a region in which the heat conduction to the wall is linearly dependent on pressure. From an equivalent of Fig. 2.3 for heat conduction the relation between the diameter of the pressure-gauge tube and the range of the gauge can be observed. The tube size must be reduced to increase the high pressure limit of operation.

In SI Λ has units of W-m^{-2}-K^{-1}-Pa^{-1}, whereas E_o has units of W/m^2. Tables of the accommodation coefficient are given elsewhere [3,6]. The accommodation coefficient of a gas is not only dependent on the material but on its cleanliness, surface roughness, and gas absorption as well.

2.2.3 Diffusion

The general phenomenon of diffusion is complex. This discussion has been simplified by restricting it to the situation in a vessel that contains two gases whose compositions vary slightly throughout the vessel but whose total concentration is everywhere the same. The coefficient of diffusion D, of the two gases is defined in terms of the particle fluxes $\Gamma_{1,2}$:

$$\Gamma_1 = -D\frac{dn_1}{dx} \qquad \Gamma_2 = -D\frac{dn_2}{dx} \qquad (2.23)$$

These fluxes are due to the partial pressure gradients of the two gases. The result from kinetic theory, when corrected for the Maxwellian distribution of velocities and for velocity persistence, is

$$D_{12} = \frac{8\left(\frac{2kT}{\pi}\right)^{\frac{1}{2}}\left(\frac{1}{m_1}+\frac{1}{m_2}\right)^{\frac{1}{2}}}{3\pi n(d_{01}+d_{02})^2} \qquad (2.24)$$

where D_{12} is the constant of interdiffusion of the two gases and, in SI, has units of m²/s. For the special case of similar molecules the coefficient of self-diffusion is

$$D_{11} = \frac{4}{3\pi n d_o^2}\left(\frac{kT}{\pi m}\right)^{\frac{1}{2}} \qquad (2.25)$$

If the density n is replaced by P/kT, it becomes apparent that the diffusion constant is approximately proportional to $T^{3/2}$ and P^{-1}.

To gain some insight into the usefulness of the diffusion coefficient examine the problem of out-diffusion of a system of $2N$ molecules distributed uniformly throughout the x-y plane at the location $z = 0$. At the time $t = 0$ the molecules start to diffuse through the gas, half in the $-z$-direction, half in the $+z$-direction. At any time t the number of molecules dn located in the region between z and $z + dz$ is given by

$$dn = \frac{Ndz}{(\pi Dt)^{\frac{1}{2}}}e^{-z^2/4Dt} \qquad (2.26)$$

From this, the fractional number of molecules f, located in the region between z_o and ∞,

$$f = erfc\frac{z_o}{2(Dt)^{\frac{1}{2}}} \qquad (2.27)$$

A plot of (2.27) is given in Fig. 2.4. From this solution we see that the fractional number of particles between some point z_o and $+\infty$ increases with time. Furthermore, if the value of z_o is increased, more time will elapse before the arrival of particles from the source. Anoth-

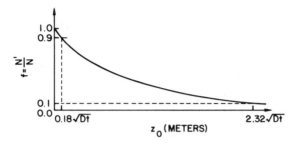

Fig. 2.4 The fractional number of particles located between z_0 and $+\infty$ that have diffused from a plane source.

er way of viewing this solution is to calculate the values of z_0, t, such that, say, 10% of the particles will have diffused beyond z_0. That condition is satisfied when the argument of the complementary error function (*erfc*) is 1.16:

$$\frac{z_0}{2(Dt)^{\frac{1}{2}}} = 1.16 \qquad\qquad (2.28)$$

or

$$z_0 = 2.32(Dt)^{\frac{1}{2}}$$

Equation 2.28 is basic to the solutions of many types of diffusion problem in that it contains a critical distance proportional to $(Dt)^{1/2}$. The diffusion "front" moves through the gas as $(Dt)^{\frac{1}{2}}$. For $(Dt)^{\frac{1}{2}} <<$ z_0 the diffusion front will not have arrived at z_0, whereas for $(Dt)^{\frac{1}{2}} >>$ z_0 the diffusing species will have equilibrated with the carrier at z_0. To illustrate the role collisional processes play in diffusion consider the diffusion of H_2 in air at atmospheric pressure and room temperature. In 60 s only one-third of the gas will have moved a distance greater than 70 mm, whereas the average H_2 molecule will have traveled a total distance of 10^5 m. Values of viscosity, thermal conductivity and diffusion in air of several common gases at $0°C$ are given in Appendix B.2.

2.2.4   Thermal Transpiration

When two chambers of different temperatures are connected by a tube or orifice, their relative pressures are a function of the ratio of mean free path to diameter (λ/d) in the connecting tubing. For $\lambda \ll d$ the relative pressures are given by the ideal gas law (2.9)

$$\frac{P_1}{P_2} = \frac{T_1}{T_2} \tag{2.29}$$

When, however, the orifice diameter is such that $\lambda \gg d$, the flux of gas through the orifice is given by (2.5)

$$\Gamma = \frac{n}{4}\left(\frac{8kT}{\pi m}\right)^{\frac{1}{2}} = \frac{P}{(2kT\pi m)^{\frac{1}{2}}}$$

In equilibrium the flux from each chamber must be equal, with the result that

$$\frac{P_1}{P_2} = \left(\frac{T_1}{T_2}\right)^{\frac{1}{2}} \tag{2.30}$$

Thermal transpiration was discovered by Neumann [8] and studied by Maxwell [9], who predicted the square-root dependence given in (2.30). Deviations are introduced by the geometry and reflectivity of the walls. Siu [10] has studied these effects theoretically and has shown that the ratio predicted in (2.30) is obtained in short tubes only for specular reflection and in long tubes only for diffuse reflection.

2.3   GAS FLOW

The flow of gases in pipes is generally a complex problem in which the nature of the solution depends on the flow rate and properties of the gas as well as on the geometry and surface properties of the duct. To simplify the discussion this section considers several classes of flow and geometry independently. More complex problems such as gas flow in ducts that contain entrance and exit orifices, aperture plates, and other geometrical shapes are usually treated by approximation or numerical methods. These methods are reviewed and their ranges of validity delineated in terms of certain dimensionless numbers.

   Gas flow at high pressures and velocities, where the flow is neither laminar nor ordered, is termed turbulent. Viscous flow commences when the velocity and surface irregularities are small enough to characterize it by laminar streamlines which flow gently around obstructions.

The boundary between turbulent and viscous flow can be expressed in terms of the Reynolds number **R** for round pipes:

$$\mathbf{R} = \frac{U\rho d}{\eta} \qquad (2.31)$$

where ρ is the mass density of the gas (kg/m^3). The Reynolds number, which is dimensionless, can be used to characterize the nature of the gas flow and may be considered a ratio of the shear stress due to turbulence to the shear stress due to viscosity. Alternatively, it tells something about the forces necessary to drive a gas system in relation to the forces of dissipation due to viscosity. Reynolds [11] found two flow situations were dynamically similar when this dimensionless number was the same. When **R** > 2200, the flow will always be turbulent and when **R** < 1200 the flow will always be viscous [12]. In the region 1200 < **R** < 2200 the flow can be viscous or turbulent, depending on the geometry of the inlet and outlet and on the nature of the piping irregularities.

Viscous flow, the ordered flow of a gas in streamlines, occurs in the region bounded by a Reynolds number lower than 1200 and a Knudsen number lower than 0.01. Knudsen's number Kn, also dimensionless, is the ratio of the mean free path to a characteristic dimension of the system, say, the diameter of a pipe:

$$Kn = \frac{\lambda}{d} \qquad (2.32)$$

In viscous flow the diameter of the pipe is much greater than the mean free path; therefore the character of the gas flow is determined by gas-gas collisions. The flow has a maximum velocity in the center of the channel and zero velocity at the wall. When the mean free path is equal to or greater than the pipe diameter, say Kn > 1, the flow properties are determined by gas-wall collisions. This region is referred to as the molecular flow region. The nature of its flow is quite different from that of the viscous region. Gas-wall collisions predominate and the concept of viscosity has no meaning. For most surfaces diffuse reflection at the wall is a good approximation; that is, each particle arrives, sticks, and is reemitted in a direction independent of its incident velocity. Thus there is a chance that a particle entering a pipe in which $\lambda \gg d$ will not be transmitted, but will be returned to the entrance. Because gas molecules do not collide with one another, two gases can diffuse in opposite directions and neither is affected by the presence of the other. In the region $1 > Kn > 0.01$ the flow is neither viscous nor molecular and is not well understood. In this range, called the transition flow range, where the pipe is several mean free paths wide, the velocity at the wall is not zero as in viscous flow and the reflection is not diffuse as in free molecular flow.

2.3.1 Throughput and Conductance

Throughput is the volume of gas at a known pressure and temperature that passes a plane in a known time. In SI throughput has units of Pa-m^3/s. Because 1 Pa=1 N/m^2, and 1 J=1 N-m, the units may be more simply expressed as J/s or watts; 1 Pa-m^3/s = 1 W. Thus throughput is the *energy* per unit time crossing a plane. The power flow is equivalent to the mass flow rate only if the system is everywhere at a constant known temperature. A spatial change in the temperature can alter the energy flow without altering the mass flow. We discuss this in greater detail in Section 9.3, where we calculate the speed of a cryogenic pump.

The flow of gas in a channel is dependent on the pressure drop across the tube as well as the geometry of the channel. Division of the throughput (Q) by the pressure drop across a channel held at constant temperature yields a property known as the intrinsic conductance of the channel.

$$C = \frac{Q}{P_2 - P_1} \qquad (2.33)$$

In SI the units of flow are Pa-m^3/s and the units of conductance or pumping speed are m^3/s; however, flow units of Pa-L/s and conductance units of L/s are also used. Unless explicitly stated, all formulas in this chapter use m^3 as the volumetric unit. For those whose first introduction to flow was with electricity (2.33) is analogous to an electrical current divided by a potential drop. As in electrical charge flow, there are situations in which the gas conductance is nonlinear, for example, a function of the pressure in the tube.

In the remaining parts of this section gas flow in the various regimes is characterized in terms of the intrinsic conductance of a channel. Reasonably good theoretical treatments exist for infinitely long cylindrical channels but for short channels, and channels of noncircular cross section empirical or numerical techniques are used. The rules for combining conductances are discussed. The definition of conductance assumes certain boundary conditions, therefore, when combinations of various orifices and pipes are connected, the usual rules of combining conductances, particularly in molecular flow, may not hold.

2.3.2 Turbulent Flow

In the region $\mathbf{R} > 2200$ the flow of gas will always be turbulent. It is helpful to replace the stream velocity in (2.31) by

$$U = \frac{Q}{AP} \qquad (2.34)$$

and if we replace the mass density, using the ideal gas law, (2.31) then becomes

$$\mathbf{R} = \frac{4m}{\pi k T \eta} \frac{Q}{d} \qquad (2.35)$$

For air at 22°C this reduces to

$$\mathbf{R} = 8.53 \times 10^{-4} \frac{Q}{d} \frac{(\mathrm{Pa-L/s})}{(\mathrm{m})} \qquad (2.36)$$

In ordinary vacuum practice turbulent flow occurs only infrequently; for example, at the entrance to a 6-in. diffusion pump ($d = 200$ mm, $S_p = 2000$ L/s) the gas flow at the highest pressure of interest, 0.13 Pa, is a product of the pressure P and the pump speed S_p, and $\mathbf{R} = 1.12$, a value far below the turbulent limit. The Reynolds number can reach high values in the piping of a large roughing pump during the initial pumping phase. For a pipe 250 mm in diameter connected to a 47 L/s pump $\mathbf{R}$ at atmospheric pressure is 1.6×10^4; turbulent flow will exist whenever the pressure is greater than 1.5×10^4 Pa (100 Torr). In practice these roughing lines are usually throttled during the initial portion of the roughing cycle to prevent the sudden out-rush of gas from scattering any process debris that may reside on the work chamber floor. Turbulent flow is encountered for a short time in the throttling line.

In the high flow limit of the turbulent flow region the velocity of the gas may reach the velocity of sound in the gas. This presents a situation in which further reduction of the downstream pressure cannot be sensed at the high-pressure side so that the flow is choked or limited to a maximum or critical value of flow. The value of critical flow depends on the geometry of the element, for example, orifice, short tube, or long tube, and in the case of tubes the shape of the tube entrance. A detailed discussion of critical flow has been given by Shapiro [13].

2.3.3 Viscous Flow

A general mathematical treatment of viscous flow results in the Navier-Stokes equations, which are most complex to solve. The simplest and most familiar solution for flow through long straight tubes is the Poiseuille equation

$$Q = \frac{\pi d^4}{128\eta\ell}\frac{(P_1 + P_2)}{2}(P_1 - P_2) \tag{2.37}$$

The conductance for air at 0°C is

$$C(\text{L/s}) = 1.38 \times 10^6 \frac{d^4}{\ell}\frac{(P_1 + P_2)}{2} \tag{2.38}$$

This specific solution is valid when four assumptions are met: (1) fully developed flow—the velocity profile is not position-dependent, (2) laminar flow, (3) zero wall velocity, and (4) incompressible gas. Assumption 1 holds for long tubes; corrections for short tubes are discussed in ref. 14. Assumptions 2 and 3 are satisfied if **R** < 1200 and Kn < 0.01, respectively. The assumption of imcompressibility holds true, provided that the Mach number **U**, the ratio of gas-to-sound velocity, is less than 0.3.

$$\mathbf{U} = \frac{U}{U_{sound}} = \frac{4Q}{\pi d^2 P U_{sound}} < \frac{1}{3} \tag{2.39}$$

For air at 22°C

$$Q(\text{Pa} - \text{L/s}) < 7.9 \times 10^5 d^2 P \tag{2.40}$$

This is a value of flow that may be encountered in vacuum practice and would render the results of the Poiseuille equation incorrect. In fact, for short pipes the flow may switch from molecular to transition to critical flow without there being any pressure region in which the Poiseuille equation would be valid.

The Poiseuille formula applies only to straight tubes of circular cross section.   Only a limited amount of work has been done for other cross sections, and that for long tubes; examples are given in ref. 15.   For tubes of zero length, for example, a small thin aperture, the conductance is a rather complicated function of the pressure.   Consider a fixed high pressure on one side of the aperture with a variable pressure on the downstream side.   As the downstream pressure is reduced, the gas flowing through the aperture will increase until it reaches a maximum.   If the downstream pressure is further reduced, the gas flow will not increase because the gas is traveling at the speed of sound and cannot communicate further with the high-pressure side of the orifice to tell it that the pressure has changed.   Guthrie and Wakerling [16] give the following formula for the conductance of a thin aperture for air at 22°C:

$$C(\text{L/s}) = 76.6 \times 10^4 \frac{P_2^{0.712}}{P_1}\left(1-\frac{P_2^{0.288}}{P_1}\right)\frac{A(\text{m}^2)}{(1-P_2/P_1)}$$

for $1 < P_2/P_1 \leq 0.52$

$$C(\text{L/s}) \sim 2 \times 10^5 \frac{A(\text{m}^2)}{(1-P_2/P_1)} \tag{2.41}$$

for $P_2/P_1 \leq 0.52$

2.3.4   Molecular Flow

Pure molecular flow begins for Kn > 1.0 or Pd < 6.6 Pa-mm and is theoretically the best understood of any flow regime.   The effects of pipe cross section and surface properties on the flow have not, however, been completely developed.   This discussion focuses on apertures, infinite tubes, finite tubes, and other geometries and includes a discussion of combinations of conducting elements.

Orifices

If two large vessels are connected by an orifice of area A and the diameter of the orifice is such that Kn > 1, then the gas flow from one vessel (P_1, n_1) to the second vessel (P_2, n_2) is given by

$$Q = \frac{KT}{4}vA(n_2-n_1) = \frac{v}{4}A(P_2-P_1) \qquad (2.42)$$

and the conductance of the orifice is

$$C = \frac{v}{4}A \qquad (2.43)$$

which for air at 22°C has the value

$$C(m^3/s) = 116A \ (m^2) \qquad (2.44)$$

or

$$C(L/s) = 1.16 \times 10^5 A(m^2) = 11.6A(cm^2) \qquad (2.45)$$

From (2.42) it is seen that in the molecular flow regime gas is flowing from vessel 2 to vessel 1; at the same time gas is flowing from vessel 1 to vessel 2 without either vessel knowing the concentration in the other.

Infinitely Long Tubes

The diffusion method of Smolochowski [17] and the momentum transfer method of Knudsen [18] and Loeb [19] were the first used to describe gas flow through infinite tubes in the free molecular flow region. For circular tubes both derivations yield conductances of

$$C = \frac{\pi}{12}v\frac{d^3}{\ell} \qquad (2.46)$$

For air 22°C this becomes

$$C \ (m^3/s) = 121\frac{d^3}{\ell} \qquad (2.47)$$

For rectangular ducts

$$C = \frac{2}{3}v\frac{a^2b^2K'}{(a+b)\ell} \qquad (2.48)$$

where K' is tabulated as [20]

b/a	1.0	0.8	0.6	0.4	0.2	0.1
K'	1.1	1.12	1.13	1.17	1.29	1.44

Guthrie and Wakerling [20] have tabulated the conductance of other infinitely long pipes with cylindrical cross sections like annular rings, triangles, and thin slitlike tubes.

These results for infinite tubes are valid when the pipe is long in comparison to its diameter, say $l > 20\ d$.

Tubes of Arbitrary Length

The flow equations for long tubes (2.46, 2.48) indicate that the conductance becomes infinite as the length tends toward zero, whereas the conductance actually becomes $vA/4$. Dushman [21] developed a solution to the problem of short tubes by considering the total conductance to be the sum of the conductances of an aperture and a section of tube of length l.

$$\frac{1}{C_{total}} = \frac{1}{C_{tube}} + \frac{1}{C_{ap}} \qquad (2.49)$$

As $l/d \rightarrow 0$, this equation reduces to (2.43) and as $l/d \rightarrow \infty$ it reduces to (2.47). Although this equation gives the correct solution for the extremal cases, it does not give the correct answer to the intermediate. There is little theoretical justification for its use.

The difficulty in performing calculations for the short tube lies in the nature of the gas wall interaction. A gas molecule that collides with a wall is reemitted in a cosine distribution and after a collision has as much likelihood of going forward through the tube as it has of going backward toward the source vessel. Clausing [22] approached this problem by calculating the probability that a molecule entering the pipe at one will escape at the other end. Clausing's solution to his integral equation is accurate to about 1% and is given by

$$C = \frac{a'v}{4}A \qquad (2.50)$$

For air at 22°C

$$C(\text{L/s}) = 1.16 \times 10^5 a' A \qquad (2.51)$$

where a', as found by Clausing, is given in Table 2.2. By comparison (2.49) is accurate to about 5%, provided that the pipe is connected between two large vessels, and no other surfaces can further modify the conductance. Although not strictly correct, the conductance of

Table 2.2   Tables of Clausing's Factor a'
for Various Values of l/d

l/d	a'	l/d	a'
0.01	.00	1.6	0.4062
0.05	0.965	1.7	0.3931
0.1	0.931	1.8	0.3809
0.15	0.899	1.9	0.3695
0.1	0.870	2.0	0.3589
0.25	0.8342	2.5	0.3146
0.3	0.7711	3.0	0.2807
0.35	0.7434	3.5	0.2537
0.4	0.7177	4.0	0.2316
0.45	0.6950	4.5	0.2131
0.5	0.6720	5.0	0.1973
0.55	0.6514	6.0	0.1719
0.6	0.6320	7.0	0.1523
0.65	0.6139	8.0	0.1367
0.7	0.5970	9.0	0.1240
0.75	0.5810	10.0	0.1135
0.8	0.5659	15.0	0.0797
0.85	0.5518	20.0	0.0613
0.9	0.5384	25.0	0.0499
0.95	0.5256	30.0	0.0420
1.0	0.5136	35.0	0.0363
1.1	0.4914	40.0	0.0319
1.2	0.4711	45.0	0.0285
1.3	0.4527	50.0	0.0258
1.4	0.4395	500.0	0.0026
1.5	0.4205	∞	$4d/3l$

Source. Reprinted with permission from *Scientific Foundations of Vacuum Technology*, J. M. Lafferty, Ed., p. 94. Copyright 1962, Wiley, New York.

short ducts of cross sections other than circular is often approximated by (2.49) with the appropriate entrance factor.

Complex Structures

The calculation of the molecular flow conductance of complex structures results in equations for transmission probability that do not yield to analysis in closed form. The statistical Monte Carlo methods which were applied to the calculation of molecular flow conductance by Davis [23], and Levenson, Milleron, and Davis [24] were therefore a major breakthrough in the calculation of complex, but practical, vacuum system elements such as elbows, traps, and baffles. The Monte Carlo technique uses a computer to follow the individual trajectories of a large number of randomly chosen molecules. Figure 2.5 is a computer graphical model of the trajectories of 15 random molecules entering an elbow [25]. It yielded a transmission probability of 0.222. When a large number of particles was used, the transmission probability of 0.31 was calculated. This points out one difficulty in the Monte Carlo technique; its accuracy depends on the number of molecular trajectories used in the calculation. A great deal of computational time is required for accurate solutions to complex problems. Figures 2.6 through 2.11 contain examples of the Monte Carlo technique for some structures of interest [23,24]. The molecular conductance is the product of the probability and conductance of an aperture identical in shape to the entrance of the structure under consideration. Because of the great computational time required to perform a transmission probability calculation by the Monte Carlo technique, methods have been developed to approximate complex systems by combining cylindrical tubes, orifices, and baffle plates, see, for example, Füstöss and Tóth [26].

Combinations of Conductances

Implicit in the definition of conductance (2.33) is the understanding that molecules will find the entrance to the pipe in a truly random manner. This is possible only if there are no other walls in the vicinity of the entrance (or exit) of the tube or component. It can be accomplished by connecting the component between two large reservoirs so that the pressures in the vessels will be unaffected by the flow through the component. In practice this condition is rarely met. More typically, traps, baffles, or elbows, whose length is of the order of the pipe

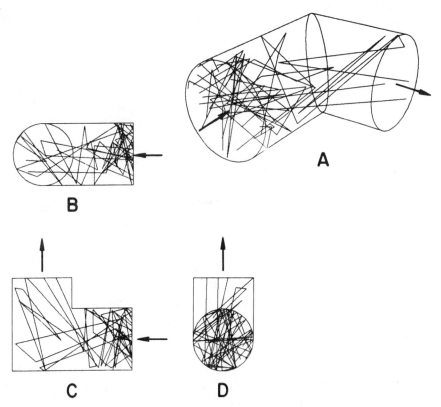

Fig. 2.5 A computer graphical display of the trajectories of 15 molecules entering an elbow in free molecular flow.   Courtesy of A. Appel, IBM T. J. Watson Research Center.

diameter, are interconnected by the shortest possible lengths of pipe that will connect the vacuum chamber and pump.

For independent elements with truly random entrance conditions the net conductance of a series of elements is

$$\frac{1}{C_T} = \frac{1}{C_1} + \frac{1}{C_2} + \frac{1}{C_3} + \dots \qquad (2.52)$$

and, for elements in parallel,

$$C_T = C_1 + C_2 + C_3 + \dots \qquad (2.53)$$

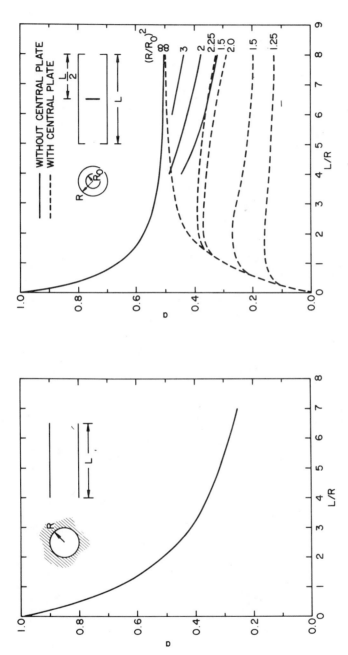

Fig. 2.6 Molecular transmission probability of a round pipe. Reprinted with permission from *Le Vide*, No. 103, p. 42, L. L. Levenson et al. Copyright 1963, Societe Francaise des Ingenieurs et Techniciens du Vide.

Fig. 2.7 Molecular transmission probability of a round pipe with entrance and exit apertures. Reprinted with permission from *J. Appl. Phys.*, **31**, p. 1169, D. H. Davis. Copyright 1960, The American Institute of Physics.

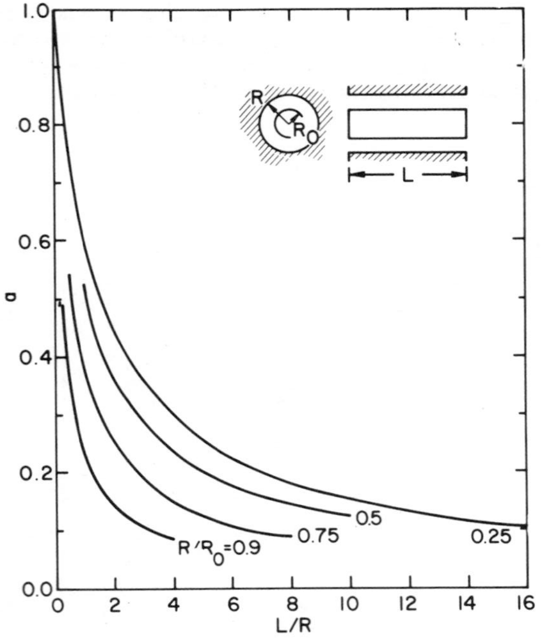

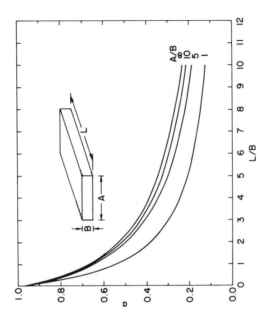

Fig. 2.9 Molecular transmission probability of a rectangular duct. Reprinted with permission from *Le Vide*, No. 103, p. 42, L. L. Levenson et al. Copyright 1963, Societe Francaise des Ingeneiurs et Techniciens du Vide.

Fig. 2.8 Molecular transmission probability of an annular cylindrical pipe. Reprinted with permission from *Le Vide*, No. 103, p. 42, L. L. Levenson et al. Copyright 1963, Societe Francaise des Ingeneiurs et Techniciens du Vide.

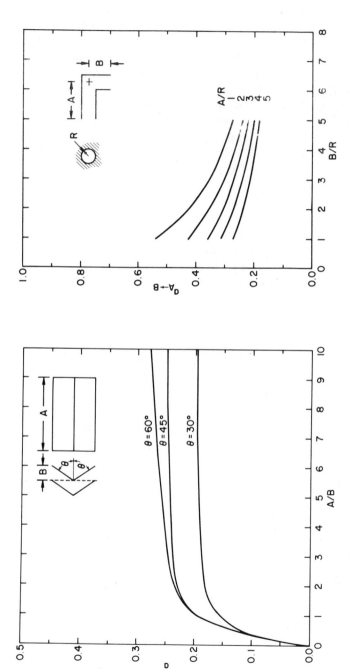

Fig. 2.10 Molecular transmission probability of a chevron baffle. Reprinted with permission from *Le Vide*, No. 103, p. 42, L. L. Levenson et al. Copyright 1963, Societe Francaise des Ingeniurs et Techniciens du Vide.

Fig. 2.11 Molecular transmission probability of an elbow. Reprinted with permission from *J. Appl. Phys.*, **31**, p. 1169, D. H. Davis. Copyright 1960, The American Institute of Physics.

33

The value of conductance given in (2.52) would be measured if the elements were connected as shown in Fig. 2.12, where each element is unaffected by the presence of others.

Problems arise because (2.52), which is valid for independently defined C's, is indiscriminately applied to series elements not isolated from one another, for example, in which the entrance conditions are not truly random. The most serious errors that can be introduced in series conductance calculations develop from the failure to account for interactions between series elements. The choice between an *exact* or *approximate* formula for an elemental conductance is usually less important than the correction for nonisolated entrance conditions.

Harries [27], Stekelmacher [28], and Oatley [29] have each used the concept of a probability factor a to calculate correctly the conductance of a series combination of vacuum elements in free molecular flow. Figure 2.13 illustrates the concept with a single component; Γ molecules per second enter at the left hand side, $a\Gamma$ molecules per second exit at the right-hand side, and $(1 - a)\Gamma$ molecules per second are returned to the source vessel. The conductance is expressed by

$$C = v\frac{A}{4}a \tag{2.54}$$

For two tubes in series Oatley [27] developed a technique for calculating a combined probability, the results of which are illustrated in Fig. 2.14. Among the Γ molecules per second that enter the first tube $a_1\Gamma$ enter the second; $\Gamma(1 - a_2)a_1$ of these are returned to the first tube and $\Gamma a_1 a_2$ enter the second. From the group $\Gamma (1 - a_2)a_1$ molecules re-

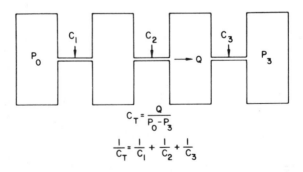

Fig. 2.12 The series conductance of isolated elements.

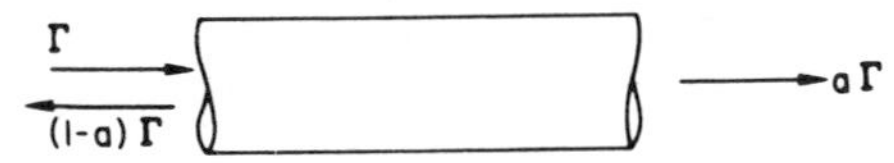

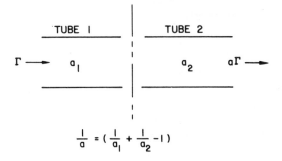

Fig. 2.13 Model for calculating the transmission probability of a single element.

turned the first tube $\Gamma a_1(1 - a_2)(1 - a_1)$ are returned to the second, and so on, until an infinite series expression is developed that simplifies to

$$\frac{1}{a} = \frac{1}{a_1} + \frac{1}{a_2} - 1 \qquad (2.55)$$

This expression, when generalized to several elements in series, was

$$\frac{1-a}{a} = \frac{1-a_1}{a_1} + \frac{1-a_2}{a_2} + \cdots \qquad (2.56)$$

Equation 2.56 may be used with care on series elements with no pronounced beaming effects. Components like chevron baffles and elbows tend to randomize the gas flow; for example, the conductance of a nondegenerate elbow can be calculated by using the conductances of the individual arms, obtained from Fig. 2.6, and summing these conductances by Oatley's method, using (2.56). If a great disparity in diameters exists, beaming effects become pronounced and (2.56) will not hold. In general, the conductance of the actual configuration

Fig. 2.14 Model for calculating the transmission probability of two series elements. Reprinted with permission from *Brit. J. Appl. Phys.*, **8**, p. 15, C. W. Oatley. Copyright 1957, The Institute of Physics.

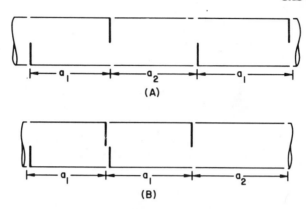

Fig. 2.15 An illustration of the position dependent nature of series vacuum elements. Interchanging elements a_1 and a_2 yields a net conductance that is strongly dependent on their relative positions. Adapted with permission from *J. Vac. Sci. Technol.*, 7, p. 237, M. H. Hablanian. Copyright 1970, The American Vacuum Society.

should be calculated, because errors can creep in when components are mathematically rearranged [30, 31]; for example, Fig. 2.15 shows two possible connections of three elements—a pipe and two baffles. It should be obvious without further explanation that the individual conductances in this extreme example are not commutative.

2.3.5 Transition Flow

The theory of gas flow in the transition region is not well developed; however, the state of the theory has been reviewed by Thomson and Owens [32]. The simplest treatment of this region, due to Knudsen and discussed in many texts, states that

$$Q = Q_{viscous} + Z'Q_{molecular} \qquad (2.57)$$

where for long circular tubes Z' is given by

$$Z' = \frac{1 + 2.507\left(\frac{d}{2}\lambda\right)}{1 + 3.095\left(\frac{d}{2}\lambda\right)} \qquad (2.58)$$

An example of this gas-flow calculation in the transition region is given in Fig. 2.16 for a tube 1 m long and 10 mm in diameter, with $P = 0$ at one end. It is seen that the transition from completely free

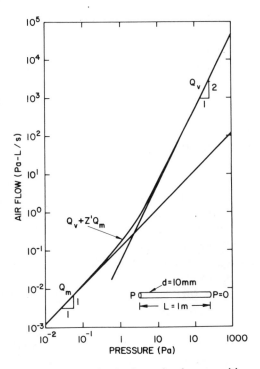

Fig. 2.16 Air flow through a long pipe in the molecular-, transition-, and viscous-flow region.

molecular flow to completely viscous flow takes place over two to three decades in pressure.

2.4 PUMPING SPEED

Pumping speed is defined as the volumetric rate at which a gas is transported across a plane. In mathematical terms speed is the gas throughput across the plane divided by the pressure at that point.

$$S = \frac{Q}{P} \tag{2.59}$$

Like conductance, it has dimensions of volume per unit time, and in SI it is expressed in units of m^3/s; units of L/s, or m^3/h are also used.

Unlike conductance, pumping speed is not a property of a passive component like a length of pipe or a baffle. Recall from (2.33) that conductance of a component is defined as the gas throughput divided by the pressure drop across a component, whereas the pumping speed is defined at a point or plane within the conductance.

In a vacuum system it is necessary to know the pumping speed at the inlet of each pump and at the end of each pumping line or manifold. The value of the pumping speed at the inlet of a pump is usually measured by the manufacturer or the user; the pumping speeds at the ends of the foreline, roughing line, and high vacuum manifold are calculated from the geometry of the ductwork and the measured pumping speed with (2.52) or by the Oatley method. Knowledge of the pumping speed may be necessary to determine whether the pump is indeed functioning as described by or to provide more information than was available from the manufacturer; for example, the pumping speed may be needed for a pump that uses a specific gas, pump fluid, cold cap, or cooling water temperature. Equation 2.59 implies independent measurements for the throughput Q and pressure P. The types of vacuum gauge used for pressure measurement are described in Chapter 3.

Several flow-measuring devices are needed to span the range of flows necessary to characterize low-, medium-, and high-vacuum pumps. At atmospheric pressure a 50 L/s (100 cfm) mechanical pump has a gas flow of 5×10^6 Pa-L/s; a 100 L/s ion pump operating at 10^{-5} Pa pumps 10^{-3} Pa-L/s—a range of more than nine orders of magnitude. No one flow-measuring instrument can cover this range. Large radius orifices are used to measure flows of more than 2×10^6 Pa-L/s, rotameters are used in the range 5×10^3 to 5×10^6 Pa-L/s, and inverted burettes are used in the range of 10^{-1} to 10^4 Pa-L/s. A full discussion of these flow-measuring instruments is given in Van Atta [33].

The AVS standard test dome [34] used for mechanical pump speed measurements is shown in Fig. 2.17. For inlet pressures below 10^{-2} Pa calibrated, trapped ionization gauges are used to measure the pressure, McLeod gauges are used in the 10^{-2} to 100 Pa range and mercury manometers are used for pressures above 100 Pa. After gas has been flowing into the test dome for at least 3 min the equilibrium pressure is recorded and the speed is calculated from $S = Q/P$.

Measurement of the pumping speed of a high-vacuum pump at pressures above 10^{-4} Pa makes use of the test dome sketched in Fig. 2.18 [35]. The pressure in the dome is first reduced to a value at least

10 times smaller than that at which the speed is to be measured. The speed is then found from the relation $S = Q/(P - P_o)$.

Determination of high-vacuum pumping speeds at inlet pressures lower than 10^{-4} Pa requires flow measurements considerably below the range of a burette. It is further complicated by the fact that the measurements are made in the molecular flow region where the pressure drop in the metering dome is large. The pressure at the entrance to the pump is also a function of the Ho coefficient [36], which may be defined as the true measured pumping speed divided by the maximum calculated speed of the pump throat area. Alternatively, it is the probability that a molecule that enters the pump body will be captured by the pump and not recoil into the test dome. The Ho factor for most pumps is generally in the range 0.2 to 0.6.

The volumetric pumping speed of a high vacuum pump at pressures lower than 10^{-4} Pa is determined with the assistance of a metering dome of standard design. In the molecular flow region the pressure within the test dome is a function of the geometry of the metering dome and the location of the inlet gas line; standardization of the geometry and the locations of the inlet gas line and pressure gauges therefore becomes a necessity. The three-gauge method of Fig. 2.19a, the large test dome method of Fig. 2.19b, and the Fischer-Mommsen method shown in Fig. 2.19c are used for measuring pumping speeds at inlet pressures lower than 10^{-4} Pa [37]. The three-gauge method is the American Vacuum Society Tentative standard [35]. Gas flow is measured by the the pressure drop along a tube with conductance C_{12} calculated by using the physical dimensions of the tube and the Claus-

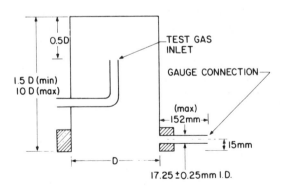

Fig. 2.17 Test dome for the measurement of mechanical vacuum pumping speed. Reprinted with permission from *J. Vac. Sci. Technol.*, **5**, p. 92. Copyright 1968, The American Vacuum Society.

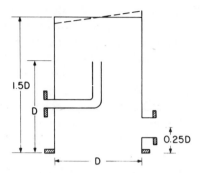

Fig. 2.18 Test dome for the measurement of high vacuum pumping speeds at pressures greater than 10^{-4} Pa. Reprinted with permission from *J. Vac. Sci. Technol.*, **8**, p. 664. Copyright 1971, The American Vacuum Society.

ing factor (2.50) determined from the length-to-diameter ratio. The gas flow is therefore $C_{12}(P_1 - P_2)$ and the speed at the entrance to the pump is

$$S = C_{12}\frac{(P_1 - P_2)}{P_3} \tag{2.60}$$

The large-test-dome method requires that the dimensions of the dome be at least three times the diameter of the pump entrance [38]. The gas flow is measured by filling a pipette of known volume V_o to a known pressure P_o. A leak valve meters the gas into the test dome in a manner that keeps the chamber pressure constant and the time t is recorded when the gas in the pipette is exhausted. The time-average gas flow is $P_o V_o / t$ and the speed is given by

$$S = \frac{P_o V_o}{P_1 t} \tag{2.61}$$

A third method developed by Fischer and Mommsen [39] has been proposed as a standard of the International Union of Vacuum Science

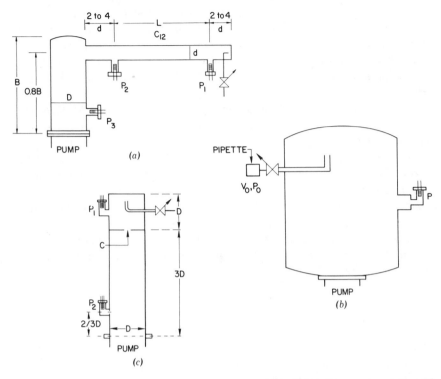

Fig. 2.19 Methods for measuring the pumping speed of pumps at pressures below 10^{-4} Pa.: (a) three-gauge, (b) large test dome, and (c) Fischer-Mommsen. Reprinted with permission from: (a) *J. Vac. Sci. Technol.*, **8**, p. 664, Copyright 1971, The American Vacuum Society; (b) *J. Vac. Sci. Technol.*, **11**, p. 337, D. R. Denison and E. S. McKee, Copyright 1974, The American Vacuum Society; (c) *Vacuum*, **17**, p. 309, E. Fischer and H. Mommsen. Copyright 1967, Pergamon Press, Ltd.

Techniques and Applications. The gas flow $C(P_1 - P_2)$ is obtained from the pressure drop across an orifice of known dimensions and the resulting speed measurement is given by

$$S = C\left(\frac{P_1}{P_2} - 1\right) \qquad (2.62)$$

Some time may be required to attain the equilibrium pressure after admission of gas at a fixed flow rate [35]. This is particularly true for sputter-ion pumps at low pressures [37, 40]. If pumping-speed data are recorded too quickly, data taken in order of decreasing pressure will yield an incorrectly small value of speed, whereas those taken in order of increasing pressure will yield an incorrectly large value of speed. It may also be necessary to erase the pump's memory for one or more gases pumped before measurement. This can be accomplished by pumping for 1 h at a pressure of 10^{-3} Pa with the gas under study [35].

In measuring the speed of an ion pump Denison and McKee [38] found that the large-test-dome method and the Fischer-Mommsen technique were in agreement, whereas the three-gauge method gave results about 8% higher than these techniques. They also discuss the advantages and disadvantages of measuring pumping speeds with air and pure gases.

An approximate measure of the pumping speed can be gained without the trouble of attaching an elaborate test dome and gas-metering system. The pumping speed can be deduced approximately if it is assumed that the pumping speed is independent of pressure in the region of interest. If this is true, then it follows that

$$S = \frac{(Q_2 - Q_1)}{(P_2 - P_1)} \tag{2.63}$$

where Q_1 is the flow that results in P_1, and so on. It can be shown that this flow is equivalent to

$$S = V\frac{\left(\frac{dP_2}{dt}\right) - \left(\frac{dP_1}{dt}\right)}{(P_2 - P_1)} \tag{2.64}$$

To measure the pumping speed the system is first pumped to its base pressure P_1 and the high vacuum valve is closed. At this time the pressure rise dP_1/dt is plotted over at least one decade pressure increase. The high vacuum value is opened and the system pumped to its original base pressure. A gas is then admitted through a leak valve until the pressure rises to a value P_2 which is several times that of the base pressure. The high vacuum value is closed and dP_2/dt is recorded. The system volume is then estimated and the speed is calculated by use of (2.64). This method, called the rate of rise or constant volume method, is only approximate, because the gas flow at the base pressure Q_1 is, in general, background desorption and not the same gas species as that admitted through the leak. If the pressure P_2 were at least 10 times P_1, the result would be be more meaningful than if it were only a few times the value of P_1.

REFERENCES

1. E. H. Kennard, *Kinetic Theory of Gases,* McGraw-Hill, New York, 1938, p. 9.
2. R. W. Moore, Jr., *8th Nat. Vac. Symp. 1961*, 1, Pergamon, New York, 1962, p. 426.
3. E. H. Kennard, *Kinetic Theory of Gases*, McGraw-Hill, New York, 1938, Chapter 4, pp. 135-205, Chapter 8, pp. 291-337.
4. S. Dushman, *Scientific Foundations of Vacuum Technique*, 2nd ed., J. M. Lafferty, Ed., Wiley, New York, 1962, pp. 35.
5. I. Langmuir, *Phys. Rev.*, 1, 337 (1913).
6. Ref. 4, p.6.
7. Ref. 4, p. 68.
8. C. Neumann, *Math Phys. K.*, 24, 49 (1872).
9. J. C. Maxwell, *Philos. Trans. R. Soc. London*, 170, 231 (1879).
10. M. C. I. Siu, *J. Vac. Sci. Technol.*, 10, 368 (1973).
11. O. Reynolds, *Philos. Trans. R. Soc. London*, 174, (1883).
12. A. Guthrie and R. K. Wakerling, *Vacuum Equipment and Techniques*, McGraw-Hill, New York, 1949, p.25.
13. A. H. Shapiro, *Dynamics and Thermodynamics of Compressible Fluid Flow*, Ronald, New York, 1953.
14. H. L. Langhaar, *J. Appl. Mech.*, 9, A-55 (1942).
15. Ref. 4, p. 85-87.
16. Ref. 12, p. 17.
17. M. von Smolochowski, *Ann. Phys.*, 33, 1559, (1910).
18. M. Knudsen, *Ann. Physik*, 28, 75 (1909); 35, 389 (1911).
19. L. B. Loeb, *The Kinetic Theory of Gases*, McGraw-Hill, New York, 2nd ed., 1934, Chapter 7.
20. Ref. 12, p. 39.
21. Ref. 4, p. 91.
22. P. Clausing, *Ann. Phys.,*, 12, 961 (1932), English Translation in *J. Vac. Sci. Technol.*, 8, 636 (1971).
23. D. H. Davis, *J. Appl. Phys.*, 31, 1169 (1960).
24. L. L. Levenson, N. Milleron, and D. H. Davis, *Le Vide*, 103, 42 (1963).
25. Courtesy of A. Appel, IBM T. J. Watson Research Center.
26. L. Füstöss and G. Tóth, *J. Vac. Sci. Technol.*, 9, 1214 (1972).
27. W. Harries, *Z. Angew. Phys.*, 3, 296 (1951).
28. W. Stekelmacher, *Proc. 6th Int. Vacuum Cong.*, Kyoto, *Japan. J. Appl. Phys.*, Sup.2, Pt.1, 117 (1974).
29. C. W. Oatley, *Brit. J. Appl. Phys.*, 8, 15 (1957).

30. D. J. Santeler et al., *Vacuum Technology and Space Simulation*, NASA SP-105, National Aeronautics and Space Administration, Washington, DC, 1966, p.115.

31. M. H. Hablanian, *J. Vac. Sci. Technol.*, **7**, 237 (1970).

32. S. L. Thomson and W. R. Owens, *Vacuum*, **25**, 151 (1975).

33. C. M. Van Atta, *Vacuum Science & Engineering*, McGraw-Hill, New York, 1965, Chapter 7.

34. Apparatus of AVS Tentative Standard 5.3, *J. Vac. Sci. Technol.*, **5**, 1968.

35. Apparatus of AVS Tentative Standard 4.1, 4.7, and 4.8, *J. Vac. Sci. Technol.*, **8**, 664 (1971).

36. T. L. Ho, *Physics*, **2**, 386 (1932).

37. D. R. Denison and E. S. McKee, *J. Vac. Sci. Technol.*, **11**, 337 (1974).

38. A. Venema, *Vacuum*, **4**, 272 (1954).

39. E. Fischer and H. Mommsen, *Vacuum*, **17**, 309 (1967).

40. D. Andrew, *Vacuum*, **16**, 653 (1966).

Measurement

In these three chapters we discuss the tools used to measure pressure in a vacuum system: the total pressure gauge and the residual gas analyzer. With the total pressure gauge routine system performance is monitored and many problems are discovered. The residual gas analyzer adds a degree of sophistication to our analytical skills. Its ability to single out the gas or vapor that is limiting the system pressure or causing a process problem greatly reduces the difficulty in trouble shooting large and complex systems. Chapter 3 is devoted to a discussion of the commonly used pressure gauges. Chapter 4 discusses the operation and installation of residual gas analyzers on vacuum systems, while Chapter 5 describes qualitative and quantitative methods of interpreting the data obtained from an RGA spectrum.

CHAPTER 3

Total Pressure Gauges

This chapter discusses some of the common total pressure gauges available to the vacuum industry—in particular those that are frequently encountered by users of modern vacuum equipment. To discuss in detail each of the 20 or more pressure measuring instruments would result in a reduced presentation of those gauges that are the most important. Descriptions of the less frequently used gauges are contained in several texts [1–3].

Over the last 50 years many techniques have been developed for the measurement of reduced pressures. Gauges are either direct- or indirect-reading. Those that measure pressure by calculating the force exerted on the surface by incident particle flux are called direct reading gauges. Indirect gauges record the pressure by measuring a gas property which changes in a predictable manner with gas density. Figure 3.1 sketches a way of classifying many pressure gauges; their operating ranges are illustrated in Fig. 3.2.

3.1 DIRECT-READING GAUGES

The diaphragm, Bourdon, and capacitance manometers are the most frequently used direct-reading gauges. Two rather well known gauges which have a necessary place in pressure measurement, the U-tube manometer and the McLeod gauge, are not described in detail because they are not routinely used by the average vacuum-system operator. In its simplest form a manometer consists of a U-tube that contains a

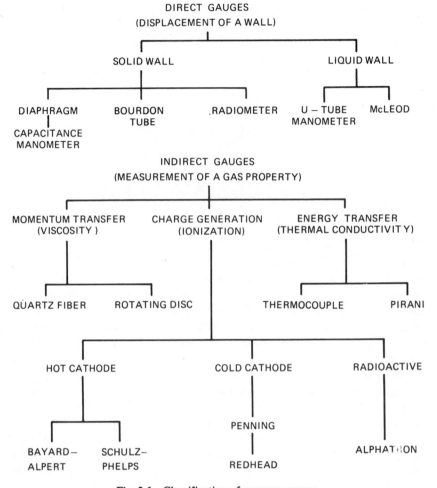

Fig. 3.1 Classification of pressure gauges.

low vapor pressure fluid such as mercury or oil. One arm is evacuated and sealed; the other is connected to the unknown. The pressure is read as the difference in the two liquid levels. The McLeod gauge is a mercury manometer in which a volume of gas is compressed before measurement; for example, precompressing a small volume of gas at 10^{-2} Pa by 10,000 times results in a measurable pressure of 100 Pa. The U-tube manometer, which is used in the 10^2 to 10^5 Pa range, and the McLeod gauge, which is a primary standard in the 10^{-3} to 10^2 Pa pressure range, are described in many texts [1–4].

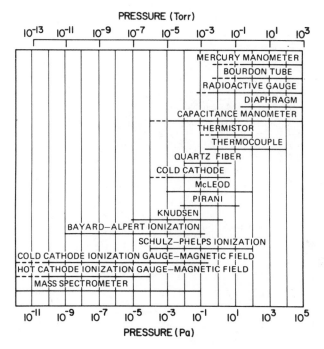

Fig. 3.2 Pressure ranges for various gauges. Adapted with permission from *Scientific Foundations of Vacuum Technique*, 2nd ed., J. M. Lafferty, Ed., p. 350. Copyright 1962, John Wiley & Sons.

3.1.1 Diaphragm and Bourdon Gauges

The simplest mechanical gauges are the diaphragm and Bourdon gauges. Both are operated by a system of gears and levers to transmit the deflection of a solid wall to a pointer. The Bourdon tube (Fig. 3.3) is a coiled tube of elliptical cross section, fixed at one end and connected to the pointer mechanism at the other. Evacuation of the gas in the tube causes rotation of the pointer. The diaphragm gauge contains a pressure-sensitive element from which the gas has been evacuated. By removing gas from the region surrounding the element, the wall is caused to deflect, and in a manner similar to the Bourdon tube the linear deflection of the wall is converted to angular deflection of the pointer.

Simple Bourdon or diaphragm gauges, for example those of the 50-mm-diameter variety, will read from atmospheric pressure to a minimum pressure of about 10^3 to 5×10^3 Pa. They are inaccurate

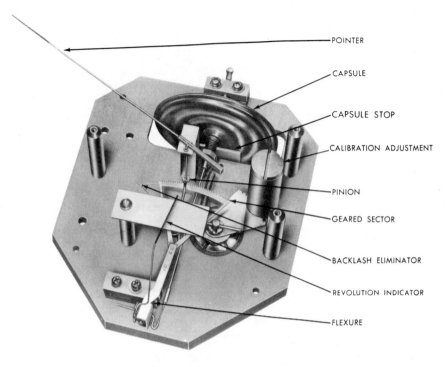

POINTER

CAPSULE

CAPSULE STOP

CALIBRATION ADJUSTMENT

PINION

GEARED SECTOR

BACKLASH ELIMINATOR

REVOLUTION INDICATOR

FLEXURE

Fig. 3.3 Diaphragm mechanism for absolute pressure measurement. Reprinted with permission from the Wallace and Tiernan Division, Pennwalt Corp., Newark, NJ.

and used only as a rough indication of pressure. They are commercially available with 316 stainless steel tubes by which they may be attached to clean systems.

Diaphragm and Bourdon gauges, which are more accurate than those described above, are available in a variety of ranges extending from 10^3 to 2×10^5 Pa and with sensitivities of an order of 25 Pa. Figures 3.3 and 3.4 illustrate two types of gauge. The gauge described in Fig. 3.3 is a diaphragm in which the entire instrument case is attached to the vacuum vessel and evacuated. The case is protected from possible overpressure damage by a blow-out plug. This gauge with its large internal volume, brass parts, and high vapor pressure lubricating materials is not the type to be appended to a clean system. The Bourdon gauge, especially the differential tube type illustrated in Fig. 3.4, is quite suitable for attachment to clean systems; only the small interior volume of the tube is added to the system. Because tubes can be fabricated from many materials, the gauge can be designed to handle

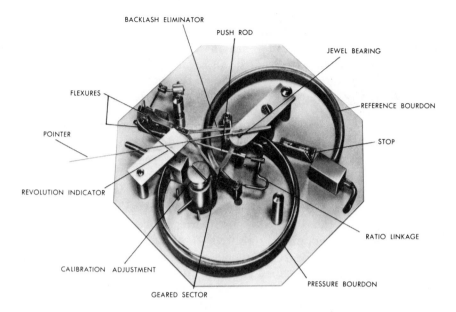

BACKLASH ELIMINATOR
PUSH ROD
JEWEL BEARING
FLEXURES
REFERENCE BOURDON
POINTER
STOP
REVOLUTION INDICATOR
RATIO LINKAGE
CALIBRATION ADJUSTMENT
PRESSURE BOURDON
GEARED SECTOR

Fig. 3.4 Bourdon tube mechanism for absolute pressure measurement. Reprinted with permission from the Wallace and Tierman Division, Pennwalt Corp., Newark, NJ.

corrosive gases. The differential gauge adds only a small surface area to the system, and when fabricated from 316 stainless steel, is excellent for use on clean chambers.

3.1.2 Capacitance Manometers

A capacitance manometer is simply a diaphragm gauge in which the deflection of the diaphragm is measured by observing the change in capacitance between it and a fixed counter electrode. The first gauge was described in 1951 by Alpert, Matland, and McCoubrey [5]. They used a differential gauge head as a null reading instrument between the vessel of unknown pressure and another whose pressure was independently adjustable and monitored by a U-tube manometer.

The capacitance of the diaphragm-counter electrode structure is proportional to geometry (area/gap) and to the dielectric constant of the gas being measured in relation to that of air. Except for the few gases that have relative dielectric constants significantly different from air, for example, certain conductive and heavy organic vapors or gases ionized by radioactivity, the use of capacitance change to measure pressure represents a true, absolute-pressure measurement: that is, the pressure may be calculated from the geometry and the observed capacitance change. A 1% difference in the dielectric constant of the measured gas and air will result in an error of 1/2% of reading. A single-sided structure is not dependent on the dielectric constant of the measuring gas because both electrodes are in the vacuum, or reference side. Modern capacitance manometers consist of two components, a transducer and an electronic sense unit that converts the membrane position to a signal linearly proportional to the pressure. A common design for a transducer is shown in Fig. 3.5. The flexible metal diaphragm, which has been stretched and welded in place, is located between two fixed electrodes. The differential transducer shown in Fig. 3.5 may be a null detector or a direct reading gauge. When used as a null detector, the pressure at the reference side P_r is adjusted until the diaphragm deflection is zero. In this mode a second gauge is necessary to read pressure. To use as a direct reading gauge the reference side must be pumped to about 10^{-5} Pa. After calibration the instrument may be used directly over the pressure range for which it was designed. Transducers are available with the reference side open for evacuation, or evacuated and permanently sealed. The permanent seal is usually a copper pinch seal; a getter is activated inside the tube at the time of manufacture.

Care should be taken in attaching the transducer to a system. It is generally advisable to have a bellows section in series with one tubulation if the transducer is to be permanently welded to a system with only short tube extensions. Even though some transducers contain filters to prevent particulates from entering the space between the diaphragm and the electrodes, it is advisable to force argon through a small diameter tube placed inside the tubulation or bellows extension to be welded on the sensor. The end of this small tube should be pushed in to a point beyond the weld location to allow the flow of argon to flush particulates out of the tube and away from the sensor during the welding operation. This procedure also stops the formation of oxide on the tube's interior walls.

Because the transducer contains ceramic insulators, cleanliness is in order; a contaminated head is hard to clean. Cleaning solvents are difficult to remove from the ceramic and may cause contamination of the system at a later time. To avoid this problem one transducer has been designed with a single sided sensor. Both electrodes are on the reference side (see Fig. 3.6). One electrode is placed at the center of the diaphragm and the second is an annular ring located around the center electrode. For zero deflection of the membrane the circuit is

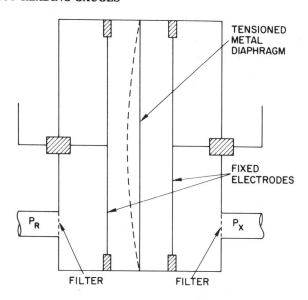

Fig. 3.5 Double-sided capacitance manometer head assembly. Reprinted with permission from *Industrial Research/Development*, January 1976, p. 41. Copyright 1976, Technical Publishing Co.

adjusted for zero output signal. The deflection or bowing of the diaphragm causes a capacitive imbalance, which is converted to a voltage proportional to pressure. A proper choice of materials results in a transducer suitable for service in corrosive environments without head damage or in extremely clean environments without contamination by the head.

The electronic sensing unit applies an ac signal to the electrodes. The changes in signal strength produced by the diaphragm are amplified and demodulated in phase to minimize the noise level. The dc output is then used to drive an analog or digital read-out. Because the resolution of the instrument is limited by system noise, the system bandwidth must be stated when specifying resolution; noise is proportional to the half-power of the bandwidth. Typical capacitance manometer systems have a resolution of 1 part in 10^6 full scale at a bandwidth of 2 Hz.

Just as low electronic noise is of prime importance in obtaining high resolution, thermal stability of the head is necessary for stable, accurate, and drift-free operation. The diaphragm deflection in the transducer can be as low as 10^{-9} cm; therefore motion of parts due to

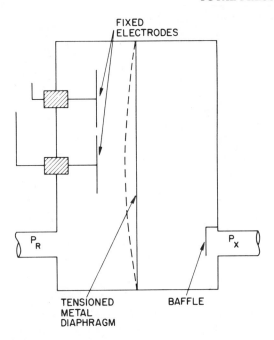

Fig. 3.6 Single-sided capacitance manometer head assembly. The outer electrode is an annular ring. Reprinted with permission from *Industrial Research/Development*, January 1976, p. 41. Copyright 1976, Technical Publishing Co.

temperature change becomes a large source of error. Transducers are available with heaters that maintain the ambient temperature at about 50°C and avoid some of the problems of ambient temperature change. Many transducers can be operated at temperatures as high as 250°C. The readings, however, must be corrected for thermal transpiration (see Section 2.2.4). Stable operation of a transducer requires that the thermal expansion coefficients of the diaphragm and electrode assemblies be well matched, but in practice designs must make a trade-off between expansion coefficient and corrosion resistance. Without proper temperature regulation a transducer may have zero and span coefficients of 5 to 50 ppm full scale and 0.004 to 0.04% of reading per degree Celsius, respectively, at ambient temperature [6]. Proper temperature regulation can result in an order of magnitude improvement in the zero and span coefficients.

Capacitive manometers can be operated over a large dynamic range, a factor of 10^4 to 10^5 for most instruments, but the overall system accuracy deteriorates at small fractions of full head range, as illustrated in Fig. 3.7 for the 1.3 × 10⁵ Pa head. Transducers with a full-scale deflection of 130 Pa have been checked in the 2.5×10^{-2} to 6.5×10^{-4} Pa pressure range by volumetric division and have been found to be linear to the lowest pressure and in agreement within 0.6% plus 5.3×10^{-5} Pa [7].

3.2   INDIRECT-READING GAUGES

In this section the most familiar indirect reading gauges are discussed. Indirect gauges calculate pressure by measuring a pressure dependent property of the gas. In the pressure range above 0.1 Pa, energy and momentum transfer techniques can be used for pressure measurement. Thermal conductivity gauges measure the heat transfer between two surfaces at different temperatures. A Pirani or a thermocouple gauge is found on every vacuum system for measuring pressure in the medium vacuum region. Ionization gauges, which measure gas density, have found wide acceptance. Hot cathode gauges are used in the Schulz-Phelps and the Bayard-Alpert geometries; together they span the pressure range 100 Pa to 10^{-9} Pa. Systems operating in the 10^{-3} to 1 Pa range often use the simpler Penning cold cathode gauge. Hot and cold cathode magnetron gauges which are capable of operation at pressures as low as 10^{-11} Pa are found on some ultrahigh-vacuum systems, but are not used on ordinary high-vacuum systems.

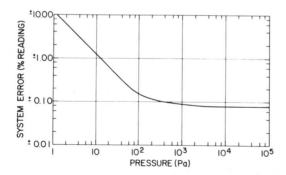

Fig. 3.7   Performance curve for a capacitance manometer with a 3×10^5 Pa full-scale transducer. Reprinted with permission from *Industrial Research/Development*, January, 1976. Copyright 1976, Technical Publishing Co.

3.2.1   Thermal Conductivity Gauges

Thermal conductivity gauges are a class of pressure-measuring instruments that operates by measuring in some way the rate of heat transfer between a heated wire and its surroundings. The heat transfer between a heated wire and a nearby wall is pressure dependent in the $0.01 <$ Kn < 10 range, where Kn is Knudsen's number. For $d = 10$ mm the pressure range is about 66 to 0.06 Pa, although the sensitivity of heat transfer with pressure is highly nonlinear at each end of the scale. The heat transfer regimes in a thermal conductivity gauge are illustrated in Fig. 3.8. At high pressures where Kn < 0.01 the heat flow is given by (2.20) and is independent of pressure except for a small convection effect. In the $0.01 <$ Kn < 10 region the *free molecular* heat flow is given by (2.22). In this region the heat flow is linearly proportional to the pressure, provided that the accommodation coefficient and the temperature difference between the heated wire and the case remain constant. In the lowest pressure region the heat flow is predominantly accounted for by radiation and conduction through the wire to the supports, as expressed in (3.1)

$$H = A\sigma\varepsilon_1(T_2^4 - T_1^4) + \text{end losses} \qquad (3.1)$$

To extend the range of a gauge to its lowest possible pressure limit it is necessary to reduce the radiation and end conduction losses. The end losses are predominant only when the length of the wire is short. Leck [1] shows that for a wire 100 mm long the increase in end losses for a pressure change from 13 to 130 Pa is about 5% of the increase in heat conductivity, whereas for a wire 10 mm long the equivalent figure is about 50%. The radiant heat losses can be minimized by reducing the diameter and the emissivity of the hot wire. The emissivity of a clean tungsten wire is about 0.1, but in practice most are not clean. The upper pressure limit of a thermal conductivity gauge is determined by the saturation pressure of the thermal conductivity. This occurs at a Knudsen number of about 0.01. The two most commonly found gauges have upper pressure limits in the 15 to 150 Pa range, but tubes which read to 10^5 by taking advantage of pressure-dependent convection losses[8] are available.

The sensitivity of the gauge is determined by tube construction and the gas as well as by the technique for sensing the change in heat flow with pressure. Tungsten is commonly used for the heater wire because

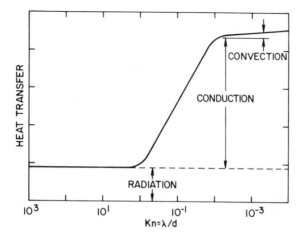

Fig. 3.8   Heat transfer regimes in a thermal conductivity gauge.

it has a large thermal resistance coefficient. (When a semiconductor is the heat sensitive element, the device is referred to as a thermistor gauge, even though it is strictly speaking a Pirani gauge). Equation 2.22 which we repeat here, describes the sensitivity of heat flow to a change in hot-wire temperature T_2

$$E_o = \frac{\alpha_1 \alpha_2}{\alpha_1 + \alpha_2 - \alpha_1 \alpha_2} \frac{1(\gamma + 1)v_1}{8(\gamma - 1)T_1}(T_2 - T_1)P \qquad (3.2)$$

The ratio of specific heats and thermal velocity depend on the gas species and in combination can produce as much as a fivefold difference in sensitivity between two gases. The accommodation coefficient for clean materials can be of an order 0.1 but for contaminated surfaces it can be as high as unity. For most cases α is stable but not known. With all other factors well controlled changes in emissivity and accommodation coefficient are large enough to allow thermal conductivity gauges to be used as only rough indicators of vacuum.

The change in temperature can be detected by monitoring the resistance of the heated wire. When a Wheatstone bridge circuit is used to measure the resistance change, the device is termed a Pirani gauge.

Alternatively, the temperature change can be measured directly with a thermocouple, in which case it is called a thermocouple gauge.

Pirani Gauge

The term *Pirani gauge* is given to any type of thermal conductivity gauge in which the heated wire forms one arm of a Wheatstone bridge. A simple form of this circuit is shown in Fig. 3.9. The gauge tube is first activated to a suitably low pressure, say 10^{-4} Pa, and R_1 is adjusted for balance. A pressure increase in the gauge tube will unbalance the bridge because the increased heat loss lowers the resistance of the hot wire. By increasing the voltage more power is dissipated in the hot wire, which causes it to heat, increase its resistance, and move the bridge toward balance. In this method of gauge operation, called the constant temperature method and the most sensitive and accurate technique for operating the bridge, each pressure reading is taken at a constant wire temperature. To correct for changes that ambient temperature would have on the zero adjustment an evacuated and sealed compensating gauge tube is used adjacent to the active gauge tube in another arm of the bridge. Bridges with a compensating tube can be used to 10^{-3} Pa.

The constant voltage and constant current techniques were devised to simplify the operation of the Pirani gauge. In each case the total bridge voltage or current is kept constant. The constant voltage me-

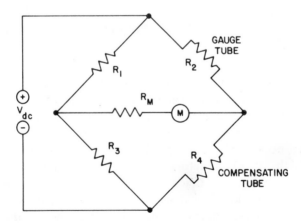

Fig. 3.9 Basic Pirani gauge circuit. Adapted with permission from *Vacuum Technology*, A. Guthrie, p. 163. Copyright 1963, John Wiley & Sons.

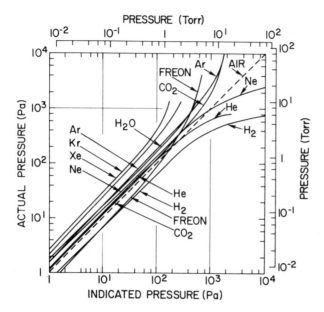

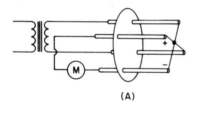

Fig. 3.10 Calibration curves for the Leybold TR201 Pirani gauge tube. Reprinted with permission from Leybold-Heraeus G.m.b.H., Postfach 51 07 60, 5000 Köln, West Germany.

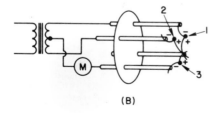

Fig. 3.11 Thermocouple gauge tubes for the 0-100 Pa range. (a) uncompensated gauge tube, (b) compensated gauge tube.

thod is widely used in modern instruments because no additional adjustments need to be made after the bridge is nulled at lower pressures. The out-of-balance current meter is simply calibrated to read the pressure.

The constant temperature method is the most sensitive and accurate because at constant temperature the radiation and end losses are constant. Because the wire temperature is constant, the sensitivity is not diminished in the high-pressure region (3.2). This method does not lend itself to easy operation; a balancing act is required before each measurement. A sudden drop in pressure can also cause over-heating of the wire if the bridge is not immediately rebalanced. Direct-reading, constant-temperature bridges that need only a zero adjustment are now commercially available, although at somewhat greater expense than a constant voltage or current bridge. Modern circuitry has eliminated tedious bridge balancing. Because the heat conductivity varies

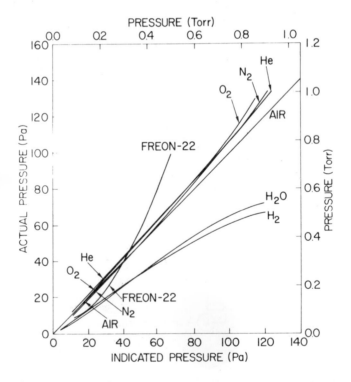

Fig. 3.12 Calibration curves for the Hastings DV-6M thermocouple gauge tube. Reprinted with permission from Hastings Instruments Co, Hampton, VA.

considerably among gases and vapors, the calibration of the gauge is dependent on the nature of the gas. Most instruments are calibrated for air; therefore a chart like the one shown in Fig. 3.10 is needed when the pressure of other gases is measured.

Thermocouple gauges

The thermocouple gauge measures pressure-dependent heat flow. Constant current is delivered to the heated wire and a tiny thermocouple, perhaps iron- or copper-constantan, is carefully spot welded to its midpoint. As the pressure increases heat flows to the walls and the temperature of the wire decreases. A low-resistance dc microammeter is connected to the thermocouple and its scale is calibrated in pressure units.

Figure 3.11 shows the four-wire and three-wire versions of the gauge tubes. The four-wire gauge tube uses a dc meter to read the temperature of the thermocouple, whereas the power supply is regulated to deliver a constant current to the wire. The current can be ac or dc. The three-wire gauge circuit reduces the number of leads between the gauge tube and controller and the number of vacuum feed-throughs by using ac to heat the wires and a dc microammeter to read the voltage between one thermocouple wire and the center tap of the transformer, which is a dc connection to the other junction. In both tubes the power delivered is not constant; instead the wire current is constant. Because the resistance of the wire is temperature-dependent, the actual power delivered decreases slightly at high pressures. Both gauge forms are rugged and reliable but inaccurate. Calibration curves for one thermocouple gauge are given in Fig. 3.12.

3.2.2 Ionization Gauges

In the high and ultrahigh vacuum region where the particle density is extremely small it is not possible, except in specialized laboratory situations, to detect the minute forces that result from the direct transfer of momentum or energy between the gas and a solid wall; for example, at a pressure of 10^{-8} Pa the particle density is only $2.4 \times 10^{12}/m^3$. This may be compared with a density of $3 \times 10^{22}/m^3$ at 300 K which is required to raise a column of mercury 1 mm. Even a capacitance manometer cannot detect pressures lower than 10^{-4} Pa. The basic principle used for the measurement of pressures lower than

10^{-3} Pa is the ionization of gas molecules, and the collection of the ions and their subsequent amplification by sensitive and stable circuitry.

Each ionization gauge has its own lower pressure limit at which the ionized particle current is equal to a residual or background current. The best of these gauges have lower limits of an order of 10^{-11} to 10^{-12} Pa. In special research environments, where pressures far below 10^{-12} Pa may be encountered, the pressure is considerably below the limit of current ionization gauge technology. At a pressure of 5×10^{-15} Pa and a temperature of 4.2 K there are only 100 (nitrogen) molecules per cubic centimeter. Even with the most efficient ionization schemes available the ion current would be lost in the system noise. In those situations adsorbed gas can be collected on a particular surface for an extremely long time, after which the pressure pulse that results from flash desorption of the surface can be recorded [9].

In routine operation of high vacuum systems in the 10^{-1} to 10^{-7} Pa range the Bayard-Alpert and Schulz-Phelps hot-cathode ionization gauges or the Penning cold cathode gauge are used. Each has its own pressure range, advantages, and disadvantages.

Hot Cathode Gauges

The operation of the ion gauge is based on ionization of gas molecules by electron impact and the subsequent collection of these ions by an ion collector. This positive ion current is proportional to pressure, provided that all other parameters, including temperature, are held constant. The number of positive ions formed is actually proportional to the number density, not the pressure; the ion gauge is not a true pressure-measuring instrument but rather a particle-density gauge. It is proportional to pressure only if the temperature is known and constant.

The earliest form of ion gauge, the triode gauge, consisted of a filament surrounded by a grid wire helix and a large diameter, solid cylindrical ion collector. This gauge, which is not illustrated here, looks a lot like a triode vacuum tube. Electrons emitted by the heated filament were accelerated toward the grid wire which was held at a positive potential of about 150 V. The external collector was biased about -30 V with respect to the filament and could collect the positive ions generated in the space between the filament and the ion collector. This gauge measured pressures as low as 10^{-6} Pa but would not give a lower reading even if indirect experimental evidence indicated the existence of lower pressures. Further progress was not made until after 1947, when Nottingham [10] suggested that the cause of this effect was an x-ray-generated photo current. Nottingham proposed that soft x-rays generated by the electrons striking the grid wire collided with the ion collector cylinder and caused photo-electrons to flow from the collector to the grid. Some photoemission is also caused by ultraviolet radiation from the heated filament. As they leave the collector these photo-electrons produce a current in the external circuit which is not distinguishable from the positive ion flow toward the ion collector, and mask the measurement of reduced pressures.

In 1950 Bayard and Alpert [11] designed a gauge in which the large area collector was replaced with a fine wire located in the center of the grid (Fig. 3.13). Because of its smaller area of interception of x-rays, this gauge could measure pressures as low as 10^{-8} Pa. Today this gauge is the most popular design for the measurement of high vacuum pressures. A more efficient design increased the sensitivity of the gauge tube by capping the end of the grid to prevent electron escape (Nottingham [12]) and reduced the x-ray limit even more by use of a fine collector wire. This efficient design can measure pressures as low as 2×10^{-9} Pa. Figure 3.14 qualitatively illustrates the x-ray limits of the triode gauge and the efficient Bayard Alpert-gauge.

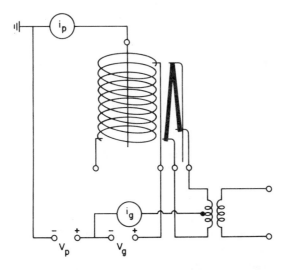

Fig. 3.13 External control circuit for a Bayard-Alpert ionization gauge tube.

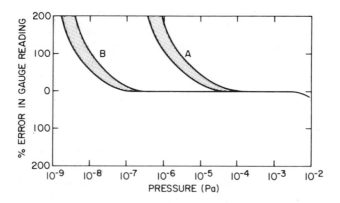

Fig. 3.14 Qualitative x-ray generated error in ion gauge tube readings. (a) triode gauge tube, (b) efficient Bayard-Alpert tube.

The proportionality between the plate current and the pressure is given by

$$i_p = S' i_e P$$

or

$$P = \frac{1}{S'} \frac{i_p}{i_e} \tag{3.3}$$

where i_p and i_e are the plate and emission currents, respectively, and S' is the sensitivity of the gauge tube. This sensitivity has dimensions of reciprocal pressure, which in SI is Pa^{-1}, and is dependent on the tube geometry, grid and plate voltages, the type of control circuitry, and the nature of the gas being measured. For the standard-design Bayard-Alpert tube with external control circuitry, a plate voltage of +150 V, and a grid voltage of 30 V the sensitivity for nitrogen is typically 0.07/Pa. Variations in tube design, voltage, and control circuitry can cause it to range from 0.05 to 0.15/Pa. The tube's sensitivity for other gases varies with the ionization probability. Alpert [13] has suggested that the relative sensitivity, (e.g., the ratio of the absolute sensitivity of a gas to that of nitrogen) should be independent of structural and electronic variations and thus be more meaningful to tabulate.

With the help of (3.4) and Table 3.1 [14] the pressure of gases other than nitrogen can be measured with an ion gauge, even though all ion

Table 3.1 Approximate Relative Sensitivity of Bayard–Alpert Gauge Tubes to Different Gases.[a]

Gas	Relative Sensitivity
H_2	0.42 - 0.53
He	0.18
H_2O	0.9
Ne	0.25
N_2	1.00
CO	1.05 - 1.1
O_2	0.8 - 0.9
Ar	1.2
Hg	3.5
Acetone	5

Source: Adapted with permission from *J. Vac. Sci. Technol.*, **8**, p 661, T. A. Flaim and P. D. Owenby. Copyright 1971, The American Vacuum Society.

[a] The pressure of any gas is found by dividing the gauge reading by the relative sensitivity.

gauges are calibrated for nitrogen. This is done by dividing the gauge reading by the relative sensitivity of the gas of interest. The relationship between the gauge pressure and the unknown pressure is

$$P(x) = \frac{S(N_2)}{S(x)} P(N_2) \qquad (3.4)$$

or because the sensitivity has been normalized to nitrogen, $S(N_2) = 1$,

$$P(x) = \frac{P(\text{meter reading})}{\text{Relative sensitivity of gas}(x)} \qquad (3.5)$$

Blears [15] observed that a nude gauge gave pressure readings much higher than a tubulated gauge when the dominant species was a pump oil vapor. He found that the vapors pumped by the walls of the tube caused the vapor pressure to be much lower inside the tubulated gauge than in the vicinity of the nude gauge. Some of the apparent difference in pressure at the two gauge locations was due to the higher sensitivity for hydrocarbons ($S' = 5$–10). Not only are hydrocarbon molecules prevented from reaching the electrodes of a tubulated gauge but those that reach either gauge cause a deflection 5 to 10 times greater than an equal number of nitrogen molecules.

Gauge sensitivity is often given in units of microamperes of plate current per unit of pressure per manufacturer's specified emission current; for example, a typical nitrogen sensitivity is (100 μA/mTorr)/10 mA. This is a confusing way of saying the sensitivity is 10/Torr, but it does illustrate an important point; not all gauge controllers have the same calibration value of emission current. A quick check in the instruction manual can avoid potential embarrassment.

Both internal and external control circuits have been developed for use with ion gauge tubes. The internal control circuit biases the plate at about $+125$ V with respect to the filament and collects the electrons while biasing the positive ion collecting grid at -20 V with respect to the filament. The external control circuit is illustrated in Fig. 3.13. The plate is held at virtual ground and collects positive ions. Most modern ion gauge controllers use the external control circuit because of its high sensitivity. Another advantage of external control is that variations in plate and grid voltages produce only small changes in plate current. At pressures of 10^{-2} Pa the emission current is usually lowered automatically to reduce space-charge buildup.

The classical control circuit is designed to stabilize the potentials and emission current while measuring the plate current. The plate current meter is then calibrated in appropriate ranges and units of pressure. The accuracy of the gauge is dependent in part on moderately costly, high quality emission current regulation. One gauge controller [16] avoids the problem of close regulation of the emission current by use of an integrated circuit to take the ratio of plate to emission current. Examination of (3.3) shows that except for a constant scale factor this current ratio is indeed proportional to pressure.

Tungsten and thoriated iridium (ThO$_2$ on iridium) are two commonly used filament materials. Thoriated iridium filaments are not destroyed when accidentally subjected to high pressures—an impossible feat with fine tungsten wires—but they do poison in the presence of some hydrocarbon vapors. The remarks contained in Section 4.2 about filament reactivity with gases in the ionizer of a residual gas analyzer also pertain to the ion gauge.

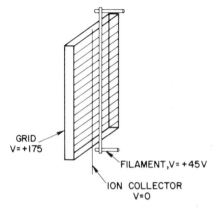

GRID
V=+175

FILAMENT,V= +45V

ION COLLECTOR
V=0

Fig. 3.15 Schulz-Phelps type ion gauge tube for operation at high pressures. Reprinted with permission from Varian Associates, 611 Hansen Way, Palo Alto, CA.

Ion gauge outgassing is accomplished by direct or electron bombardment heating. Either the grid wire is heated directly by connecting it to a low-voltage high-current transformer or the grid and plate wire are connected to a high-voltage transformer and heated by electron bombardment. It is best to wait until the pressure is on a suitably low scale ($<10^{-4}$ Pa) before outgassing. An unbaked tubulated gauge should be outgassed until the walls have desorbed. (The pressure may be monitored during outgassing on gauges that use resistance-heated grids.) The time for this initial outgassing is variable but 15–20 min is typical. After the initial outgassing the tube should be left on. Subsequently only short outgassing times, say 15 s, are periodically needed to clean the electrodes. It is useful to operate the gauge at reduced emission (0.1 mA) because it will pump the least when the emission current is the lowest.

At pressures greater than 10^{-2} Pa space charge reduces the number of electrons capable of producing ionizing collisions and the apparent sensitivity is reduced. In addition, the mean free path becomes small and ions are scattered before reaching the collector. A high-pressure gauge has been designed by Schulz and Phelps [17], versions of which are marketed by several manufacturers (Fig. 3.15). The close spacing of the electrodes allows this tube to be used at high pressures. Ion generation however, is reduced because the chance for an ionizing collision is proportional to the path length. A typical sensitivity for a Schulz-Phelps tube is 4×10^{-3} /Pa and a typical pressure range is 10^{-4} to 100 Pa. The ability to read lower pressures is again limited by x-ray generated electrons. These tubes are excellent for monitoring chamber pressure during sputtering, reactive-ion etching, and other plasma processes. It is necessary to mount the Schulz-Phelps tube in a way that will prevent it from being affected by optical or other electromagnetic energy radiating from the plasma. This is accomplished by mounting it on an elbow and placing a piece of stainless screen over the end of the elbow at its entrance to the process chamber.

Cold Cathode Gauge

The cold cathode gauge developed by Penning [18] some 40 years ago provides an alternative to the hot cathode gauge which in some respects is superior, but in other respects more limited. The gauge tube illustrated in Fig. 3.16 uses a wire anode loop maintained at a potential of 2–10 kV and grounded cathode electrodes. Surrounding the tube is a permanent magnet of about 0.1–0.2 Tesla.

The arrangement of the electric and magnetic fields causes electrons to travel long distances in spiral paths before finally colliding with the anode. These long trajectories considerably enhance the ionization probability and result in a gauge with a much higher ionization efficiency than the hot cathode gauge. The total current, which is the sum of the electron and positive ion currents, is so much greater than in a hot cathode gauge that a current amplifier is not needed. Output currents of 10–50 mA/Pa are typical.

The range of operation of the cold cathode gauge is $1-10^{-4}$ Pa; gauge operation becomes erratic at low pressures because of the difficulty of maintaining the discharge. Penning and Nienhuis [19] were able to overcome some of the problems in this design by using a cylin-

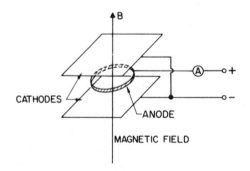

Fig. 3.16 Penning type cold cathode discharge tube.

drical anode with cathode plates at each end and a cylindrical magnet. Cold cathode gauges like hot cathode gauges have sensitivities which vary with gas species and in a similar manner. One advantage of the cold cathode gauge is that it overlaps the range of the hot cathode and thermal conductivity gauges but has the disadvantage of not operating below 10^{-4} Pa. Because of sputtering the nitrogen pumping speed of these tubes is typically in the 0.1–0.5 L/s range—a factor of 10–100 times greater than a hot cathode. Because of this high pumping speed, gauges are fabricated with a large diameter entrance tubulation, typically 30 mm; they should not be connected to a system with tubulation of a smaller diameter or a considerable pressure drop will result. Cold cathode gauges should be mounted in a way that will not allow metal particles to fall inside the tube.

3.2.3 Ultrahigh Vacuum Gauges

The realization that photo electrons emitted from the ion collector are indistinguishable from the positive ions incident on the ion collector led to the development of the Bayard-Alpert ionization gauge and a reduction of almost three orders of magnitude in the minimum detectable pressure. This, in turn, led to the development of a number of gauges with even lower x-ray, field-emission, or photon-limiting currents. Both hot and cold cathode gauges have been developed. Each gauge was designed to screen the ion collector from the pressure-independent current or to increase the electron ionization path length.

Redhead investigated the possibility of using a Penning-type cold cathode gauge for pressures below 10^{-8} Pa because the electron current that produced x-rays was proportional to pressure (no x-ray limit). He developed the inverted magnetron gauge [20,21] which contained auxiliary cathode electrodes. These electrodes shielded the ion collector from the field emission current that limited the minimum detectable pressure of the Penning gauge. The inverted magnetron gauge and the magnetron gauge developed by Redhead [22] have low pressure limits of 10^{-11} Pa. Young and Hession [23] developed the trigger gauge, a cold cathode gauge that is started by momentary operation of a hot cathode. The trigger gauge has a low pressure limit of 5×10^{-10} Pa.

Several hot cathode gauges have also been developed. The modulated Bayard-Alpert gauge designed by Redhead [24] contains an extra electrode wire adjacent to the ion collector. By measuring the ion current at two modulator potentials the effects of the x-ray current may be subtracted. With careful calibration this gauge can measure pressures as low as 10^{-12} Pa. Lafferty [25] devised a high sensitivity hot cathode magnetron gauge with an x-ray limit of 5×10^{-12} Pa. Other hot cathode gauges, which shield the ion current from the x-ray-generated photo current with supressor grids [26], extracting electrodes [27], and bent beams, have also been developed.

The ultrahigh vacuum gauges we have discussed are usually calibrated or referenced to a mass spectrometer which, in turn, is calibrated by a Bayard-Alpert gauge. The Bayard-Alpert gauge is referenced to the primary standard, the McLeod gauge. Not all the ultrahigh vacuum gauges we have described are in widespread use today. Many workers choose to use the nude, enclosed grid ion gauge to its limiting pressure of 2×10^{-9} Pa, and a residual gas analyzer for pressures below that value because a knowledge of the spectrum is usually desired.

3.2.4 Accuracy of Indirect-Reading Gauges

It has already been noted that changes in emissivity and accommodation coefficient in thermal conductivity gauges are large enough to allow these gauges to be used only to indicate the degree of vacuum. Ionization gauges also suffer from inaccuracy but for different reasons.

Redhead [28] has shown that the sensitivity of a Bayard-Alpert gauge varied as much as a factor of 2.5 when the filament-to-grid spacing was varied from 0.5 to 6 mm. This variation resulted from the changing electron orbit length, hence the total ionization was a function of the filament-to-grid spacing. Redhead also observed that a number of nominally identical Bayard-Alpert gauges had sensitivity variations as large as $\pm 15\%$. The grid-to-filament spacing appeared to be the most inadequately controlled dimension in most Bayard-Alpert structures. For this reason a sagging filament induces error. Electron-induced ion desorption is another common source of error in pressure measurement. The high collector current, observed with a Bayard-Alpert gauge after it had been exposed to a relatively high gas pressure, results from desorption of gases chemisorbed on the grid. Some gas molecules are desorbed as ions that strike the collector and result in artificially high collector currents [29–33]. Ion gauges may often be a dominant source of gas in ultrahigh vacuum systems. Ion desorption causes major errors in pressure measurement and residual gas analysis because the conditions of measurement reflect the conditions in the ion gauge or RGA and not the system as a whole. In spite of the problems that relate to construction tolerances, pumping, dissociation, gas generation, and temperature effects, the Bayard-Alpert gauge can give readings that are accurate within 25%; the various ultrahigh vacuum versions are accurate within an order of magnitude. Under suitable laboratory conditions, in which a gauge has been calibrated for a known gas composition, accuracies of a few percent can be obtained [34]. The average gauge is performing well, however, if it reads within 25%. The calibration of high and ultrahigh vacuum gauges has been reviewed by Sellenger [35] and Fowler and Bock [36]. Procedures for hot-filament gauge calibration are given in AVS Tentative Standard 6.4 [37].

REFERENCES

1. J. H. Leck, *Pressure Measurement in Vacuum Systems*, 2nd ed. Chapman and Hall, London, 1964.
2. J. P. Roth, *Vacuum Technology*, North Holland, Amsterdam, 1976.
3. S. Dushman, *Scientific Foundations of Vacuum Technique*, 2nd ed., J. M. Lafferty, Ed., Wiley , New York, 1962, p. 220.
4. A. Guthrie, *Vacuum Technology*, Wiley, New York, 1963, p. 150.
5. D. Alpert, C. G. Matland, and A. C. McCoubrey, *Rev. Sci. Instrum.*, **22**, 370 (1951).
6. J. J. Sullivan, *Ind. Res. Dev.*, January 1976, p. 41.
7. G. Loriot and T. Moran, *Rev. Sci. Instrum.*, **46**, 140 (1975).
8. For example, the Granville-Phillips Convectron Gauge, Series 275, Granville Phillips Co, Boulder CO., or Leybold-Heraeus TR201 gauge, Leybold-Heraeus G.m.b.H., Köln, West Germany.
9. W. Thompson and S. Hanrahan, *J. Vac. Sci. Technol.*, **14**, 643 (1977).
10. W. B. Nottingham, *7th Ann. Conf. on Phys. Electron.*, M.I.T., 1947.
11. R. T. Bayard and D. A. Alpert, *Rev. Sci. Instrum.*, **21**, 571, (1950).
12. W. B. Nottingham, *1954 Vacuum Symp. Trans.*, Comm. Vacuum Techniques, Boston, 1955, p. 76.
13. D. Alpert, *J. Appl. Phys.*, **24**, 7 (1953).
14. T. A. Flaim and P. D. Owenby, *J. Vac. Sci. Technol.*, **8**, 661 (1971).
15. J. Blears, *Proc. R. Soc. London*, **188A**, 62 (1946).
16. Ratio-matic Gauge,® Varian Associates, 611 Hansen Way, Palo Alto, CA.
17. G. J. Schulz and A. V. Phelps, *Rev. Sci. Instrum.*, **28**, 1051 (1957).
18. F. M. Penning, *Physica*, **4**, 71 (1937).
19. F. M. Penning and K. Nienhuis, *Philips Tech. Rev.*, **11**, 116 (1949).
20. P. A. Redhead, *Can. J. Phys.*, **36**, 255 (1958).
21. J. P. Hobson and P. A. Redhead, *Can. J. Phys.*, **36**, 271 (1958).
22. P. A. Redhead, *Can. J. Phys.*, **37**, 1260 (1959).
23. J. R. Young and F. P. Hession, *Trans. 9th Nat. Vacuum Symp.*, Macmillan, New York, 1963, p.234.
24. P. A. Redhead, *Rev. Sci. Instrum.*, **31**, 343 (1960).
25. J. M. Lafferty, *J. Appl. Phys.*, **32**, 424 (1961).

26. W. L. Schuemann, *Rev. Sci. Instrum.*, **34**, 700 (1963) .
27. P. A. Redhead, *J. Vac. Sci. Technol.*, **3**, 173 (1966).
28. P. A. Redhead, *J. Vac. Sci. Technol.*, **6**, 848 (1969).
29. P. A. Redhead, *Vacuum*, **12**, 267 (1962).
30. P. A. Redhead, *Vacuum*, **13**, 253, (1963).
31. T. E. Hartman, *Rev. Sci. Instrum.*, **34**, 1190 (1963).
32. G. Rettinghaus and W. K. Huber, *J. Vac. Sci. Technol.*, **6**, 89 (1969).
33. H. F. Winters, *J. Vac. Sci. Technol.*, **7**, 262 (1970).
34. J. M. Lafferty, *J. Vac. Sci. Technol.*, **9**, 101 (1971).
35. F. R. Sellenger, *Vacuum*, **18**, 645 (1969).
36. P. Fowler and F. J. Bock, *J. Vac. Sci. Technol.*, **7**, 507 (1970).
37. AVS Tentative Standard 6.4, *J. Vac. Sci. Technol.*, **7**, (1970).

CHAPTER 4

Residual Gas Analyzers

Residual gas analyzers have become so common that almost every vacuum technologist has seen one in the laboratory or attached to an occasionally troublesome system on the manufacturing line. In 1960 Caswell [1] used the RGA to study the residual gases in vacuum evaporators, and demonstrated that the performance of a conventional system could be improved by Viton gaskets, Meissner traps, and getters. For thin-film deposition the importance of thoroughly outgassing the source material was stressed. Caswell also used the RGA to study the effects of residual gases on the properties of tin and indium films [2, 3]. Since that time the instrument has proved its value in the solution of device-fabrication problems by measuring other properties such as gas purity, gas loads evolved from films, and background gas composition during sputtering and other plasma processes.

This chapter is concerned with the theory of the operation of magnetic sectors and RF quadrupoles and the methods of installation of and data collection on vacuum and sputtering chambers. The interpretation of data is discussed in Chapter 5.

4.1 INSTRUMENT DESCRIPTION

RGAs and mass spectrometers are used to measure the ratio of mass-to-electric charge of a molecule or atom. First the molecules are ionized, then directed through a mass separator, and finally detected.

See Fig. 4.1. A variety of methods has been developed for each of the three stages of particle identification. Some approaches are sophisticated and are applicable to analytical laboratory mass spectrometers that are capable of differentiating small fractional mass differences; for example, carbon monoxide ions (M = 27.9949 AMU) and molecular nitrogen ions (M = 28.0061 AMU) are easily separable on a double-focusing magnetic sector. (The weight of ^{12}C is 12 AMU.) Other methods have been developed into portable instruments that scan the mass range 1 to 50 AMU or perhaps 1 to 300 AMU and are able to resolve adjacent peaks 1 AMU apart. The latter instruments are collectively referred to as residual gas analyzers. This section reviews briefly the most commonly used ionization, separation, and detection methods in RGAs.

4.1.1   Ionization Sources

Virtually the only technique applicable to the production of positive ions in commercial residual gas analyzers is electron impact ionization. Other techniques such as field ionization and chemical ionization are

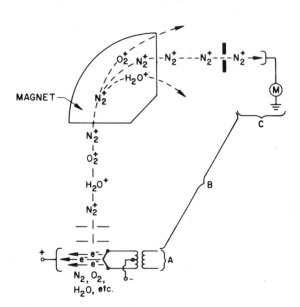

Fig. 4.1 Three stages of partial pressure analysis: (a) Ionization—hot filament illustrated; (b) mass separation—magnetic sector illustrated; and (c) detection—Faraday cup illustrated.

useful in some research applications. Figure 4.2 is an ionization chamber that might be used in a residual gas analyzer. The electrons from the filament are drawn across the chamber to the anode. While crossing this space some of the electrons collide with gas molecules, strip off one or more of their electrons, and create positive ions. Not all ionization chambers are geometrically similar to the one sketched in Fig. 4.2. One instrument looks very much like a Bayard-Alpert ionization gauge except for the absence of the wire collector, and the addition of an electron reflector outside the filament. These and other ionizers were designed to maximize ion production and instrument sensitivity.

As in the ion gauge, positive ion production is not the same for all gases. The RGA differs from the ion gauge in that it sorts ions by their mass-to-charge ratio and counts each ratio separately. Thus for nitrogen the ion gauge makes no distinction between a current due to N^+ ($M/z = 14$) or N_2^+ ($M/z = 28$), whereas the RGA does distinguish the two ion currents. Table 4.1 gives the total positive ion cross sections relative to N_2 for several common gases at an ionizing energy of 70 eV [4]. Although the ionization cross section does not peak at the same energy for all gases, it is generally greatest for most gases somewhere in the 50-to-150-eV range. For this reason most ionizers operate at a potential of 70 eV. Some instruments make provision for the adjustment of the ionizing voltage because it is sometimes desirable to reduce the potential in order to reduce the dissociation of complex molecules. This is essential in qualitative analysis.

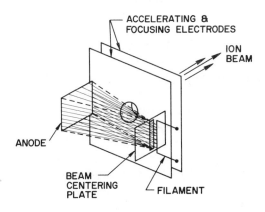

Fig. 4.2 One form of an ionizing source used in a residual gas analyzer.

Table 4.1   Experimental Total Ionization
Cross Sections (70 V) for Selected
Gases Normalized to Nitrogen

Gas	Relative Cross Section
H_2	0.42
He	0.14
CH_4	1.57
Ne	0.22
N_2	1.00
CO	1.07
C_2H_4	2.44
NO	1.25
O_2	1.02
Ar	1.19
CO_2	1.36
N_2O	1.48
Kr	1.81
Xe	2.20
SF_6	2.42

Source. Reprinted with permission from *J. Chem. Phys.*, **43**, p. 1464, D. Rapp and P. Englander-Golden, Copyright 1965, The American Institute of Physics.

The ion production of each species is proportional to its density or partial pressure. Consider a sample of a gas mixture containing only equal portions of nitrogen, oxygen, and hydrogen, whose total pressure is 3×10^{-5} Pa. A mass scan of this mixture would show three main peaks of unequal amplitudes. All other factors being equal, the main oxygen peak would be slightly larger and the hydrogen peak about half as large as the nitrogen peak because of differences in relative sensitivity or ionizer yield. If however, the total pressure of the gas mixture were increased to 6×10^{-5} Pa, the amplitudes of each of the three main peaks would double. In other words, the instrument is linear with pressure. Linearity of the ionizer extends to a maximum total pressure of order 10^{-3} Pa. At higher pressures space charge effects and gas collisions become important. The ions produced in the space between the filament and anode are drawn out of that region, focused, and accelerated toward the mass separation stage. The acceleration energy depends on the type of mass analyzer that follows, and in a magnetic sector instrument ion acceleration is really a part of the mass separation stage.

4.1.2 Mass Separation

Almost a dozen techniques have been developed for mass separation of ions generated by a method like the one just described. For various reasons only two techniques, the magnetic sector and the RF quadrupole have survived the test of widespread commercial development. Those who are interested in a thorough discussion of all types of mass separation schemes are referred to other sources [5]. The more common methods are outlined in Fig. 4.3. This section discusses the quadrupole and magnetic-sector mass separation methods as they are commonly used in residual gas analyzers.

Magnetic Sector

The magnetic sector analyzer, which was developed 60 years ago [6], separates ions of different mass-to-charge ratios by first accelerating the ions through a potential V_a and then directing them into a uniform

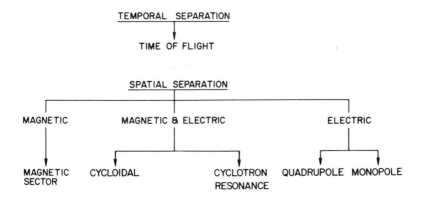

Fig. 4.3 Mass separation methods.

magnetic field perpendicular to the direction of the ion motion. While under the influence of this magnetic field the ions are deflected in circular orbits of radii r given by

$$r = \frac{1}{B}\left(\frac{2mV_a}{ze}\right)^{\frac{1}{2}} \tag{4.1}$$

If B is given in units of teslas, the ion accelerating energy V_a, in volts, the mass M, in mass units, and z is the degree of ionization, the radius of curvature will be

$$r = \frac{1.44 \times 10^{-4}}{B}\left(\frac{MV_a}{z}\right)^{\frac{1}{2}} \text{ meters} \tag{4.2}$$

A practical mass analyzer that uses magnetic separation is shown in Fig. 4.4 for a 60° magnetic sector. In principle, any angle will work, but angles of 180, 90, and 60° are common. The 60° sector is a common filter for RGA applications. It provides sufficient separation between source and collector, good focusing for divergent ions, and requires a minimum amount of magnetic material.

As illustrated in Fig. 4.4, the location of the exit and entrance slits determine the radius r at which the beam will be properly focused. With the radius so specified, the mass-to-charge ratio M/z of the beam in focus is determined by the accelerating potential and the magnetic field-strength. In the example shown a singly ionized molecule $z = 1$ of mass M_2 is focused on the exit slit for $V_a = V_2$ and $B = B_2$. Mas-

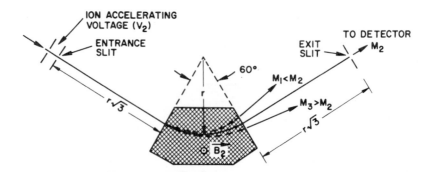

Fig. 4.4 A magnetic sector mass separator (60°) with symmetrical entrance and exit slits. Adapted with permission from *Mass Spectroscopy in Science and Technology*, p. 18, F. A. White. Copyright 1968, John Wiley & Sons.

ses $M_1 < M_2$ and $M_3 > M_2$ will be deflected through greater and lesser angles, respectively, than M_2. To focus mass M_3 on the detector, B must be increased or V_a must be decreased. Commercial RGAs generally use permanent magnets and a variable acceleration voltage. Electromagnets are available to extend the range and provide magnetic scanning.

Equation 4.1 states that for constant r and B, the quantity MV_a/z is a constant. From this it can be seen that sweeping a large mass range, say, $1 \leq M \leq 300$, requires prohibitively large linear sweep voltages. Because of this limitation, permanent magnetic sector analyzers divide the instrument range into at least two scales by changing the magnet; for example, one scale might cover the mass range $2 \leq M < 50$ with a magnet of about 0.1 tesla and a second mass range of $12 \leq M \leq 300$ with a magnet of 0.25 tesla. Traditionally, but not exclusively, such instruments sweep the voltage linearly with time. Because MV_a/z is a constant, the resulting mass scan is not linear with time; as the mass number increases the peak separation decreases. Somewhat more expensive instruments allow the accelerating potential to be held constant while an electromagnet of 0 to .25 tesla sweeps the range $1 \leq M \leq 100$.

Differentiation of (4.1) reveals that the mass dispersion Δx of the instrument is mass dependent. For ions of equal energy traversing a uniform magnetic sector the mass dispersion, or spatial separation between adjacent peaks of mass m and $m + 1$ has been found to be [5],

$$\Delta x \propto \frac{r}{m} \tag{4.3}$$

Equation 4.3 illustrates why instruments of small radii cannot effectively separate adjacent heavy mass peaks. The resolution and sensitivity of a magnetic sector are dependent on mass and exit slit width.

Figure 4.5 shows two idealized mass peaks being scanned with slits of different widths and indicates that wide slits are efficient collectors of ions (high sensitivity) but poor resolvers of adjacent mass peaks (low resolving power). To first order there are no mass-dependent transmission losses in a fixed-radius, magnetic-sector instrument. This means that if equal numbers of, say, hydrogen and xenon molecules pass through the entrance slit, equal numbers will pass through the exit slit. It will be noted later that this is not always true of the quadrupole. Figure 4.6 shows an RGA trace taken with a small sector instrument [7] on a 35-in. oil diffusion pumped system. This trace clearly illustrates the nonlinear nature of the sweep and the mass-dependent resolution. Because the slit width at the detector is fixed and the distance between adjacent mass peaks varies as $1/m$, the valley between two peaks of unit mass difference is not so pronounced at high mass numbers as at low mass numbers. Figure 4.7 was taken on an instrument in which the magnetic field was varied to produce a linearized scan. The same kind of mass dependent resolution is also evident here. More detailed discussions of magnetic sectors are found in a number of texts [8, 9].

RF Quadrupole

The rf quadrupole, developed by Paul [10] and coworkers, is the most popular nonmagnetic mass filter in modern RGAs. Its acceptance has been due, in part, to the development of the necessary stable, high-power quadrupole power supplies. Figure 4.8 illustrates the mass filter geometry and the path of a filtered ion. The ideal electrodes are hyperbolic in cross section. In practice they are realized by four rods of cylindrical cross section located to provide the optimum approximation to the hyperbolic fields. Each of the rods is spaced a distance r_o from the central axis. Mosharrafa [11] has provided a nonmathematical explanation of quadrupole operation. The two rods with positive

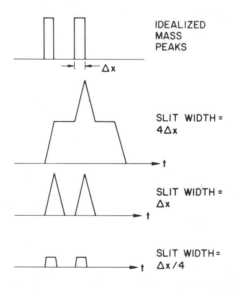

Fig. 4.5 Idealized mass peaks illustrate the trade off between sensitivity and resolution.

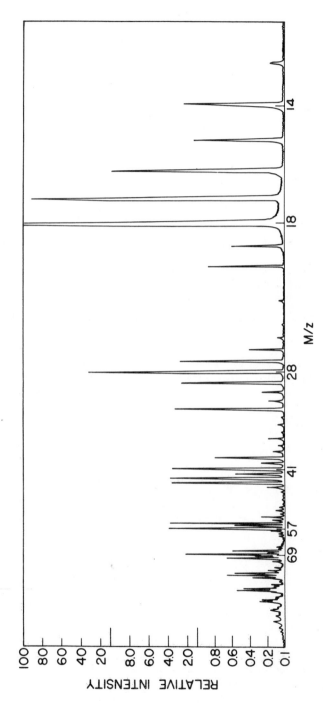

Fig. 4.6 Mass scan from a large diffusion pumped evaporator taken with a magnetic sector instrument with a permanent magnet and voltage sweep.

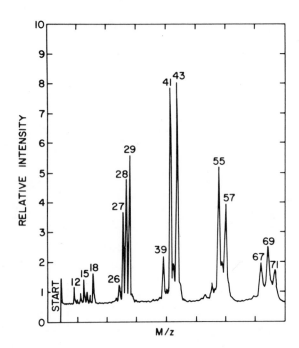

Fig. 4.7 Linear mass scan taken with an Aero Vac 700 series magnetic sector instrument. Reprinted with permission from High Voltage Engineering Corp., Burlington, MA.

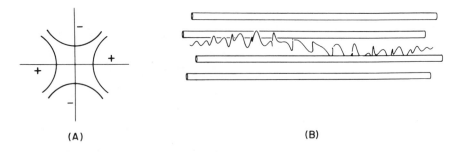

(A)   (B)

Fig. 4.8 Quadrupole mass filter: (a) idealized hyperbolic electrode cross section; (b) three-dimensional computer generated representation of a stable ion path. Courtesy of A. Appel, IBM T. J. Watson Research Center.

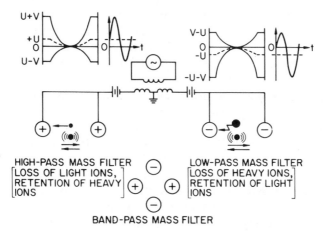

Fig. 4.9   Electric fields in a quadrupole mass filter.   Reprinted with permission from *Industrial Research/Development*, (March 1970), p. 24, M. Mosharrafa. Copyright 1970, Technical Publishing Co.

dc potential, $+U$ in Fig. 4.9, create a potential valley near the axis in which positive ions are conditionally stable.   The potential is zero along the axis of symmetry, shown in the dotted curve of Fig. 4.9. This field is zero only if the potential $-V$ is simultaneously applied to the other pair of quadupole rods.   It is a property of a quadrupole, not a dipole, field.   The addition of an RF field of magnitude greater than the dc field $(U + V \cos \omega t)$ creates a situation in which positive ions are on a potential "hill" for a small portion of the cycle.   Heavy ions have too much inertia to be affected by this short period of instability, but light ions are quickly collected by the rods after a few cycles.   The lighter the ion, the fewer number of cycles required before ejection from the stable region.   This rod pair acts as a "high-pass" filter.

The rod pair with the negative dc potential $-U$ creates a potential "hill" that is unstable for positive ions.   However, the addition of the RF field creates a field $-(U + V \cos \omega t)$ which allows a potential "valley" to exist along the axis of the quadrupole for a small portion of the cycle, provided $V > U$.   In this field light ions are conditionally stable and heavy ions drift toward the electrodes because the potential hill exists for most of the cycle.   This half of the quadrupole forms a "low-pass" filter.

Together the high- and low-pass filters form a band-pass filter that allows ions of a particular mass range to go through a large number of stable, periodic oscillations while traveling in the z-direction.   The width of the pass band, or resolution, is a function of the ratio of dc to RF potential amplitudes U/V, whereas the "sharpness" of the pass band is determined by the electrode uniformity, electrical stability, and ion entrance velocity and angle. A detector is mounted on the z-axis at the filter's exit to count the transmitted ions. Ions of all other M/z ratios will follow unstable orbits and be collected by the rods before exiting the filter. The stability limits for a particular M/z ratio are determined from the solutions of the equations of motion of an ion through the combined RF and dc fields and involve ratios of ω, M/z, r_o^2 and the potentials U and V. A thorough discussion of the RF quadrupole has been given by Dawson [12]. By sweeping the RF and dc potentials linearly in time the instrument can be made to scan a mass range. Scan times as slow as 10 to 20 min and as fast as 80 ms are typically attainable in commercial instruments with a range of $1 \leq M/z \leq 300$. One noticeable distinguishing feature of the quadrupole is that no additional restriction other than linear sweeping of the RF and dc potentials is needed to obtain a graphical display that is linear in mass scan.

Although the stability of the trajectory of an ion may be calculated without consideration of the z-component of the ion velocity or the beam divergence, experimentally the situation is more complicated. There is a reasonable range of velocities and entrance angles that does yield stable trajectories. In the magnetic sector both the ion energy and the magnetic field determine the focus point of an ion. One of the advantages of the quadrupole is that ions with a range of energies or entrance velocities will focus even though not with the same resolution. The slow ions are resident in the filter for a longer time and therefore are subjected to a greater number of oscillations in the RF field than are those ions with larger z-components of velocity. As a result the slow ions are more finely resolved but suffer more transmission losses than the light ions. For this reason quadrupole transmission usually decreases with increasing mass. In a typical instrument adjusted for unity resolution the gain is constant to about $20 < M/z < 50$, after which it decays at the rate of approximately a decade per 150 AMU. See Fig. 4.10. This is only typical; there is considerable instrumental and manufacturer variation. By proper choice of the potentials U and V the mass dependence of the transmission can be considerably improved at the expense of resolution. Experimentally mass independent transmission can be achieved by adjusting the sensitivity for two gases, say, argon and xenon, to be the same. If accurate knowledge of the transmission versus mass is desired, it must be measured for the particular filter and potentials in question. A typical mass scan taken

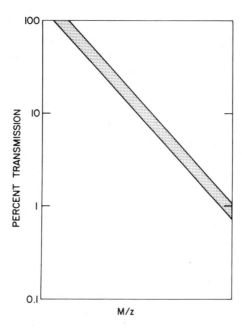

Fig. 4.10 Relative transmission of a typical RF quadrupole as a function of charge-to-mass ratio when adjusted for unity absolute resolution.

on a small oil diffusion pumped system with a quadrupole adjusted for constant absolute resolution is displayed in Fig. 4.11.

Resolution and Resolving Power

The absolute resolution of an RGA is a measure of the ion-separating ability of the instrument of a given mass M and is given by the peak width ΔM. The American Vacuum Society's tentative standard for absolute resolution [13] specifies that the peak width shall be measured at a point equal to 10% of the peak height, (see Fig. 4.12a). The resolving power of an RGA is the ratio of mass to resolution, $M/\Delta M$. In addition to the AVS definition of ΔM, Fig. 4.12 provides some other common definitions. It is evident that the numerical value of the resolving power of a given instrument will vary with the definition. A difference of a factor of two in numerical values between these definitions is not uncommon.

The reason that myriad definitions exist is partly historical and partly because no one definition seems to be suitable for all situations.

Definition (b) in Fig. 4.12 is acceptable for two peaks of equal size but does not cover the trace peak next to a main peak. Definition (c) in the same figure is adequate for peaks of unequal magnitude, whereas definitions a and d do not require adjacent mass peaks to compute resolving power. In RGA work it is necessary to resolve adjacent peaks separated by one mass unit so that the minimum absolute resolution needed is unity. Analytical spectroscopy necessitates the discrimination of mass peaks separated by small fractional mass units. According to definition b in Fig. 4.12 a resolving power of 2000 is needed to distinguish $^{32}S^+$ ($M/z = 31.9720$) from $^{16}O_2^+$ ($M/z = 31.9898$). One important aspect of the definitions of resolving power and resolution has not been adequately emphasized; that is, the definitions are incomplete unless the sensitivity is specified. See (4.4). As illustrated in Fig. 4.5, the sensitivity and resolution of a magnetic sector vary with the slit width, whereas the same parameters are electronically controlled in the RF quadrupole. Because the resolving power can be adjusted over a wide range at the expense of sensitivity, it is misleading to quote the value of only one parameter.

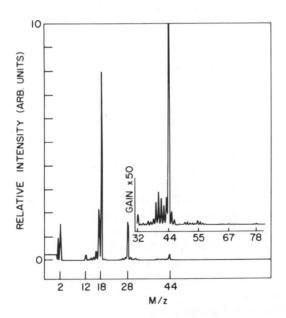

Fig. 4.11 Mass scan taken on a small oil diffusion pumped chamber with an rf quadrupole instrument.

4.1.3   Detection

The ion current detector located at the exit of the mass filter stage must be sensitive to small ion fluxes.   The ion current at mass n is related to the pressure in the linear region by

$$i_n = S'_n P_n \qquad (4.4)$$

where i_n, S'_n, and P_n are, respectively, the ion current, sensitivity of the ionizer and filter, and partial pressure of the nth gas.   A typical sensitivity for nitrogen is 5×10^{-6} A/Pa to 2×10^{-5} A/Pa.   We might ask why the sensitivity is defined with dimensions of current per unit pressure instead of reciprocal pressure as in ion gauge tubes.   The

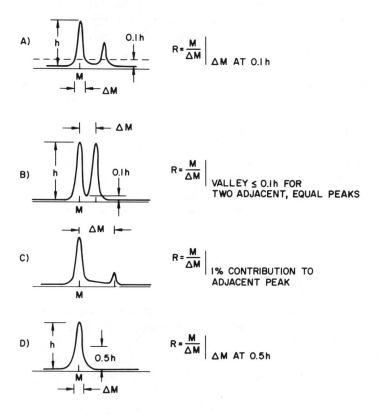

Fig. 4.12   Four definitions of resolving power.

answer is that the ion sources used in RGAs are sometimes space-charge controlled, and their ion current is not linearly proportional to their emission current. Some instruments with high sensitivity use very high emission currents, up to 50 mA, but a typical ionizer with a nitrogen sensitivity of 7×10^{-6} A/Pa will have an emission current of 1 to 5 mA. For an emission current of 1 mA the sensitivity, defined as an ion gauge, would be 7×10^{-3} Pa^{-1}, which is an order of magnitude smaller than that of an ion gauge. The reasons for this low "ion gauge" sensitivity are rooted in the design of the mass analyzer. Not all the ions generated in the ionizer are extracted through the drawing-out electrode nor do all the extracted ions traverse the mass filter.

If we assume an average sensitivity of 10^{-5} A/Pa and a dynamic pressure range of 10^{-1} to 10^{-12} Pa, the ion current at the entrance to the detector can range from 10^{-6} to 10^{-17} A. For the upper half of this range a simple Faraday cup detector, followed by a stable, low-noise, high-gain FET amplifier will suffice, but below 10^{-12} A an electron multiplier is needed. Figure 4.13 illustrates a typical installation in which high gain is required—a combination Faraday cup–electron multiplier. When the Faraday cup is in operation, the first dynode is grounded to avoid interference. When the electron multiplier is used, the Faraday cup is grounded or connected to a small negative potential to improve the focus of the ions as they make a 90° bend toward the first dynode. In quadrupole analyzers the first dynode is generally located off-axis to avoid x-ray and photon bombardment.

Amplification in an electron multiplier is achieved when positive ions incident on the first dynode generate secondary electrons. The secondary electrons are amplified as they collide with each succeeding dynode. The multipliers are usually operated with a large negative voltage (-1000 to -3000 V) on the first dynode. The gain of the multiplier is given by

$$G = G_1 \cdot G_2^n \tag{4.5}$$

where G_1 is the number of secondary electrons generated on the first dynode per incident ion and G_2 is the number of secondary electrons per incident electron generated on each of the n succeeding dynodes. The values of G_1 and G_2 depend on the material, energy, and nature of the incident ion. For a 16-stage multiplier, whose overall gain G is 10^6, $G_2 = 2.37$; for $G = 10^5$, G_2 would be 2.05.

The electron multiplier sketched in Fig. 4.13 typically uses a Cu-2 to -4% Be alloy as the dynode material and when suitably heat treated to

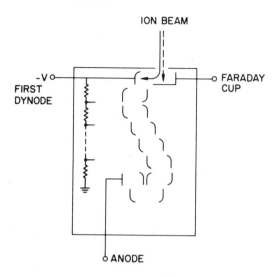

Fig. 4.13 Combination Faraday cup–electron multiplier detector. Reprinted with permission from Uthe Technology Inc, 325 N. Mathilda Avenue, Sunnyvale, CA, 94086.

form a beryllium oxide surface will have an initial gain as high as 5×10^5. These tubes should be stored under vacuum at all times, because a continued accumulation of contamination will cause the gain to decrease slowly. If the tube has been contaminated with vapors, such as halogens or silicone-based pump fluids, the gain will drop below 10^3; at this point the multiplier must be cleaned or replaced. If the multiplier has had only occasional exposure to air or water vapor and has not been operated at high output currents near saturation for prolonged times, its gain can usually be restored with successive ultrasonic cleanings in toluene and acetone, followed by an ethyl alcohol rinse, air drying, and baking in air or oxygen for 30 min at 300°C. Tubes contaminated with silicones cannot be reactivated. Tubes contaminated with chemisorbed hydrocarbons or fluorocarbons often form polymer films on the final stages in which the electron current is great. Such contamination can be quickly removed by plasma ashing.

A channel electron multiplier is illustrated in Fig. 4.14. The structure consists of a finely drawn PbO–Bi$_2$O$_3$ [14] tube typically 5 mm OD, and 1 mm ID. A high voltage is applied between the ends of the tube. The high resistivity of the glass makes it act like a resistor chain which causes secondary electrons generated by incoming ions to be

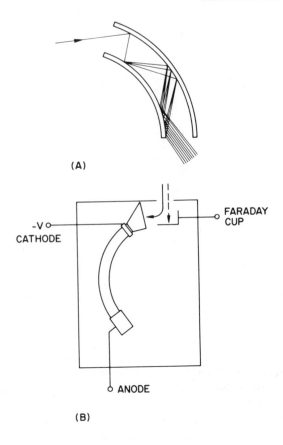

(A)

(B)

Fig. 4.14 Channeltron® electron multiplier: (a) schematic detail of capillary; (b) incorporation into mass analyzer. Reprinted with permission from Galileo Electro-Optics Corp., Galileo Park, Sturbridge, MA 01518.

deflected in the direction of the voltage gradient. The tubes are curved to prevent positive ions generated near the end of the tube from traveling long distances in the reverse direction, gaining a large energy, colliding with the wall, and releasing spurious, out of phase secondary electrons [15, 16]. If the tube is curved, the ions will collide with a wall before becoming energetic enough to release unwanted secondary electrons. An entrance horn can be provided if a larger entrance aperture is needed. A channel electron multiplier can be operated at pressures up to about 10^{-2} Pa and at temperatures up to 150°C; it has a distinct advantage over the Be-Cu type multiplier in

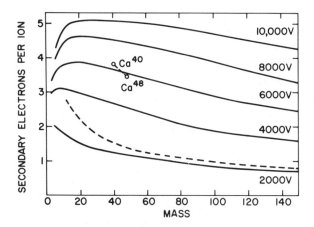

Fig. 4.15 Secondary electron yield of an AgMg dynode as a function of the mass and energy of the impinging ion. Reprinted from *Mass Spectroscopy* (1954), p. 43, with the permission of the National Academy of Sciences, Washington, DC.

that it is relatively unaffected by long or repeated exposure to atmosphere. Because the secondary emission material is a suboxide, its emission is adversely affected by prolonged operation at high oxygen pressures. It seems to be less affected by some contaminants than a Be-Cu multiplier. Under normal operating conditions the gain of either multiplier will degrade in one or two years to a point at which replacement or rejuvination is necessary. Little increase in gain will be realized by increasing the voltage on a multiplier whose gain is less than 1000.

Although channel electron multipliers offer a significant advantage in their ability to accept air exposure, they saturate at a lower current than Be-Cu multipliers and cannot be operated with a linear output at high pressures unless the operating voltage or the emission current is reduced. The manufacturer's literature should be consulted for the precise values of saturation current and range of linearity before any electron multiplier is used. The gain of a multiplier is not a single-valued function. Ions illuminate the first dynode and electrons collide with each succeeding dynode. The gain of the latter dynodes G_2 is dependent on the interstage voltage and the dynode material, whereas the gain of the first dynode G_1 is more complex. The gain of the first stage is primarily dependent on the material, mass, and the energy of the impinging ion, as illustrated in Fig. 4.15 for an AgMg dynode [17]. Ideally, the gain is proportional to m^k, the dotted curve in Fig. 4.15, but this is valid only for ions of low energy and high mass. The gain of any multiplier is dependent on several other factors. Stray magnetic fields can distort the path of the electrons and complex molecules may dissociate on impact, to produce more electrons than a simple compound or element of the same mass. Isotopes and doubly ionized molecules also react somewhat differently than the curve in Fig. 4.15; for example, the gains of ^{40}Ca and ^{48}Ca are different because of the variations in binding energy of the outer electron [17].

If the gain of a specific electron multiplier needs to be accurately known, a calibration curve of gain versus mass must be experimentally measured for all gases and vapors of interest. The gain of an electron multiplier is never constant. It may be measured by taking the ratio of the currents from the Faraday cup and the electron multiplier output, a tedious process, done only when semiquantitative analysis is required. Even then the gain of the multiplier must be periodically checked with one major gas to account for day-to-day aging of the tube. Only periodic checking of the gain at mass 28 with N_2 is necessary for ordinary residual gas analysis work.

4.2 INSTALLATION AND OPERATION

4.2.1 High Vacuum

For routine gas analysis work the RGA head is mounted on a work chamber port. A valve is usually placed between the head and the chamber to keep the electron multiplier clean when the chamber is frequently vented to air. This is especially important with a Be-Cu electron multiplier. The head should be positioned to achieve maximum sensitivity in the volume being monitored. If the instrument is to monitor beams or evaporant streams, a line-of-sight view is necessary. It is often worthwhile to mount the head inside the chamber and shield its entrance so that the beam impinges on a small portion of the ionizer. This shielding will reduce the contamination of the ionizer and associated ceramic insulators.

Bayard-Alpert ion gauge tubes should be turned off when the RGA is operated. In the high vacuum region surface desorption by ion bombardment near the ion gauge will cause increases in all background gas levels. In the ultrahigh vacuum region inert gases are ionically pumped, whereas hydrogen and mass 28 are desorbed by electron bombardment [18]. If the system is to be baked or operated at an elevated temperature, the mass head should be baked at the same temperature or higher to avoid contamination from the condensable vapors that collect on the coolest surfaces. The necessity of baking will affect the choice of electron multiplier. Mass heads that contain channel electron multipliers can be baked to 320°C and operated at temperatures up to 150°C, whereas those with a Be-Cu multiplier can usually be baked to about 400°C and operated at temperatures up to 185°C. These temperature limitations are due to the materials used in channel electron structures and the glass encapsulated resistors in Be-Cu multiplier chains. Further contamination of the head and multiplier can be prevented in diffusion pumped systems by avoiding silicone-based pump fluids in favor of polyphenylethers or perfluoro-polyethers. Silicone-based fluids will polymerize under electron bombardment to form insulating layers; therefore the absence of fluid fragments from the spectrum does not imply the absence of polymer films on a surface. During the initial operation of an RGA in an unbaked system heat dissipation by the filament will warm nearby surfaces and cause them to outgas. After 30 to 60 min of operation these surfaces will equilibrate thermally. Ceramic insulators will often exhibit a "memory effect" in which they continuously evolve fragments of hydrocarbons, fluorides, or chlorides after having been exposed to their vapors. This memory effect often confuses analysis when a mass head is moved from one system to another. Ion sources are easily cleaned by vacuum firing at 1100°C.

Filaments made from tungsten, thoriated iridium, and rhenium are available for RGA use. Tungsten filaments are recommended for general purpose work, although they generate copious amounts of CO and CO_2. Rhenium filaments do not consume hydrocarbons. Thoriated iridium filaments are used for oxygen environments although their emission characteristics are easily changed after contamination by hydrocarbons or halocarbons. As in the ion gauge, the thoriated iridium or "non-burnout" filaments are advantageous when a momentary vacuum loss occurs. Table 4.2 gives some of the properties of three filament materials as tabulated Raby [19]. Molecular hydrogen dissociates at a temperature of 1100 K [20]. This can be avoided by the use of a lanthanum hexaboride filament operating at 1000 K [21].

An oscilloscope and a chart recorder simplify data taking. The oscilloscope is handy for observing the desired mass range and gain before plotting and necessary for using the instrument to detect transient leaks. It is important to remember that the electron multiplier may saturate at pressures above 10^{-3} Pa if the applied voltage is too large. The gain of the multiplier should be reduced by decreasing the voltage so that the largest ion current is in the multiplier's linear range.

Table 4.2   Some Properties of RGA Filament Materials

Property	ThO_2	W/3% Re	Re
CO production	Unknown	High	High
CO_2 production.	High	Mod.	Mod
O_2 consumption	Low	High	High[a]
H-C consumption	Unknown	High	Low
Water entrapment	Maybe high	Low	Low
Volatility in O_2	Low	High	V. high[b]
Good filament for ...	Nitrogen oxides	Hydrogen	Hydrogen
	Oxygen	Hydrogen halides	Hydrocarbons
	Sulfur oxides	Halogens	Hydrogen halides
		Halocarbons[c]	Halocarbons[c]
			Halogens

Source. Reprinted with permission from Uthe Technology Inc., 325 N. Mathilda Ave., Sunnyvale, CA 94086.
[a] Loss of one filament caused O_2/N_2 ratio to increase 17.6%.
[b] Exposure to air at 10^{-4} Pa caused failures of the filament pair at 30 and 95 h.
[c] Freons, etc.

Because of the long time constants used in low-current, high-gain amplifiers, measurements of extremely small signals must be made slowly on a chart recorder at speeds of about 1 AMU/s or less. The turbomolecular pump should be included in the usual grounding procedures used with all high-gain amplifiers.

4.2.2   Differentially Pumped

It is sometimes desired to sample the gas in a system operating at a pressure higher than can be tolerated by the RGA, one in which sputtering or ion etching is performed. This can be done by differentially pumping the RGA. Honig [22] has given three primary conditions for gas analysis with a differentially pumped RGA: (1) The beam intensity should be directly proportional to the pressure of the gas in the sample chamber but should be independent of its molecular weight. (2) In a gas mixture the presence of one component should not affect the peaks due to another component. (3) The gas flow through the orifice should be constant during the scan. These conditions are satisfied if there is good mixing in the chamber, if the leaks are in molecular flow and if the pump is throttled.

Examine the model of the differentially pumped system sketched in Fig. 4.16. In this model P_c is the pressure in the sputtering system, P_s is the pressure in the spectrometer chamber, and P_p is the pressure in the auxiliary pump whose speed is S_p. The conductances C_1 and C_2 which connect these chambers are schematically shown as capillary tubes, but in practice they may be tubes or small diameter holes in thin plates. The gas flow through the auxiliary system which is everywhere the same leads to the following equation.

$$C_1(P_c - P_s) = C_2(P_s - P_p) = S_p P_p \qquad (4.6)$$

When P_p is eliminated, we obtain

$$P_s = \frac{P_c}{1 + \left(\dfrac{C_2}{C_1}\right)\left(\dfrac{S_p}{S_p + C_2}\right)} \qquad (4.7)$$

Now consider two cases: Case 1: let the conductance C_2 be much larger than the speed of the pump. The auxiliary pump is located in or immediately adjacent to the spectrometer chamber with no interconnecting conductance. Equation 4.7 then reduces to

$$P_s = \frac{P_c}{1 + (S_p/C_1)} \qquad (4.8)$$

The conductance C_1 has a mass dependence which varies as $m^{\frac{1}{2}}$, whereas the pumping speed has a mass dependence that is a function of the pump type; for example, an ion pump has different pumping

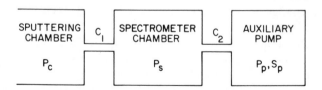

Fig. 4.16 Analysis of a differentially pumped residual gas analyzer.

speeds for noble and reactive gases and diffusion pump speeds increase somewhat at low mass numbers but not as $m^{1/2}$. The important result of this calculation is that the ratio of the gas pressure in the spectrometer to the pressure in the chamber, P_s/P_{c}, is mass dependent when the auxiliary pump is appended directly onto the spectrometer chamber. Gases will not exist in the spectrometer region in the same proportion as in the chamber. Honig's first criterion will therefore not be satisfied.

Case 2 considers the situation in which a small conductance C_2 is placed between the spectrometer and pump such that $C_2 \ll S_p$. For this condition (4.7) becomes

$$P_s = \frac{P_c}{1 + (C_2/C_1)} \tag{4.9}$$

Light gases will still pass from the sputtering chamber to the spectrometer chamber more rapidly than heavy gases, but so will they exit to the pump. Stated another way, the mass dependences of C_1 and C_2 are the same and negate each other so that (4.9) is mass independent. Therefore to sample the ratios of gases in the sputter chamber accurately, the auxiliary pump should be throttled to about $1/10$ of its speed. If this is done, the pressure ratio of two gases in the chamber (P_{AC}/P_{BC}) will be the same as their ratio in the spectrometer (P_{AS}/P_{BS}).

The orifice molecular flow ratio is given by

$$\frac{N_A}{N_B} = \left(\frac{M_B}{M_A}\right)^{1/2} \frac{P_A}{P_B} \tag{4.10}$$

Therefore the ratio of the two gases in the sputtering chamber will eventually change as the lighter gas is selectively removed at a faster rate unless the gas in the sputtering chamber is in dynamic equilibrium. If the gas throughput to the main sputtering pump is much larger than the flow through the aperture, the steady-state composition of the gas in the sputtering chamber will not be affected by the auxiliary pump. Sullivan and Busser [23] note that the time constant of the spectrometer chamber (V_s/C_2) must be equal to or less that the reaction time of the process to obtain time dependent information concerning the process.

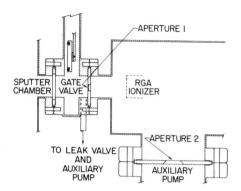

Fig. 4.17 Installation of a differentially pumped RGA on a sputtering chamber.

The RGA may be connected to the working chamber with a valve that will permit monitoring the background when the system is evacuated to the high vacuum range and a parallel leak valve for use when the chamber is being operated at sputtering pressures. In this manner the RGA can be used to monitor rough gas composition in the chamber during the process. For those situations in which high sensitivity is desired the leak valve may take the form of a capillary tube or a small hole in a plate located in front of a clear aperture gate valve, (Fig. 4.17). The size of the aperture should be small enough to keep the gas in molecular flow: for example, at a work chamber pressure of 2 Pa (15 mTorr), the mean free path is 3 mm. If the aperture diameter is made to be equal to or less than 0.1 λ (0.3 mm), the flow of gas will be completely molecular. To maintain the spectrometer chamber at a pressure of 1.3×10^{-3} Pa (10^{-5} torr), and the pump at a pressure of 1.3×10^{-4} Pa (10^{-6} Torr) the conductance C_2 must be 17 L/s, which is equivalent to a 13.5-mm diameter aperture; the pump speed required is 150 L/s. For these values (4.7) is essentially independent of mass. The flow from the sputtering chamber may be increased by drilling several holes to allow the flow to remain molecular in each hole. A reduced flow is possible for use with small auxiliary pumps by choosing an orifice or a capillary tube for C_1 that is smaller than that in the above example. The capillary tube has the advantage of beaming the sample gas to the ionizer and increasing the signal-to-noise ratio. For high chamber pressures (atmospheric) the aperture is replaced by a porous plug. Small triode ion pumps are sometimes used for differentially pumping the RGA. They are convenient but not suitable.

One problem is that the amount and kind of gas desorbed from the walls of an ion pump is a function of the quantity of sputtering gas being pumped; for example, the amount of H_2 backstreaming is a function of sputter chamber pressure. This creates a high background for hydrogen and some other gases. Diffusion pumps suffer from the problem that some gases, caught and slowly desorbed on the cold trap yield a steady state background level that is neither constant nor reliable. Argon is one of the gases subject to this memory effect, but turbomolecular pumps show no evidence of it and are recommended for this application.

Residual gas analysis is then performed by recording a process and a background spectrum. The background spectrum is obtained under the conditions of gas flow in the sputtering chamber and into the RGA at the desired sputtering pressure. The process spectrum is taken with the discharge operating. Because the pressures in the RGA are lower than in the work chamber, the sensitivity of the instrument is similarly reduced. The large pressure reduction factor (1500 for the above example) and the ever-present background gases reduce the detection limit of the RGA. If, in this example, the minimum detectable partial pressure of nitrogen were 10^{-6} Pa in the spectrometer after subtracting the background, the minimum detectable pressure of nitrogen in the sputter chamber would be 1.5×10^{-3} Pa, or 0.1% of the sputter gas pressure. The minimum detectable partial pressure is usually limited to $\sim$ 0.2 to 0.5% of the sputter gas pressure for an unbaked auxiliary chamber pumped by an ion pump. The ultimate detection sensitivity is determined by the pressure reduction factor and the minimum detectable pressure in the analyzer region. If the RGA were capable of detecting a partial pressure of 10^{-9} Pa and the pressure reduction factor between the sputtering chamber and the ionizing chamber were 1500, a partial pressure of 1.5×10^{-7} Pa (1 part in 10^6) would be detectable in the sputtering chamber. The unbaked ion pumped system described above cannot reach this limt for several reasons. Visser [24] showed that vapors such as pump fluids or water cannot be detected easily because they sorb on the walls of the leak valve and analyzer chamber. Heating the walls will stop this sorption but will increase the background pressures of other gases. Gases that do not sorb also present problems. Argon ionized in the analyzer will collide with nearby surfaces and release gases and vapors—even those surfaces that have been thoroughly baked. These problems can be avoided by using a turbomolecular pump, an aperture, a chopper, a synchronous detector or energy filter, and a line-of-sight path to the ionizer. If necessary, a hole may be cut in the ionizer grid. In this manner only those molecules traversing the space between the aperture and ionizer without collision are counted. With a phase sensitive or energy selective technique such as this it is possible to achieve a detection sensitivity of $1:10^4$ for a gas like hydrogen and a sensitivity of $1:10^7$ for a gas or vapor with no background contribution.

Differentially pumped gas analysis is a useful technique for determining the time constants of contaminant decay during presputtering and for sampling the background gases during sputtering. Coburn [25] has used a carefully constructed differentially pumped mass spectrometer to sample the ionization in a sputtering environment. A small hole in the anode allowed study of the energy distribution of ionized plasma particles and the thickness profile of multilayered films located on the cathode. Visser [24] demonstrated an alternative to differentially pumped gas analysis. A mini-quadrupole was operated directly in the glow discharge; this technique however is applicable only at pressures of ≤ 0.5 Pa.

4.3 INSTRUMENT SELECTION

The instrument chosen for residual gas analysis work must be simple and reliable and have adequate resolution, sensitivity, and mass range. A variety of instruments is available from the elementary sector or quadrupole, with a mass range of 1 to 50 AMU, a Faraday cup detector, and a resolving power of 1 to 3 mass units. Other instruments have mass ranges up to 800 AMU. The only differences are complexity and cost.

For simple monitoring of background gases in regularly cycled, nonbakable high-vacuum systems the simplest of instruments with a mass range of 1 to 50 AMU and a resolving power of 1 mass unit is adequate. With such an instrument the dominant fixed gases up to mass 44 and hydrocarbons at mass numbers 39, 41, and 43 may be monitored. Units with Faraday cup detectors and a sensitivity of 10^{-8} Pa are commercially available. Some instruments have an alarm that alerts the operator when a particular mass number has exceeded a preset ion current.

More detailed residual gas analysis with a sector or a quadrupole requires both more sensitivity and a greater mass range. Partial pressures of 10^{-12} Pa are detectable with an electron multiplier. A mass range of at least 1 to 80 AMU is necessary to distinguish most pump oils, but some solvents such as xylene have major peaks in the mass range 104 to 106 AMU. A mass range of 1 to 200 AMU permits identification of many heavy solvents. A resolving power of one mass unit is desired. The ability to detect partial pressures in the 10^{-10} to 10^{-12} Pa range is needed to leak-check and analyze the background of ultrahigh vacuum systems and to do serious semiquantitative analysis. An electron multiplier is necessary for differentially pumped systems. Instruments for these purposes need both the Faraday cup and an electron multiplier to measure the gain and to work over a wide pressure range.

The recent microprocessor revolution has resulted in a variety of instruments that provides automatic control and information display. The operator convenience is a definite asset, but the accuracy with which data are often displayed can seduce the user into believing that the data are indeed that well known. An RGA is limited by the stability of the ionizer, mass filter, and detector, as well as the history and cleanliness of the instrument, and *not* by the accuracy to which data can be presented.

REFERENCES

1. H. L. Caswell, *J. Appl. Phys.,* **32,** 105 (1961).
2. H. L. Caswell, *J. Appl. Phys.,* **32,** 2641 (1961).
3. H. L. Caswell, *IBM J. of Res. and Dev.,* **4,** 130 (1960).
4. D. Rapp, and P. Englander-Golden. *J. Chem. Phys.,* **43,** 1464 (1965).
5. F. A. White, *Mass Spectrometry in Science and Technology,* John Wiley, New York, 1968, p. 13.
6. A. J. Dempster, *Phys. Rev.,* **11,** 316 (1918).
7. Aero Vac Model 610 Magnetic Sector, Aero Vac Products, High Voltage Engineering Corp., Burlington, MA.
8. G. P. Barnard, *Modern Mass Spectrometry,* The Institute of Physics, London, 1953.
9. C. A. McDowell, *Mass Spectrometry,* McGraw-Hill, New York, 1963.
10. W. Paul and H. Steinwedel, *Naturforsch,* **8A,** 448 (1953).
11. M. Mosharrafa, *Industrial Research/Development,* (March 1970), p. 24.
12. P. H. Dawson, Ed., *Quadrupole Mass Spectrometry and its Applications,* Elsevier,n New York, 1976.
13. American Vacuum Society Standard (tentative), AVS 2.3-1972, *J. Vac. Sci. Technol.,* **9,** 1260 (1972).
14. U.S. Patent #3, 492,523.
15. J. G. Timothy and R. L. Bybee, *Rev. Sci. Instrum.,* **48,** 292 (1977).
16. E. A. Kurz, *Am. Lab.,* March 1979, p. 67.
17. M. G. Inghram and R. J. Hayden, National Academy of Science, National Research Council, Report #14, Washington, D.C., 1954.
18. G. A. Rozgonyi and J. Sosniak, *Vacuum,* **18,** 1 (1968).

19. B. Raby, UTI Technical Note 7301, May 18, 1977.
20. T. W. Hickmott, *J. Vac. Sci. Technol.,* **2,** 257 (1965).
21. T. W. Hickmott, *J. Chem. Phys.,* **32,** 810 (1960).
22. R. E. Honig, *J. Appl. Phys.,* **16,** 646 (1945).
23. J. J. Sullivan and R. G. Busser, *J. Vac. Sci. Technol.,* **6,** 103 (1969).
24. J. Visser, *J. Vac. Sci. Technol.,* **10,** 464 (1973).
25. J. Coburn, *Rev. Sci. Instrum.,* **41,** 1219 (1970).

CHAPTER 5

Interpretation of RGA Data

The method of interpreting mass scans like those in Figs. 4.6, 4.7, and 4.11 is based on detailed knowledge of the cracking or fragmentation patterns of gases and vapors found in the system. The degree of difficulty encountered depends on the type and quality of information sought. After becoming familiar with an RGA, qualitative analysis of many major constituents is rather straightforward, whereas precise quantitative analysis requires careful calibration and complex analysis techniques. This section discusses cracking patterns, some rules of qualitative analysis, and methods of determining the partial pressures of the gases and vapors in the spectrum.

5.1 CRACKING PATTERNS

When molecules of a gas or vapor are struck by electrons whose energy can cause ionization, fragments of several mass-to-charge ratios are created. The mass-to-charge values are unique for each gas species, whereas the peak amplitudes are dependent on the gas and instrumental conditions. This pattern of fragments, called a cracking pattern, forms a fingerprint that may be used for absolute identification of a gas or vapor. The various peaks are primarily created by dissociative ionization, isotopic mass differences, and higher ionization states. The cracking pattern of methane CH_4, illustrated in Fig. 5.1, shows that in addition to the parent ion CH_4^+ the energetic electrons crack the molecule into lighter fragments, CH_3^+, CH_2^+, CH^+, C^+, H_2^+, H^+.

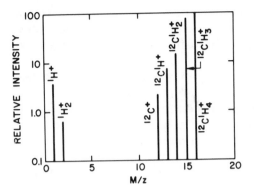

Fig. 5.1 Cracking pattern of methane illustrates the dissociative ionization peaks.

Fragments containing ^{13}C are not shown. This cracking pattern is produced by dissociative ionization. Figure 5.2 illustrates the multiplicity of isotopic peaks in the mass spectrum of singly ionized argon. By comparing the peak heights of the isotopes with the relative isotopic abundances given in Appendix D, we see that this is a spectrum of naturally occurring argon. The relative isotopic peak heights observed on the RGA will generally mirror those given in Appendix D unless the source was enriched or the sensitivity of the RGA was not constant over the isotopic mass range. Higher degrees of ionization are also

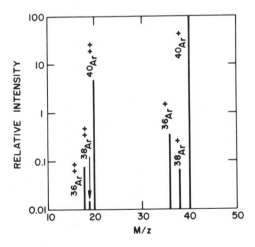

Fig. 5.2 Cracking pattern of argon, adapted from Appendix E.2.

visible in the argon spectrum. Argon has three isotopes of masses 36, 38, and 40; the doubly ionized peaks Ar^{++} show up at $M/z = 18$, 19, and 20. The cracking patterns of heavy metals may show triply ionized states.

The cracking pattern of a somewhat more complex molecule CO, given in Appendix E.2, is illustrated in Fig. 5.3 to show the combined effects of isotopes, dissociation, and double ionization. The amplitude of the largest line $^{12}C^{16}O^{+}$ has been normalized to 100, whereas the amplitude of the weakest line shown, $^{16}O^{++}$, is 10^6 times smaller. The spectrum, as sketched in Fig. 5.3, can be observed only under carefully controlled laboratory conditions. During normal operation of an RGA the spectrum will be cluttered with other gases and only the four or five most intense peaks will be identifiable as a part of the CO spectrum. Even then some of the peaks will overlap the cracking patterns of other gases. Under the assumption that this clean spectrum has been obtained, it is instructive to classify the eight lines according to their origin. The main peak at mass 28 is due to the single ionization of the dominant isotope $^{12}C^{16}O$. Energetic electrons decompose some of these molecules into two other fragments $^{12}C^{+}$, and $^{16}O^{+}$. Because carbon has two isotopes and oxygen three, several isotopic combina-

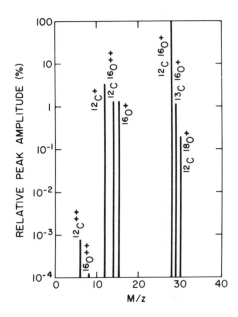

Fig. 5.3 Cracking pattern of CO, adapted from Appendix E.2.

tions are possible. The two most intense lines are $^{13}C^{16}O^+$ and $^{12}C^{18}O^+$. Isotopic fragments such as $^{13}C^+$ and $^{18}O^+$ are too weak to be seen here. Other complexes, for example, $^{12}C^{16}O^{++}$ ($M/z = 14$), $^{16}O^{++}$ ($M/z = 8$), and $^{12}C^{++}$ ($M/z = 6$), have high enough concentrations to be seen.

The relative amplitudes of the eight lines of CO are determined by the source gas and instrument. If isotopically pure source gas were used (^{12}C and ^{16}O), only six peaks would be found. Changes in the operating conditions of the instrument also affect the relative peak heights. The ion temperature and electron energy affect the probability of dissociation and the formation of higher ionization states. Dissociation of a molecule into fragments also changes the kinetic energy of the ion fragment. This can be a serious problem in a magnetic sector instrument because the kinetic energy of the ion directly affects the focusing and dispersion of the instrument. Quadrupoles can focus ions with a greater range of initial energies and therefore are less sensitive to transmission losses of fragment ions than sector instruments.

Electron- and ion-stimulated desorption of atoms from surfaces [1, 2] can also add to the complexity of the cracking pattern. Oxygen, fluorine, chlorine, sodium, and potassium are some of the atoms that can be released from surfaces by energetic electron bombardment. In a magnetic sector instrument energetic oxygen or fluorine will often occur at fractional mass numbers of mass 16⅓ and 19¼ respectively [3]. These peaks are not representative of gaseous oxygen and fluorine in the chamber but of molecules desorbed from the walls. They occur at fractional mass numbers because they are formed at energies different from those corresponding to the ionized states of free molecules [1] and leave the surface with some kinetic energy [3]. The generation of gases resulting from the decarburization of tungsten filaments can add other spurious peaks to the cracking pattern.

Representative cracking patterns are given in the appendix. Appendixes E.2 and E.3 give the cracking patterns of some common gases and vapors, and Appendix E.4 contains the patterns of frequently encountered solvents. Appendix E.5 describes the patterns of gases used in semiconductor processing and partial cracking patterns of six pump fluids are given in Appendix E.1. The patterns in the appendices are intended to be representative of the substances and are not unique. It cannot be emphasized too strongly that each pattern is quantitatively meaningful only to those who use the same instrument under identical operating conditions. Nonetheless, there are enough similarities between patterns to warrant their tabulation.

5.2   QUALITATIVE ANALYSIS

Perhaps the most important aspect of the analysis of spectra for the typical user of an RGA is qualitative analysis; that is, the determination of the types of gas and vapor in the vacuum system. In many cases the existence of a particular molecule points the way to fixing a leak or correcting a process step. The quantitative value or partial pressure of the molecular species in question usually does not need to be known because industrial process control is frequently done empirically. The level of a contaminant for example, water vapor, which will cause the process to fail, is determined experimentally by monitoring the quality of the product. An inexpensive RGA tuned to the mass of the offending vapor is then used to indicate when the vapor has exceeded a predetermined partial pressure. With experience many gases, vapors, residues of cleaning solvents, and traces of pumping fluids will be recognizable without much difficulty.

The mass spectra shown in Figs. 4.6, 4.7, and 4.11 contain considerable information about the present condition as well as the history of the systems on which they were recorded. To help in their interpretation examine the hypothetical spectrum in Fig 5.4   which was constructed from five gases H_2O, N_2, CO, O_2, and CO_2 in the ratio (20:4:4:1:1) from the cracking patterns in Appendixes E.2 and E.3. It is interesting to study this pattern under the assumption that its origin and composition are not known. Examination of the cracking patterns of common gases, like those tabulated in Appendixes E.2 and E.3, quickly verifies the presence of carbon dioxide, oxygen, and water vapor. Notice that the mass 32 peak is not due to the dissociation of carbon dioxide. The presence of oxygen at 32 AMU usually indicates an air leak unless it is being intentionally introduced. Analysis of the mass 28 peak is not so clear. Some of this peak is certainly due to the dissociation of CO_2; however, if we assume that the sensitivity of the instrument is reasonably constant over the mass range in question, that contribution cannot be very great. The majority of the peak amplitude would then be attributable to N_2 or CO or both. To distinguish these gases further the amplitudes of the peaks at mass $16(O^+)$, mass $14(N^+, N_2^{++}, CO^{++})$, and mass $12(C^+)$ are examined. In practice it is difficult to conclude much from the presence of carbon because it originates from so many sources, both organic and inorganic. The mass 14 peak is largely due to N^+, therefore nitrogen is definitely present. Analysis of the mass 16 peak is complicated by the fact that there are other sources of atomic oxygen beside CO, namely O_2 and CO_2, as well as electron-stimulated desorption of O^+ from the walls. Oxygen desorp-

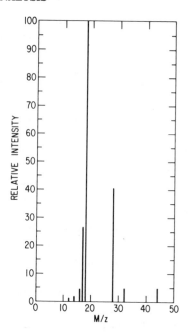

Fig. 5.4 Background spectrum constructed from a (20:4:4:1:1) mixture of H_2O, N_2, CO, O_2, and CO_2.

tion is a common phenomenon. Referring to the cracking pattern tables, we see that the mass 16 peak looks too large to be accounted for totally by the dissociation of CO_2 and O_2, whereas the mass 14 peak looks too small to be only a fragment of nitrogen. We then conclude that both nitrogen and carbon monoxide are present but in undetermined amounts.

After some familiarity with the combined effects of these common background gases they should be easily identifiable in the spectra shown in Figs. 4.6, 4.7, and 4.11. The large peaks at mass 32, 28, and 14 seen in Fig. 4.6 are an indication of an air leak, and the fact that water vapor is the dominant gas load in an unbaked vacuum system is verified by the large mass 18 peak shown in Figs. 4.6 and 4.11. These spectra also show fragments that are characteristic of organics as well as other fixed gases. One way to determine the nature of the organics in the system qualitatively is to become familiar with the cracking patterns of commonly used solvents, pump fluids, and elastomers.

The cracking patterns of several solvents are listed in Appendix E.4. One characteristic shared by all organic molecules is that the probability of fragmentation in a 70 eV ionizer is so great that the parent peak is rarely the most intense peak in the spectrum. The lighter solvents such as ethanol, isopropyl and acetone have fragment peaks that bracket nitrogen and carbon monoxide at mass 27 and 29, but each has a prominent peak at a different mass number. Methanol and ethanol both have a major fragment at mass 31; they can be distinguished by the methanol's absence of a fragment at mass number 27. Solvents that contain fluorine or chlorine have characteristic fragments at mass numbers 19, 20, and 35, 36, 37, 38, and respectively. The extra fragments at mass numbers 20 (HF) and 35–38 (HCl) seem to be present, whether or not the solvent contains hydrogen. Fragments due to CF, CCl, CF_2, and CCl_2 are also characteristic of these compounds. As with all fragments, their relative amplitudes will vary with instrumental conditions. Although the fragments of mass number 27, 29, 31, 41, and 43 are prominent in these solvents, they are also common fragments of many pump fluids, which further complicates the interpretation of a spectrum.

Appendix E.1 lists the partial cracking patterns of six common pump fluids. All the fragment peaks up to mass number 135 are tabulated. For the sector instrument the largest peak (100%) often occurs at a higher mass number and therefore is not shown. The complete spectra of the four fluids which were taken on the sector instrument were tabulated by Wood and Roenigk [4]. Most organic pump fluid molecules are quite heavy and the parent peaks are not often seen in the system because only the lighter fragments backstream through a properly operated trap. Saturated straight-chain hydrocarbon oils are characterized by groups of fragment peaks centered 14 mass units apart and coincide with the number of carbon atoms in the chain. Figure 5.1 shows the fragment peaks for group C_1 in the mass range 12 to 16. Higher carbon groups have similar characteristic arrangements; C_2 (mass numbers 24–30), C_3 (mass numbers 36–44), C_4 (mass numbers 48–58), and C_5 (mass numbers 60–72), and so on. The spectrum of Apiezon BW diffusion pump oil taken by Craig & Harden [5] and shown in Fig. 5.5 illustrates these fragment clusters. In most hydrocarbon oils the fragments at mass numbers 39, 41, 43, 55, 57, 67, and 69 are notably stable and their presence is a guarantee of hydrocarbon contamination. These odd-numbered peaks, which are more intense than the even-numbered peaks in straight-chain hydrocarbons, clearly stand out in the mass scans shown in Figs. 4.6 and 4.7. Some traces of mechanical pump oil are seen in Fig. 4.11. Hydrocarbon oils are used in most rotary mechanical pumps, except those for oxygen or corrosive gas service, and in diffusion pumps for many

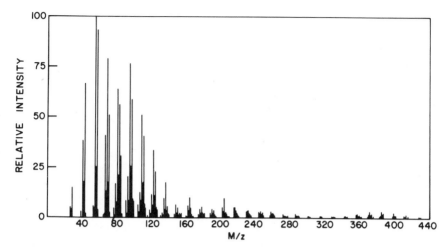

Fig. 5.5 Mass spectrum of Apiezon BW oil (obtained using MS9 sector spectrometer), source temperature 170°C. Reprinted with permission from *Vacuum*, **16**, 67, R. D. Craig and E. H. Harden. Copyright 1966, Pergamon Press, Ltd.

applications. These diffusion pump oils are not resistant to oxidation and will decompose when exposed to air while heated. Their continued popularity for certain applications is due to their low cost. See Table 8.2.

Silicones are an important class of diffusion pump fluids. Not only do they have low vapor pressures but also extremely high oxidation resistance [6]. Cracking patterns for Dow Corning DC-704 and DC-705, given in Appendix E.1, show many of the characteristic fragments of benzene. By way of illustration a partial spectrum taken in a contaminated diffusion pumped system using DC-705 fluid is compared with the cracking pattern of benzene in Fig. 5.6. The peaks labeled M are due to mechanical pump oil. Systems that have been contaminated with a silicone pump fluid will always show the characteristic groups at $M/z = 77$ and 78 and usually those at $M/z = 50$, 51, and 52. Notice the lack of these peaks in Fig 4.6, which was taken on a system with a straight chain hydrocarbon oil in the diffusion pump. Wood and Roenigk [4] observed fragments of DC-705 at $M/z = 28$, 32, 40, and 44, all of which could naturally occur in a vacuum system. They urged caution that the interpretation of these peaks as residual air could cause erroneous conclusions.

Esters and polyphenylethers are also widely used pump fluids because they polymerize to form conducting layers. They find use in

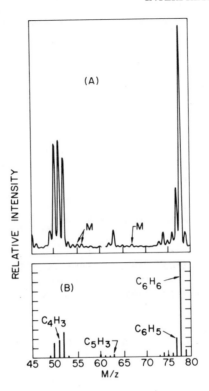

Fig. 5.6 Comparison of (*a*) Residual-gas background in a system contaminated with DC-705 fluid and (*b*) cracking pattern of benzene.

systems that contain mass analyzers, glow discharges, and electron beams. Octoil-S is characterized by its repeated C_mH_n groupings. Polyphenylether (Santovac 5, Convalex-10, and BL-10) also contains the characteristic fragments of the phenyl group which include the fragments at M/z = 39, 41, 43, 44, and 64. Cracking patterns will vary greatly from one instrument to another and the inability to match the data exactly to the patterns in any table should not be considered evidence of the absence or existence of a particular pumping fluid in the vacuum system. In fact, the cracking pattern for DC-705 given in Appendix E.1 does not show the same relative intensities at M/z = 50, 51, 52, and 78, as shown in Fig. 5.6*a* or seen by other workers. The spectrum taken from a gently heated liquid pump fluid source contains proportionally more high mass decomposition products than the spectrum of backstreamed vapors from a trapped diffusion pump because the trap is effective in retaining high molecular weight fragments. See Section 8.4. A liquid nitrogen trap is a mass-selective filter. These patterns cannot be used to differentiate between back-streamed DC-704 and DC-705. High-mass ion currents are severely attenuated when a quadrupole instrument is operated at constant absolute resolution. This built-in attenuation, sketched in Fig. 4.10, can easily lead to the conclusion that the environment is free of heavy molecules.

Elastomers are found in all systems with demountable joints except those using metal gaskets. The most notable property of all elastomers is their ability to hold gas and release it when heated or squeezed. The mass spectra of Buna-N is shown in Fig. 5.7 during heating [7]. Also shown are the initial desorption of water followed by the dissociation of the compound at a higher temperature. The decomposition temperature is quite dependent on the material. Mass spectra obtained during the heating of Viton fluoroelastomer (Fig. 5.8 [7]) show the characteristic release of water at low temperatures and the release of carbon monoxide and carbon dioxide at higher temperatures. At a temperature of 300°C the Viton begins to decompose. Silicone rubbers have a polysiloxane structure. They are permeable, and their mass spectra usually show a large evolution of H_2O, CO, and CO_2. At high temperatures they begin to decompose. Their spectra show groups of peaks at $Si(CH_3)_n$, $Si_2O(CH_3)_n$, and $Si_3O_2(CH_3)_n$; $n = 3, 5, 6$, respectively, are the largest [8]. Polytetrafluoroethylene (Teflon) is suitable for use up to 300°C, although it outgasses considerably. A spectrum taken at 360°C shows major fragments at $M/z = 31(CF)$, $50(CF_2)$, $81(C_2F_3)$, and $100(C_2F_4)$ [8].

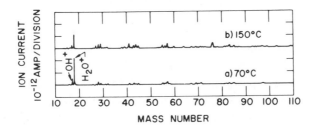

Fig. 5.7 Mass spectra obtained during the heating of Buna-N rubber. Reprinted with permission from *Trans. 7th Nat. Vacuum Symp. (1960)*, p. 39, R. R. Addis, L. Pensak and N. J. Scott. Copyright 1961, Pergamon Press, Ltd.

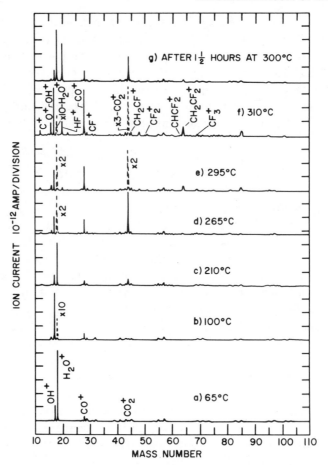

Fig. 5.8 Mass spectra obtained during the heating of Viton fluoro-elastomer. Reprinted with permission from *Trans 7th Nat. Vacuum Symp. (1960)*, p. 39, R. R. Addis, L. Pensak, and N. J. Scott. Copyright 1961, Pergamon Press, Ltd.

The potential limits of qualitative analysis become clear after some practice with an RGA. A knowledge of the instrument, the cleaning solvents, and the pumping fluids used in mechanical, diffusion, or turbomolecular pumps combined with periodic background scans will result in the effective use of the RGA in the solution of equipment and process problems.

5.3   QUANTITATIVE ANALYSIS

The RGA, with a resolving power of 50 to 150, is not intended to be an analytical instrument. It cannot eliminate overlapping peaks, nor are its cracking patterns as stable as those of an analytical instrument. Quantitative analysis of a single gas or vapor or combination of gases and vapors with unique cracking patterns is a simpler task than the analysis of combinations that contain overlapping peaks. This section demonstrates approximation techniques for quickly obtaining quantitative partial pressures within a factor of 10 but points out that the acquisition of data accurate to, say, 10% requires careful calibration. Either crude data are obtained quickly or accurate data painstakingly—there is little middle ground. We now consider techniques, both approximate, and precise, for gases with isolated and overlapping cracking patterns.

5.3.1   Isolated Spectra

Approximate analysis techniques for gases with isolated spectra will yield results of rough accuracy with minimal effort, for example, a mixture of Ar, O_2, N_2, and H_2O would be reasonably easy to examine because of the unique peaks at M/z = 40, 32, 28, and 18. One technique is the summation of the heights of all peaks of any significant amplitude which are due to these gases, followed by the division of that number into the total pressure. The resulting sensitivity factor, expressed in units of pressure per unit scale division, is then applied to all the peaks without further correction. Some improvement in accuracy can be made if the ion currents are first corrected for the ionizer sensitivity by dividing by the values given in Table 4.1. There is an alternative technique that is equally accurate and does not require the knowledge of the total pressure. It relates the partial pressure of a given gas to the ion current, sensitivity of the mass analyzer, and gain of the electron multiplier according to (5.1).

$$P(x) = \frac{\text{total ion current}(x)}{GS(N_2)} \tag{5.1}$$

The sensitivity is usually provided by the manufacturer for nitrogen and is typically of the order of 10^{-5} A/Pa. By taking the ratio of electron multiplier current to the Faraday cup current we can determine the gain G of the multiplier. From this information the partial pressure of the gas is obtained. These two techniques are accurate to within a factor of five.

A more accurate correction accounts for the gas ionization sensitivity and the mass dependencies of the multiplier gain and mass filter transmission. The mass dependence of the multiplier is often approximated as $M^{\frac{1}{2}}$, as shown on the dotted curve of Fig. 4.15, but this is not always valid. The transmission of a fixed radius sector with variable magnetic field, is independent of mass, but the most common RGA, the quadrupole, has a transmission that is dependent on the energy, focus, and resolution settings. The more accurate correction may turn out to be less accurate unless a significant amount of calibration is done to obtain the mass dependence of the mass filter and electron multiplier accurately. The time spent in applying corrections such as those shown in Figs. 4.10 and 4.15 is probably out of proportion to the information gained.

Accurate measurements of the partial pressures of gases with nonoverlapping spectra are best accomplished by calibrating the system for each gas of interest. The vacuum system must be thoroughly clean and baked if possible before the background spectrum is recorded. It is then backfilled with gas to a suitable pressure so that the cracking pattern can be recorded and the gas sensitivity, measured. The values of all the ionizer potentials and currents, the gain of the electron multiplier, and the pressure should be recorded at that time. The system should be thoroughly pumped and cleaned between each successive background scan and gas admission. This is a laborious process and is only done when precise knowledge of the partial pressure of a particular species is required. Even then periodic checks of the multiplier gain, for example, at mass 28, are still necessary.

5.3.2 Overlapping Spectra

Analysis of overlapping spectra is made more difficult by the fact that the peak ratios for a given gas may not be stable with time because of electron multiplier contamination and because trace contaminants in the system will add unknown amounts to the minor peaks of gases under study. To gain an appreciation of the problems involved in determining partial pressures two simplified numerical examples are worked out here.

A mass spectrum is taken on a system that contains peaks mainly attributable to N_2 and CO. Peaks due to nitrogen appear at mass numbers 28 and 14 and peaks due to CO appear at mass numbers 28,

16, 14, and 12. Trace amounts of carbon present in the system from other sources dictate that the amplitude of the mass 12 peak cannot be relied on for accurate determination of the CO concentration. In a similar manner the mass 16 peak is of questionable value because of the surface desorption of atomic oxygen, methane, or other hydrocarbon contamination. Therefore the analysis in this simplified example is weighted heavily in favor of using the peaks at mass numbers 14 and 28. From cleanly determined experimental cracking patterns the relative peak heights for nitrogen were found to be 0.09 and 1.00 for mass numbers 14 and 28, respectively, whereas values of 0.0154 and 1.10 were measured at the same mass numbers for CO. (The CO cracking patterns have been corrected for the difference in ionizer sensitivity; see Table 4.1). The sensitivity and multiplier gain of the instrument were $S = 10^{-5}$ A/Pa and $G = 10^{+5}$. From the mass spectrum the ion currents were $i_{14} = 10.54$ μA, and $i_{28} = 210$ μA. The individual partial pressures were then found by solving the following two equations simultaneously:

$$i_{28} = SG[a_{11}P(N_2) + a_{12}P(CO)]$$

$$i_{14} = SG[a_{21}P(N_2) + a_{22}P(CO)] \tag{5.2}$$

or

$$210 \ \mu A = 1A/Pa[1.00P(N_2) + 1.10 \ P(CO)]$$

$$10.54 \ \mu A = 1A/Pa[0.09P(N_2) + 0.0154 \ P(CO)]$$

which yielded $P(N_2) = P(CO) = 10^{-4}$ Pa.

Now consider how a change unaccounted for in the cracking pattern would affect the accuracy of this calculation. If the actual cracking pattern of nitrogen, and consequently the measured ion currents, were to change without the knowledge of the operator, because of contamination in the first dynode or a change in the temperature of the ion source, an error would be introduced into the calculation because the coefficients a_{mn} in the right-hand side of (5.2) were not altered simultaneously. Figure 5.9a shows the calculated values of $P(N_2)$ and $P(CO)$ that would be obtained for the example in (5.2) if the actual mass 14 fragment of nitrogen were changed from 9 to 5% of the mass number 28 peak without our knowledge and therefore without our having made the corresponding change in the coefficient a_{21}. It can be seen that even with moderate changes in the cracking pattern the partial pressures of the two gases can still be determined within 25% for this example in which the N_2 and CO are present in equal amounts. If greater accuracy is desired, the cracking pattern of the gases should be taken frequently and the coefficients a_{mn} adjusted to account for these instrumental changes.

For *unequal* concentrations of the two gases the errors are far greater than when the gases are present in equal proportions. Figure 5.9*b* illustrates the pressure measurement error as a function of the cracking pattern change for a 10:1 ratio of N_2:CO. Again, this represents an actual change in cracking pattern which was not accounted for by a corresponding change in the coefficient a_{21}. This demonstrates that even modest changes in the cracking pattern ratios of the major constituent can cause the error in partial pressure calculation of the minor constituent to be as great as 200 to 400% when the major and minor constituents have overlapping cracking patterns.

These two illustrations demonstrate that quantitative analysis of overlapping spectra requires accurate and often, frequent measurements of the cracking patterns and that accurate quantitative measurements of trace gases are difficult when the trace gas peaks overlap those of a major constituent.

The effects of certain cracking pattern errors in a residual gas spectrum have been illustrated by Dobrozemsky [9]. These data taken on an Orb-Ion pumped system are presented in Table 5.1. Column 1 shows the correctly analyzed partial pressures and their standard deviations, Column 2 and 3, respectively, show the effects of interchanging the peaks at mass numbers 17 and 18 and the effects of doubling the peak height at mass number 14. This analysis vividly demonstrates the effects of errors in the accuracy of calculating trace gas compositions. Column 1, the correct analysis, shows that the standard deviation for H_2O is about 0.3%, whereas for a trace gas such as hydrogen the standard deviations can be large enough to render the measurement useless. The standard deviation for hydrogen is large because the signal at $M/z = 2$ arises from many sources. Literally any hydrocarbon that is ionized in the RGA has a fragment at $M/z = 2$ and the standard deviation of each fragment is additive. The result is that the hydrogen concentration, if any, is not known. It demonstrates the ease with which false data can be generated when cracking patterns are not accurately known. Note also that the incorrect analyses shown in Column 2 and Column 3 yield unreasonably large standard deviations for all gases even when only one ion current was in error.

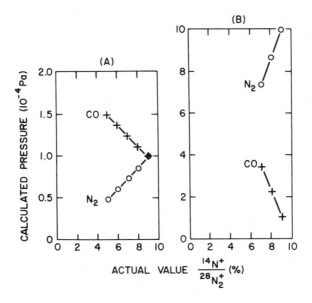

Fig. 5.9 Errors induced in the calculation of pressures of mixtures of N_2 and CO resulting from physical changes in the dissociation of nitrogen (ratio of $N^{14}:N^{28}$) which were not compensated for by appropriate changes in the coefficient a_{21} of the nitrogen cracking pattern. (*a*) 1:1 mixture of N_2:CO, (*b*) 10:1 mixture of N_2:CO.

Even though only a limited number of low molecular weight gases are present in a vacuum system, the analysis procedure is complicated by the fact that there are often several gases that produce peaks at the same mass numbers; for example, CO, N_2, C_2H_4, and CO_2 produce ion current at mass number 28 and CO_2, O_2, CH_4, and H_2O produce ion currents at mass number 16. For a system containing *n* gases and *m* ion current peaks, (5.2) becomes

$$
\begin{bmatrix} i_1 \\ i_2 \\ \cdot \\ \cdot \\ \cdot \\ \cdot \\ i_m \end{bmatrix} = \begin{bmatrix} a_{11} & \cdots & a_{1n} \\ a_{21} & & a_{2n} \\ \cdot & & \cdot \\ \cdot & & \cdot \\ \cdot & & \cdot \\ \cdot & & \cdot \\ a_{m1} & \cdots & a_{mn} \end{bmatrix} \begin{bmatrix} P_1 \\ P_2 \\ \cdot \\ \cdot \\ \cdot \\ \cdot \\ P_n \end{bmatrix} \tag{5.3}
$$

where i_m is the ion current at mass *m*, a_{mn} are the components of the cracking pattern matrix, and P_n is the partial pressure of the *n*th gas.

Table 5.1   Analysis of Background Gases in an Orb-Ion Pumped System[a]

	Partial Pressures $\times$ 10^{-8} Torr		
Gas	1	2	3
H_2	2.21 $\pm$ 3.15	2.85 $\pm$ 8.29	2.07 $\pm$ 2.67
He	15.54 $\pm$ 0.62	14.95 $\pm$ 22.6	14.94 $\pm$ 7.2
CH_4	5.37 $\pm$ 0.49	-13.5 $\pm$ 5.13	9.64 $\pm$ 1.63
NH_3	2.64 $\pm$ 0.89	49.56 $\pm$ 6.83	5.03 $\pm$ 2.17
H_2O	50.95 $\pm$ 0.15	15.4 $\pm$ 5.29	48.44 $\pm$ 1.68
Ne	4.54 $\pm$ 0.54	4.39 $\pm$ 16.5	4.38 $\pm$ 5.24
N_2	-1.99 $\pm$ 3.26	34.88 $\pm$ 57.6	-13.6 $\pm$ 18.3
CO	4.54 $\pm$ 2.91	-30.7 $\pm$ 51.4	15.38 $\pm$ 16.4
C_2H_6	3.42 $\pm$ 0.74	8.44 $\pm$ 18.4	1.89 $\pm$ 5.84
O_2	0.00 $\pm$ 0.13	-0.99 $\pm$ 4.87	0.3 $\pm$ 1.55
Cl	-0.02 $\pm$ 0.12	0.04 $\pm$ 4.64	0.02 $\pm$ 1.48
Ar	1.36 $\pm$ 0.09	1.29 $\pm$ 3.29	1.31 $\pm$ 1.04
CO_2	5.67 $\pm$ 0.14	5.3 $\pm$ 3.22	5.50 $\pm$ 1.02
C_3H_8	2.72 $\pm$ 0.56	0.27 $\pm$ 12.9	3.32 $\pm$ 4.11
Acetone	2.99 $\pm$ 0.31	7.94 $\pm$ 6.91	1.33 $\pm$ 2.2

Source. Reprinted with permission from *J. Vac. Sci. Technol.*, **9**, p. 220, R. Dobrozemsky. Copyright 1972, The American Vacuum Society.
[a] The total pressure is 1.3×10^{-7} Torr. (1) correct analysis. (2) incorrect spectrum obtained by interchanging ion currents at mass numbers 17 and 18; (3) incorrect spectrum obtained by doubling the ion current at mass number 14.

Most gases have more than one peak, so that $m > n$, and the system is overspecified. A least mean squares or other smoothing criterion is then applied to the data to get the best fit. If the cracking patterns are carefully taken and if the standard deviations are measured as well, then accuracies of 1 to 3% may be obtained for major constituents [9,10]. The matrix for a real problem would contain about 10 $\times$ 50 elements, and require the assistance of a computing machine in order to obtain a solution. In this case it is practical to use a computer for data acquisition and instrument control as well as for analysis [10]. These experiments are expensive and time consuming and are only performed in situations where such precision is required.

REFERENCES

1. P. Marmet and J. D. Morrison, *J. Chem. Phys.*, **36**, 1238 (1962).
2. P. A. Redhead, *Can. J. Phys.*, **42**, 886 (1964).
3. J. L.Robbins, *Can. J. Phys.*, **41**, 1383 (1963).
4. G. M. Wood, Jr. and R. J. Roenigk Jr., *J. Vac. Sci. Technol.*, **6**, 871 (1969).
5. R. D. Craig, and E. H. Harden, *Vacuum*, **16**, 67 (1966).
6. C. W. Solbrig and W. E. Jamison, *J. Vac. Sci. Technol.*, **2**, 228 (1965).
7. R. R. Addis, Jr., L. Pensak, and N. J. Scott, *Trans. 7th A.V.S. Nat. Vacuum Symp. 1960*, Pergamon, Oxford, 1961, p. 39.
8. A. H. Beck, *Handbook of Vacuum Physics*, Vol. 3, Macmillan, New York, 1964, p. 243.
9. R. Dobrozemsky, *J. Vac. Sci. Technol.*, **9**, 220 (1972).
10. D. L. Ramondi, H. F. Winters, P. M. Grant and D. C. Clarke, *IBM J. Res. Dev.*, **15**, 307 (1971).

Materials

Knowledge of the basic properties of the materials from which vacuum systems are fabricated is essential to operate and maintain vacuum systems properly. One of the most troublesome properties of materials used in vacuum applications is gas release from solids at low pressures. Chapter 6 focuses on this issue. The origins of this gas and methods for its removal from a variety of materials used in vacuum system construction are reviewed here. There is nothing to vacuum. It is all in the packaging.

CHAPTER 6

Materials in Vacuum

A superficial examination of a high-vacuum or ultrahigh-vacuum pumping system gives an impression of simplicity; clean, polished metal or glass surfaces, glass view ports, electrical and motion feed-throughs, piping and pumps. A close examination reveals that many requirements are placed on materials in vacuum environments, requirements that sometimes conflict. The chamber walls must support a load of 10,335 kg/m^2, a load that is present on the surfaces of all vacuum systems, even those that are merely roughed to 100 Pa. This load can be sustained by materials of high mechanical strength which have been fabricated according to sound design principles. The materials must be easy to machine and easy to join by welding, brazing, soldering or by use of demountable seals. Because view ports and electrical feed-throughs are required, methods are needed for sealing glasses, ceramics, and other insulators to metals. The importance of reducing the outgassing load from the fixturing of high vacuum chambers and tooling is discussed in Chapter 10. These systems usually contain a large internal surface area which cannot be baked. Therefore it is essential that they be fabricated from materials that can be processed otherwise to yield low outgassing rates. A further reduction of the surface outgassing rate by baking is of paramount importance in reaching the lowest possible pressure in an ultrahigh vacuum system. This places additional restrictions on the choice of materials.

This chapter reviews the sources of gas evolution from solid surfaces and describes the outgassing properties of some of the metals, glasses, ceramics, and polymeric materials that are used in the construction of vacuum equipment. Stainless steel is the dominant material of construction and its properties are discussed in some detail. A comprehensive review of materials for ultrahigh vacuum has been given by Weston [1], and Perkins [2] and Elsey [3, 4] have thoroughly reviewed the literature on outgassing of materials used in vacuum.

6.1   GAS EFFLUX FROM SOLIDS

If all the gas to be removed from a vacuum chamber were located in the volume of the chamber, it could be removed easily by a pump in a very short time. Consider a 100-L chamber previously roughed to 10 Pa and just connected to a 1000 L/s high-vacuum pump. The gas pressure is given by $P = P_o \exp(-St/V)$. According to this equation the pressure will reach 4.5×10^{-4} Pa in 1 s. In practice this will never happen; the slow evolution of additional gases and vapors from the interior surfaces will lengthen the actual pumping time. A pump-down time of 15 to 30 min is more likely.

This surface gas release, collectively referred to as outgassing, is actually a result of several processes. Figure 6.1 shows all the possible sources of gas in addition to the gas located in the volume of the chamber. Gases and vapors released from the surface are a result of vaporization, thermal desorption, diffusion, permeation, and electron- and ion-stimulated desorption.

6.1.1   Vaporization

A vapor is a gas near its condensation temperature and vaporization is the thermally stimulated entry of molecules into the vapor phase. In dynamic equilibrium the rate of molecules leaving the surface equals the rate of molecules arriving at the surface. The pressure of the vapor over the surface in dynamic equilibrium is the vapor pressure of the solid or liquid, provided that the solid or liquid and the vapor are at the same temperature. In Chapter 2 it was noted that the molecular flux of a vapor across a plane was $nv/4$. In equilibrium this is therefore the rate of molecular release from the surface. For the case of evaporation of a solid from a heated source none of the molecules return to the surface and (2.5), given here in a different form, may be used to calculate the maximum rate of evaporation of the solid from its temperature, vapor pressure, surface area, and molecular weight:

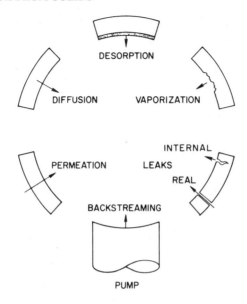

Fig. 6.1 Potential sources of gases and vapors in a vacuum system.

$$\Gamma(\text{molecules/s}) = 2.63 \times 10^{24} \frac{PA}{(MT)^{\frac{1}{2}}} \qquad (6.1)$$

The vapor-pressure–temperature curves for many gases are given in Appendix B.5. Appendix C.7 lists the vapor pressures of the solid and liquid elements.

6.1.2 Thermal Desorption

Thermal desorption is the heat-stimulated release of gases or vapors adsorbed on the interior walls of the system. These gases may develop from several sources. They may have been adsorbed on the chamber surface while it was exposed to the atmospheric environment and then slowly released as the pump removed gas from the chamber. Desorption is also the final step in the processes of diffusion and permeation. The rate of desorption is a function of the energy with which the various molecules are bound to the surface, the temperature of the surface, and the number of monolayers of surface coverage. Gas is sorbed onto surfaces by physisorption and chemisorption. Physisorbed molecules are bonded to surfaces by weak van der Waal's forces of energy less than 40 J/(kg-mole); chemical adsorption energies are greater than 60 J/(kg-mole). It can be shown that physisorbed particles may be quickly removed from solid surfaces at ambient temperatures and do not hinder system pump-down. Chemical desorption is responsible for most of the slow outgassing encountered vacuum systems.

Most room-temperature outgassing data for metals show that the outgassing rate varies inversely with time, at least for the first 10 h of pumping [5,6]. This may be expressed as

$$q_n = \frac{q_1}{t^{-\alpha}} \qquad (6.2)$$

where the subscript n denotes the time in hours for which the data apply. The slope will range from -0.7 to -2, with -1 the most common value. Several desorption models have been analyzed by Elsey [4] but are not discussed here because their assumptions of monoenergetic surfaces of less than monolayer coverage do not hold. Experimental data show that surface activation energies vary over a wide range and that adsorbed gas loads of up to 100 monolayers which are mainly water vapor are often encountered. Dayton [5] modeled this $1/t$ behavior as the sum of the successive evaporation of molecules of a range of energies. He proposed that a uniform distribution of activation energies like that found in a surface oxide containing pores of varying diameters will result in a sum of individual outgassing curves that will appear as one curve with a slope of about -1 on a log-log plot of outgassing rate versus pumping time. This dependence is sketched in Fig. 6.2.

6.1.3 Diffusion

Diffusion is the transport of one material through another. Gas diffusing from the wall of a vacuum container contributes to the outgassing of the system. The gas pressure in the solid establishes a concentration gradient that drives molecules or atoms to the surface, where they desorb. Because diffusion is a much slower process than desorption, the rate of transport through the bulk to the surface governs the rate of release into the vacuum. A concentration gradient moves the atoms

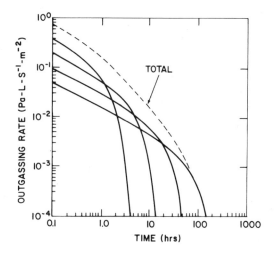

Fig. 6.2   Total outgassing rate as a sum of four rates, each resulting from a single outgassing time constant whose value depends on the shape of the surface oxide pores and the activation energy for desorption. Reprinted with permission from *Trans. 8th Vac. Symp. (1961)*, p. 42, B. B. Dayton. Copyright 1962, Pergamon Press, Ltd.

or molecules to the surface with a flux given by (2.23). The outgassing rate from a thick solid wall containing gas at an initial uniform concentration C_o may be obtained from Fick's law of diffusion and is given by [7]:

$$q = C_o \left(\frac{D}{t}\right)^{\frac{1}{2}} \left(1 + 2 \sum_{n=1}^{\infty} (-1)^n \exp\left(-n^2 d^2 / Dt\right)\right) \qquad (6.3)$$

where D is the diffusion constant in m^2/s and d is the thickness of the material. If C_o is given in units of Pa, then q will have units of W/m^2. For small times $q \sim C_o (D/t)^{\frac{1}{2}}$ or q varies as $t^{-\frac{1}{2}}$. For long times the exponential terms must be included and the outgassing decays as $t^{-\frac{1}{2}} e^{-at}$.

The diffusion constant is a function of the thermal activation energy of the diffusing species and is given by [7]:

$$D = D_o e^{-E/kT} \qquad (6.4)$$

Because of this exponential dependence on temperature, a rather modest increase in temperature will sharply increase the initial outdiffusion rate and reduce the time necessary to unload the total quantity

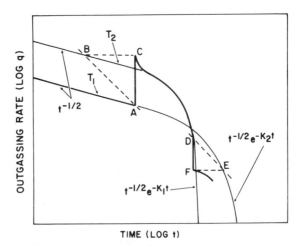

Fig. 6.3 Change in outgassing rate for an increase in temperature from T_1 to T_2 for a diffusion process. Reprinted with permission from *Vacuum Technology and Space Simulation*, D. J. Santeler et al., SP-105, 1966, Nationa Aeronautics and Space Administration, Washington, DC.

of gas contained in the solid. Figure 6.3 illustrates how vacuum baking reduces the final outgassing rate to a level far below that which is possible in the same time without baking. A solid exposed to vacuum at ambient temperature T_1 outgasses along the initial portion of the curve with a slope of $t^{-1/2}$, as given by (6.3). At a time corresponding to point A, the temperature is increased to T_2. Because the gas concentration in the solid cannot change instantaneously, the outgassing rate increases to the value given by point B on the high temperature diffusion curve but shifted in time to point C. Outdiffusion continues along the high-temperature curve but is displaced in time. At a time corresponding to point D the baking operation is terminated and the temperature is reduced to T_1. The outdiffusion rate is reduced to a value corresponding to point E on the low-temperature curve but at a point earlier in time given by point F. Outgassing now continues at a rate given by the low-temperature curve but shifted to an earlier point in time. The new pressure at point F is given by [8]:

$$P_f = (D_1/D_2)P_d \qquad (6.5)$$

where $D_{1,2}$ are the diffusion coefficients at temperatures $T_{1,2}$, respectively. The net effects of baking are the reduction in the outgassing rate and the reduction of the time required to remove the initial concentration of gas dissolved in the solid. Recalling that surface desorption was modeled by Dayton [5] as the sum of diffusive release of gas from the surface oxide pores, we find that baking will reduce the desorption time in exactly the same manner as it reduces the diffusion time.

6.1.4 Permeation

Permeation is a three-step process. Gas first adsorbs on the outer wall of a vacuum vessel, diffuses through the bulk, and finally desorbs from the interior wall. The time-dependence of the permeation rate of a gas through a solid wall is qualitatively sketched in Fig. 6.4. If the material is initially devoid of the permeating gas, a period of time will pass before the gas is observed to desorb from the vacuum wall. After equilibrium has been established, gas will desorb from the wall at a constant rate. Steady-state permeation therefore behaves like a small leak. The permeability of the wall for a gas is given by

$$K_p = DS' \qquad (6.6)$$

where D is the diffusion constant and S' is the solid solubility [4]. The permeability has an exponential dependence similar to that of the diffusion constant and consequently increases rapidly with temperature.

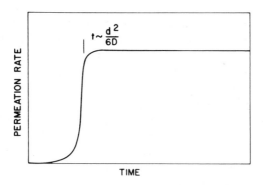

Fig. 6.4 Permeation of a gas through a solid wall.

The steady-state permeation rate of a gas that does not dissociate on adsorption can be expressed as

$$q_k = \frac{K_p P}{d} \qquad (6.7)$$

where q_k is the total flux in units of $(Pa\text{-}m^3)/(s\text{-}m^2)$ or W/m^2 and P is the pressure drop across the solid of thickness d. The permeability K_p has units of m^2/s and is the quantity of gas in m^3 at STP flowing through 1 m^2 of material that is 1 m thick when the pressure difference is 1 atm. Diatomic gases dissociate to atoms as they adsorb on and diffuse through metals, therefore the steady-state permeation rate of a gas such as hydrogen through a metal is given by

$$q_k = \frac{K_p(P_2^{1/2} - P_1^{1/2})}{d} \qquad (6.8)$$

The permeability K_p for gases that dissociate has units of $Pa^{1/2}\text{-}m^2/s$.

6.1.5   Electron- and Ion-Stimulated Desorption

Electrons and ions incident on solid surfaces can release adsorbed gases from these surfaces and generate new vapors in quantities large enough to limit the ultimate pressure in a vacuum chamber. Many reactions are possible when ions and electrons collide with surfaces [9]; here we mention electron-stimulated desorption, ion-stimulated desorption, and electron- or ion-induced chemical reactions. An energetic electron incident on the surface gas layer excites a bonding electron in an adsorbate atom to a nonbonding level. This results in a repulsive effective potential between the surface and atom which gives rise to desorption as a neutral atom or an ion [10–13]. The desorbed neutral gas flux can be as high as 10^{-2} atom per electron, whereas the desorbed ion flux is much smaller, of an order of 10^{-5} ion per electron. This desorption process is specific; that is, it depends on the manner in which a molecule or atom is bonded to the surface. Some molecules are not desorbed at all by this process. These gas-release phenomena have been shown to cause serious errors in pressure measurement with the Bayard-Alpert gauge [14, 15] and residual gas analyzer [16] and to increase the background pressure in systems that use high energy electron or ion beams.

Ion stimulated desorption has been extensively studied by Winters and Sigmund [17] and Taglauer and Heiland [18]. Winters and Sigmund show that nitrogen chemisorbed on tungsten can be desorbed by noble gas ions in an energy range up to 500 V. They also showed that the adsorbed atoms were removed by a sputtering process as a result of direct knock-on collisions with impinging and reflected noble gas ions. Ion-stimulated desorption is responsible for part of the gas release observed in sputtering systems, Bayard-Alpert gauges, and glow discharge cleaning.

Ion- and electron-stimulated chemical reactions may also occur at solid surfaces; for example, hydrogen and oxygen can react with surface carbon to produce methane and carbon monoxide. An ion stimulated chemical reaction is responsible for the rapid etch rates observed in reactive-ion etching. In the reactive-ion etching of silicon high-energy ions in a collision cascade process greatly enhance the reaction of neutral F with Si and produce a much greater yield of volatile SiF_4 than is possible without ion bombardment.

In this discussion no attempt was made to relate the various gas release rates numerically to a real vacuum system problem. The relative roles of surface desorption, diffusion, and permeation in determining the pump-down time of a vacuum chamber will vary considerably with the material used for the construction of the vacuum walls (steel, aluminum, or glass), the seal material (metal or elastomer) and the history of the system (newly fabricated, unbaked, chemically cleaned, or baked). Despite these many variables, we can still construct a composite pump-down curve that will illustrate the relative roles of

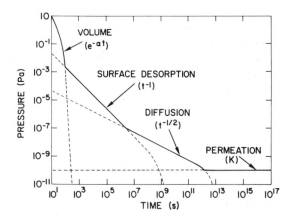

Fig. 6.5 Rate limiting steps during the pumping of a vacuum chamber.

these phenomena. Figure 6.5 shows the high-vacuum pumping portion of an unbaked, metal-gasketed system. In the initial stages the pressure is reduced exponentially with time as the volume gas is removed. This portion of the pumping curve takes only a short time because a typical system time constant is less than 1 s. It is expanded here for clarity. In the next phase, surface desorption controls the rate of pressure decrease. In a typical unbaked system most of this gas load is water vapor, however, nitrogen, oxygen, carbon oxides, and hydrocarbons are also present. The total quantity of gas released is determined by the material and its history. Glass or steel that has been exposed to room ambient for extended periods may contain up to 100 monolayers of water vapor, whereas a carefully vented chamber may contain little water vapor. If the system has a number of large elastomer O-rings or a large interior surface area, a slow decrease in pressure is to be expected. Unbaked, routinely cycled vacuum systems are never pumped below this pressure range.

If the system is allowed to continue pumping without baking, the surface gas load will ultimately be removed and the outdiffusion of gases in solution with the solid walls will be observed. The slope of the curve will change from t^{-1} to $t^{-\frac{1}{2}}$, for example, hydrogen that diffused into the steel in a short time at high fabrication temperatures diffuses out very slowly at room temperatures. If we were to continue pumping until the dissolved hydrogen was removed, the pressure would become constant even though the system was leak free. The system would now be at its ultimate pressure given by $P = Q_k/S$. Experimentally, hydrogen permeates the walls of metal systems and helium permeates glass walls. Notice that the time required to reach the ultimate pressure in this hypothetical example of an unbaked metal-gasketed system is 10^8 h. This clearly demonstrates the absolute necessity of baking in ultrahigh vacuum technology.

The order of importance of the processes shown in Fig. 6.5 is not always as given here for this example. Elastomer gaskets have a high permeability for atmospheric gases, and if the system contained a significant amount of these materials the limiting permeation rate would be several orders of magnitude higher than illustrated here. In fact, it could be large enough to mask the diffusion process.

The level of outgassing that results from the various electron- and ion-stimulated desorption processes is not shown in Fig. 6.5. These processes play a variable but important role in determining the ultimate pressure in many instances because they are able to desorb atoms that are not removed by baking. Electron-stimulated desorption in an ion gauge is easily observable during gauge outgassing. A 10-mA electron flux can desorb 6×10^4 neutrals/s. Because a gas flux of 2×10^{17} atoms/s equivalent to 1 Pa-L/s, this electron flux can initially provide up to 2.5×10^{-3} Pa-L/s of desorbed gas. If the system is pumping on the ion gauge tube at a rate of 10 L/s, this desorption peak can reach a pressure of an order of 10^{-4} Pa in the gauge tube. The scattered electrons from Auger and LEED systems can stimulate desorption in a similar manner, with the result that background pressure will increase when the beam is operating. Electron-induced desorption is a first-order process that should produce an exponential decay of the desorbed species; however, if readsorption is included, the initial rate will decay exponentially to a steady-state level [19]. Ion beams can also cause significant surface desorption. A 20-μA ion beam can desorb surface gas at an initial rate of 10^{-3} Pa-L/s, assuming a desorption efficiency of two atoms per incident ion.

6.2 METALS

Metals are used both in the vacuum chamber and to form its walls. Each metal must have a vapor pressure low enough to prevent vaporization from occurring at the highest temperature encountered. Each should also have a low outgassing rate. The particular application in the chamber, such as the filament, radiation shield, and thermal sink, will more narrowly define the selection. Metals used for vacuum chamber walls should be easily joinable and sealable to one another and to ceramics and have high strength. They should not be porous or permeable to atmospheric gases. This section presents the vaporization, permeation, and outgassing properties of several metals of interest along with the structural properties of austenitic stainless steels.

6.2.1 Vaporization

The vapor pressure of most metals is so low that it does not restrict their use for vacuum applications. The vapor pressures of the elemental metals are listed in Appendix C.7. Some commonly found metals should not be used in vacuum construction because their vapor pressures are high enough to interfere with normal vacuum baking procedures. Alloys that contain zinc, lead, cadmium, selenium, and sulfur, for example, have unsuitably high vapor pressures for vacuum applications. Zinc is a component of brass, cadmium is commonly used to plate steel screws, and sulfur and selenium are used to make the free

machining grades of 303 stainless steel.   Under no circumstances
should these materials be used in vacuum system construction.

6.2.2   Permeability

Any gas that is soluble in a metal may permeate it.   Its permeability is
directly proportional to its solid solubility [see (6.5)].   Because of its
high solubility, hydrogen is one of the few gases that permeate most
metals to a measurable extent.   Its permeation rate is proportional to
the square root of the pressure difference, given in (6.8).   Figure 6.6
shows the temperature dependence of the permeation constant of
hydrogen through several metals [20].   Hydrogen permeation is the
least in aluminum, followed by Mo, Cu, Pt, Fe, Ni, and Pd.   The
addition of chrome to iron allows the formation of a chrome oxide
barrier that reduces the hydrogen permeation rate.   Even though the
permeation rate for hydrogen through steel is small, it is a concern for
users of ultrahigh vacuum systems.   The total permeation rate of a

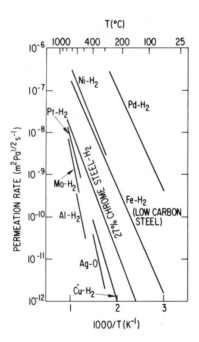

Fig. 6.6 Permeation constant for hydrogen through various metals as a function of
temperature. Reprinted with permission from *Trans. 8th Nat. Vac. Symp. (1961)*, p. 8, F.
J. Norton. Copyright 1962, Pergamon Press, Ltd.

vacuum chamber wall 2-mm-thick and 1 m² in area can be calculated from (6.8) as 5×10^{-13} Pa-m³/s at 20°C for an atmospheric hydrogen pressure of 0.05 Pa. If the chamber is pumped by a 500 L/s pump, the ultimate pressure will be 10^{-10} Pa. The influx will be greater if rusting occurs on the external wall because the reaction of water vapor with iron creates a high partial pressure of hydrogen. Perkins [2] has described the experimental techniques used to measure the permeation and diffusion constants in solids and has tabulated the permeability and diffusivity of hydrogen in paladium, stainless steels, Fe-Co-Ni alloys, copper, and nickel and the permeability and diffusivity of nitrogen in tungsten and molybdenum. Begeal [21] has measured the hydrogen permeation in copper and copper alloys.

6.2.3 Outgassing

The gas load in vacuum fixturing and chamber walls will affect the performance of high and ultrahigh vacuum systems if it is not removed. Gas is dissolved in a metal during its initial melting and casting and consists mainly of hydrogen, oxygen, nitrogen, and carbon oxides. The gas load which is physi- and chemisorbed on the interior surfaces arises from the exposure of these surfaces to ambient atmosphere. It consists typically of a large quantity of water vapor, with carbon oxides, oxygen, and some nitrogen. The nature of the adsorbed layer is also a function of the gas used to release the system to atmosphere and the time and extent that it was exposed to the surrounding air.

Efflux of gas from within the metal fixturing and walls can be eliminated by rendering it immobile, reducing its concentration, or erecting a barrier to its passage. Dissolved gas can be rendered immobile by completely immersing a system in liquid helium. At 4.2 K the diffusion constant for any gas is so small [see (6.4)], that no special system-cleaning precautions need be taken. The initial concentration can be substantially reduced by vacuum melting of the metal, by predegassing parts in a vacuum furnace, or by in-situ bakeout of the completed system. A barrier to this outgassing flux can be created by incorporating a layer of metal such as copper, which has a low permeability, or by forming an oxide barrier such as chrome oxide on stainless steel. An oxide barrier to hydrogen diffusion can be formed by an air or oxygen bake or by a multistep chemical treatment such as Diversey [22] cleaning, however, the latter treatment leaves water on the surface. Less thorough cleaning methods are needed after the system has been initially treated. Either glow discharge cleaning or a vacuum bake can be used to clean a chamber after each exposure to ambient.

The type and extent of cleaning depends on the system construction and ultimate pressure.

Vacuum melting is an excellent technique for removing dissolved gas under certain conditions. It is expensive and used mainly for specialized applications that require hydrogen and oxygen-free material in small quantities such as certain internal parts, charges for vacuum evaporation hearths, and cathode materials. It is not used for structural components because further heat treatments such as forging and welding would only recontaminate the highly purified material.

Vacuum firing of components and subassemblies will effectively remove the dissolved gas load in cleaned and degreased parts. Hydrogen firing is traditionally used for this purpose because it effectively reduces surface oxides. It has the disadvantage of incorporating considerable hydrogen into the material at the firing temperature which can slowly outdiffuse at lower temperatures. Vacuum or inert gas firing is preferred for vacuum components, especially those in ultrahigh

Table 6.1 Firing Temperatures for Some Common Metals

Material	Firing Temperature (°C)
Tungsten	1800
Molybdenum[a]	950
Tantalum	1400
Platinum	1000
Copper and alloys[b]	500
Nickel and alloys (Monel, etc.)	750-950
Iron, steel, stainless steel	1000

Source. Reprinted with permission from *Vacuum Technology*, p. 277, A. Guthrie. Copyright 1963, John Wiley & Sons, New York.

[a] Embrittlement takes place at higher temperatures. The maximum firing temperature is 1760°C.

[b] Except zinc bearing alloys which cannot be vacuum fired at high temperatures because of excessive zinc evaporation.

vacuum systems. The maximum firing temperatures are determined by the metal. Examples are given in Table 6.1. Iron and steel are usually fired at a temperature of 800 to 1000°C, whereas copper and its zinc free alloys are fired at 500°C It is important to use oxygen-free, high-conductivity copper because any copper oxide will react with hydrogen to form water vapor. This vapor can cause voids in the material when it is heated that will create a porous leaky metal. It is not good practice to fire both a screw and the part containing the tapped hole because the screw will bind when it is tightened.

The vacuum firing of metal at low temperatures will also reduce the outgassing rate, but the pumping time is prohibitively long. Because the gas depletion time is given by $t \sim d^2/(6D)$ [23], a 1 h bake of stainless steel at 1000°C is equivalent to a 2500 h bake at 300°C. Calder and Lewin [24] have calculated the time required to reach an outgassing rate of 10^{-13} W/m² for a stainless steel sample 2 mm thick with an initial hydrogen concentration of 4×10^4 Pa. Their results, shown in Table 6.2, show that the hydrogen diffusion in stainless steel at intermediate temperatures of 420 to 570°C is rapid enough to make processing at these temperatures practical.

In SI the outgassing rate (quantity of gas evolved per unit time per unit surface area) has units of $(Pa\text{-}m^3)/(m^2\text{-}s)$ or W/m². Factors for converting old units to W/m² are given in Appendix A.3. The pressure in a chamber with outgassing rate q, and area A, when pumped at a speed S, is given by

$$P = 1000 \frac{q(W/m^2)A(m^2)}{S(L/s)} \qquad (6.9)$$

The factor of 1000 needs to be included because the pumping speed is expressed in L/s rather than m³/s. The outgassing rates for various metals are given in Appendix C.1.

The procedures used for cleaning metal vacuum system parts depend on the system application. The outgassing rate of unbaked, untreated stainless steel is about 5×10^{-5} W/m² after 10 h of pumping. An unbaked system with a 0.5 m³ work chamber may have as much as 6 m² of internal tooling and wall area. If the high vacuum pump has a baseplate pumping speed of 2000 L/s, the pressure after 10 h of pumping will be 1.5×10^{-4} Pa, not a good base pressure for such a system. Clearly the outgassing rate of stainless steel must be reduced by a factor of 10 to 100 over its untreated value to be suitable for

Table 6.2   The Theoretical Time to Reach
an Outgassing Rate of 10^{-13} W/m^2
in Stainless Steel.

t (s)	D (m^2/s)	T (°C)
10^6 (11 days)	3.5×10^{-12}	300
8.6×10^4 (24 h)	3.8×10^{-11}	420
1.1×10^4 (3 h)	3.0×10^{-10}	570
3.6×10^3 (1 h)	9.0×10^{-10}	635

Source. Reprinted with permission from
Brit. J. Appl. Phys., **18**, p. 1459, R. Calder
and G. Lewin. Copyright 1967, The Institute of Physics.

high vacuum applications and by a factor of 10^4 to 10^6 for ultrahigh
vacuum applications.   Unbaked systems with base pressures of 5 ×
10^{-6} Pa are generally cleaned by chemical or glow discharge cleaning
techniques and perhaps a bake at hot water temperatures, whereas
ultrahigh vacuum chambers require some form of vacuum heat treating.
    Barton and Govier [25] have measured the outgassing rates of
stainless steel when treated by several methods that did not involve
baking.   Some of their results are shown in Fig. 6.7.   The electropolishing treatment gave an outgassing rate as low as 10^{-8} W/m^2, a value
similar to those obtained from honed and turned surfaces.   Trichloroethylene vapor degreasing was reported to be the best technique for
general use.   After 70 h of pumping the vapor-degreased specimen
showed a mass spectrum that consisted of 58% H_2O, 32% carbon
oxides, 9% hydrocarbons, and 1% trichloroethylene.   Freon and
alcohol or 1, 1, 1, trichloroethane are usually applied instead of trichloroethylene.   Diversey chemical cleaning, which produces a chrome-
oxide-rich surface, yielded an outgassing rate three times greater than
that of the vapor-degreased sample.   Water vapor was the dominant
product of desorption.   Plasma arc cleaning gave the lowest observed
outgassing rate, but the local heating outside the chamber (700°C)
makes the process impractical for general use.
    Govier and McCracken [26] have studied the cleaning of stainless
steel (AISI 321) in rare-gas glow discharges.   The total outgassing rate

was about the same after discharge cleaning as after baking, but the composition of the desorbing species was drastically altered. They found that the outgassing rate after 20 h of pumping was 1.7 × 10⁻⁷ W/m², whereas the argon glow discharge cleaned-sample measured 2 × 10⁻⁷ W/m² after the same pumping time. However, the residual gas load from the sample that was not discharge cleaned was 68% H_2O, 7% CO_2, 5% CO, and 20% H_2. The residual gas composition from the discharge cleaned sample was 98% H_2 and 2% Ar. Halama and Herrera [27] reported that argon and oxygen glow-discharge cleaning of aluminum 6061 alloy gave an outgassing rate of 5 × 10⁻¹⁰ W/m². After a 24 h, 200°C vacuum bake an outgassing rate of < 1.3 × 10⁻¹¹ W/m² was measured. The residual gas composition also consisted of more than 99% hydrogen. They reported that argon was imbedded in the metal and not released at room temperature. Its release rate reached a maximum between 280 and 300°C. At this temperature it was rapidly depleted from the metal. Halama and Herrera found that during oxygen glow-discharge cleaning oxygen combined with aluminum to form stable oxides and removed carbon from the surface by forming CO and CO_2; the hydrogen remained on the surface.

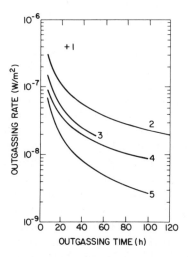

Fig. 6.7 Total outgassing per m² of sample as a function of pumping time and treatment. (1) New sample contaminated with lard oil, no cleaning. (2) New sample cleaned by the Diversey process. (3) New sample electropolished only. (4) Vapor degreased following the electropolishing. (5) New sample machined, degreased and contaminated with lard oil, then plasma arc cleaned. Reprinted with permission from *Vacuum*, **20**, p. 1, R. S. Barton and R. P. Govier. Copyright 1970, Pergamon Press, Ltd.

The effect of a vacuum bake on the outgassing rate of aluminum and stainless steel has been extensively studied because of the wide use of these two metals in vacuum-system construction. The data of Calder and Lewin [24], Young [28], Dobrozemsky and Moraw [29], Samuel [30], and Strausser [31] for stainless steel and aluminum all fall in the range of 10^{-9} to 10^{-11} W/m^2. The exact values are dependent on the effects of treatment before the vacuum bake as well as on its time and temperature. Nuvolone [32] has compared several of these treatments on 316L stainless steel under identical precleaning and measurement conditions in an attempt to separate these effects. This work, which is consistent with earlier work, is described in Table 6.3. The lowest outgassing rate was obtained with oxidation in pure oxygen at 2700 Pa. The oxide barrier reduced the hydrogen outgassing rate effectively. Most importantly, this shows that samples cleaned by an 800°C vacuum bake or a 400°C air bake can be stored for long periods before use, provided they are given a low-temperature bake (150°C) after assembly.

6.2.4 Structural Metals

Stainless steel and aluminum are the two metals most commonly used in the fabrication of vacuum chambers. Aluminum is inexpensive and easy to machine but hard to join to other metals. It is often used in the fabrication of vacuum collars for glass bell jar systems which are sealed with elastomer O-rings, and in some internal fixturing. It is more difficult to weld aluminum than stainless steel, but leak-tight seams can be made by a skilled welder. Because of these difficulties and its inability to make a seal via a metal gasket, aluminum has been largely bypassed in modern vacuum system construction. Recently, however, it has been reexamined for use in the construction of chambers for very large high-energy particle accelerators and storage rings [26]. Its high electrical and thermal conductivity and low cost are an asset in the construction of beam tubes. Because its atomic number is less than that of iron, chrome, or nickel, its residual radioactivity is less than that of stainless steel. In this application explosively bonded aluminum-to-stainless-steel sections are used to make the transition to stainless steel flanges [26]. Ishimaru [33] has attempted to eliminate stainless steel altogether by designing a system of flanges and bolts of high-strength aluminum alloy for use with aluminum O-rings.

For ordinary laboratory high vacuum systems stainless steel is the preferred material of construction. It has a high yield strength, is easy to fabricate, and is stable. The stainless steels used in vacuum systems

Table 6.3 Outgassing Rates of 316L Stainless Steel After Different Processing Conditions[a]

Sample	Surface Treatment	Outgassing Rates (10^{-10} W/m^2)				
		H_2	H_2O	CO	Ar	CO_2
A	Pumped under vacuum for 75 h.	893	573	87	-	13.
	50 h vacuum bakeout at 150°C	387	17	6	-	0.4
B	40 h vacuum bakeout at 300°C	83	0.7	2.2		0.01
C	Degassed at 400°C for 20 h in a vacuum furnace (6.5×10^{-7} Pa)	19	0.3	0.44	0.16	0.11
D	Degassed at 800°C for 2 h in a vacuum furnace (6.5×10^{-7} Pa)	3.6	-	0.07	-	0.05
	Exposed to atmosphere for 5 mo, pumped under vacuum for 24 h	-	73	67	-	13.
	20-h vacuum bakeout at 150°C	3.3	-	0.08	-	0.04
E	2 h in air at atmospheric pressure at 400°C	17	-	1.12	-	0.4
	Exposed to atmosphere for 5 mo, pumped under vacuum for 24 h	-	80	69	-	33.
	20-h vacuum bakeout at 150°C	17	0.75	0.37	-	0.17
F	2 h in oxygen at 27,000 Pa at 400°C	600	253	-	123	-
	20-h vacuum bakeout at 150°	5.2	0.09	0.4	0.51	-
G	2 h in oxygen at 2700 Pa at 400°C	-	20	13	8.7	-
	20-h vacuum bakeout at 150°C	-	0.9	0.64	0.45	-
H	2 h in oxygen at 270 Pa at 400°C	-	16	52	19	-
	20-h vacuum bakeout at 150°C	5.7	3.2	0.36	2	-

Source. Reprinted with permission from *J. Vac. Sci. Technol.*, **14**, 210, R. Nuvolone. Copyright 1977, The American Vacuum Society.
[a] All samples were first degreased in perchloroethylene vapor at 125°C, ultrasonically washed for 1 h in Diversey 708 cleaner at 55°C, rinsed with clean water, and dried.

are part of a family of steels characterized by an iron-carbon alloy that contains greater than 13% chrome. The 300 series of austenitic steels is the most frequently used in vacuum and cryogenic work because it is corrosion resistant, easy to weld, and nonmagnetic. The maintenance of corrosion resistance, and methods of increasing strength and reducing porosity in AISI 300 series stainless steels have been discussed by Geyari [34] and are summarized here.

The AISI 300 series is basically an "18/8" steel that contains 18% chrome and 8% nickel. To this basic composition additions and changes are made to improve its properties. Figure 6.8 outlines some of the 300 series alloys and their characteristics. Appendix C.8 contains a more complete description of the properties and uses of the entire series.

Carbide precipitation in stainless steel is a concern in certain applications. Carbon that has precipitated at grain boundaries as a result of welding or improper cooling after annealing removes with it a substantial fraction of the chrome from nearby regions. See Fig. 6.9. The nearby regions then contain less than 13% chrome and are no longer stainless steel. They are subject to corrosion if exposed to a corrosive atmosphere; subsequent baking in the presence of water vapor can increase hydrogen permeation through the affected region. The formation of microcracks is also a concern for stainless steel subjected to low temperatures. A crack in a carbide-rich zone can cause a leak in a cold trap or cold finger.

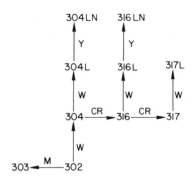

Fig. 6.8 Stainless steels used in vacuum equipment (AISI designation) CR = corrosion resistance. W = weldability, Y = yield strength, and M = machinability. Reprinted with permission from *Vacuum*, **26**, p. 287, C. Geyari. Copyright 1976, Pergamon Press, Ltd.

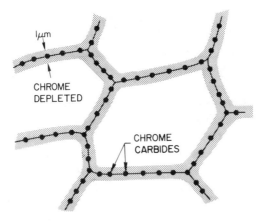

Fig. 6.9 Chrome carbide precipitation in the grain boundaries. The chrome depleted areas are shown in gray. Reprinted with permission from *Sheet Metal Industries*, February 1970, p. 93, C. H. Rosendahl. Copyright 1970, Fuel and Metallurgical Journals, Ltd.

Carbide precipitation may be prevented by the use of an alloy with a low carbon content, a stabilized alloy, or a minimum-heat welding technique. The best solution is the use of low-carbon steel alloys such as 304L and 316L, but they are not so strong as their higher carbon counterparts and may not suffice for some applications, because they require more nickel and are more expensive to manufacture. Titanium, niobium, and tantalum form carbides more easily than chrome; so that an alternative solution is the use of an alloy (321, 347, or 348) that is stabilized with one of these elements. A minimum heat weld will also reduce the time that the metal in the weld region spends in the dangerous 500 to 900°C temperature region. An 18/8 alloy of 0.06% carbon will not precipitate carbides at the grain boundaries if heated to this temperature, provided that it is cooled from 900 to 550°C in less than 5 min. A steel with a lower carbon content can remain in this temperature range for a longer time without carbide precipitation [34]. Heat reduction to the weld can also be accomplished by the use of weld relief grooves on each side of the part to be joined. It is assumed that all welding is done by the tungsten inert-gas (TIG) process to avoid oxidation. TIG welding, also known as heliarc or argon arc welding, is a technique for forming clean, oxide-free, leak-tight joints by flooding the arc with an inert gas, usually argon.

For some applications such as cryogenic vacuum vessels it is desirable to reduce the thickness of the structural steel wall to reduce heat losses. This can be accomplished without loss of any other properties by the use of a nitrogen-bearing alloy such as 304LN or 316LN or by cold stretching and annealing.

A most important concern of the high and and ultrahigh vacuum user is the proper fabrication of stainless steel flanges and components to eliminate leaks through inclusions in the metal. These minute inclusions, which occur in the process of making the steel, are masked by grease and other impurities until the steel wall is thoroughly baked out. At that time minute leaks will appear.

As a steel ingot cools the impurities distribute themselves at the top and center of the ingot (Fig. 6.10). The portion with most of the oxide and sulfide impurities is removed before rolling and any remaining impurities are stretched into long narrow leak paths. The inclusions are in the direction of rolling (see Fig. 6.10). It is important to be aware of this fact when selecting the raw stock from which compo-

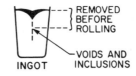

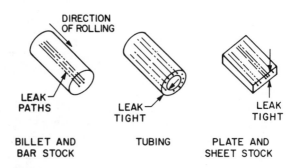

Fig. 6.10 Schematic-inclusions in steel during casting and rolling. Reprinted with permission from Varian Report VR-39, *Stainless Steel for Ultra-high Vacuum Applications*, V. A. Wright, Copyright 1966, Varian Associates, 611 Hansen Way, Palo Alto, CA.

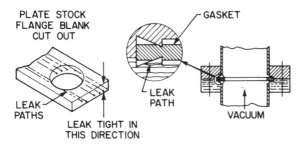

PLATE STOCK
FLANGE BLANK
CUT OUT

GASKET

LEAK
PATH

LEAK
PATHS

LEAK TIGHT IN
THIS DIRECTION

VACUUM

Fig. 6.11   Porosity in high vacuum flanges. Reprinted with permission from Varian Report VR-39, *Stainless Steel for Ultra-high Vacuum Applications*, V. A. Wright. Copyright 1966, Varian Associates, 611 Hansen Way, Palo Alto, CA.

nents are to be fabricated.   Figure 6.11 illustrates the origin of leak paths in high vacuum flanges made of plate stock.   To avoid such potential leaks modern flanges are fabricated from bar stock, as shown in Fig. 6.10.   In addition, some manufacturers make their flanges of steel that has been cross forged.   Cross forging breaks up the long filamentary inclusions and reduces the potential for leaks.   For critical applications the inclusion index is also individually measured after the ingot is rolled to the desired diameter.

6.3   GLASSES AND CERAMICS

A glass is an inorganic material that solidifies without crystallizing. The common glasses used in vacuum technology are formulated from a silicon oxide base to which other oxides have been added to produce a product with specific characteristics.   Soft glasses are formed by the addition of sodium and calcium oxides (soda-lime glass) or lead oxide (lead glass) and hard glasses are formed by the addition of boric oxide (borosilicate glass).   Table 6.4 lists the chemical composition and physical properties of glasses encountered in vacuum applications.

The physical properties of a glass are best described by the temperature dependence of the viscosity and expansion coefficient.   Because glass has no definite melting point, its important temperatures have been defined by specific viscosities.   At the strain point ($10^{15.5}$ Pa-s) stresses are relieved in hours; the annealing temperature represents a viscosity (10^{14} Pa-s) at which stresses are relieved in minutes.   At the softening point the viscosity is about $10^{8.6}$ Pa-s; the working point corresponds to a viscosity of 10^5 Pa-s.   Glass is brittle, and because of

Table 6.4   Properties of Some Glasses Used in Vacuum Applications

Property	Fused Silica	Pyrex 7740	7720[a]	7052[a]	Soda 0080	Lead 0120
Composition						
SiO_2	100	81	73	65	73	56
B_2O_3		13	15	18		
Na_2O		4	4	2	17	4
Al_2O_3		2	2	7	1	2
K_2O				3		9
PbO				6		29
LiO				1		
Other				3	9	
Viscosity characteristics						
Strain point °C	956	510	484	436	473	395
Annealing point °C	1084	560	523	480	514	435
Softening point °C	1580	821	755	712	696	630
Working point °C	-	1252	1146	1128	1005	985
Expansion Coefficient × $10^{-7}/°C$	3.5	35.0	43	53	105	97
Shock Temp., 1/4-in. plate °C	1000	130	130	100	50	50
Specific gravity	2.20	2.23	2.35	2.27	2.47	3.05

Source. Reprinted with permission from Corning Glass Works, Corning, NY.
[a] 7720 glass is used for sealing to tungsten and 7052 glass is used for sealing to Kovar.

its high thermal expansion and low tensile strength it can shatter if unequally heated. Its expansion coefficient is therefore important when selecting the components of a glass-to-metal or glass-to-glass seal; for example, Corning 7720 glass is formulated to seal to tungsten, whereas the expansion coefficient of 7052 glass matches Kovar. The expansion coefficients of soda-lime and borosilicate glasses are so different that they cannot be sealed directly to each other but only through a graded seal consisting of 5 to 7 intermediate glasses whose expansion coefficients differ successively from one another by about $10 \times 10^{-7}/°C$.

The viscosity-temperature and thermal expansion characteristics determine the suitability of a glass for a specific application. Borosilicate glasses are used whenever the baking temperature exceeds 350°C, whereas fused silica is required for temperatures higher than 500°C. The thermal expansion coefficient and strength determine the maximum temperature gradient that a glass can withstand and to what it can be sealed. Glasses are used in the production of vacuum apparatus, ion gauge tubes, view ports, metal-to-glass seals, and internal electrical and thermal insulation.

Solder glasses, a third class of glasses, have low melting points and are used to make glass-to-glass, glass-to-metal, or ceramic-to-metal seals. A review of modern seal glasses has been published by Takamori [35].

A ceramic is a polycrystalline nonmetallic inorganic material formed under heat treatment with or without pressure. Ceramics are mechanically strong, with high dielectric breakdown strength and low vapor pressures. The general class of ceramics includes glass bonded crystalline aggregates, and single-phase compounds such as oxides, sulfides, nitrides, borides, and carbides. Because ceramics are formed by processes such as sintering wetted powers, they contain entrapped gas pores and are not so dense as their crystalline counterparts. Their physical properties improve as their density approaches that of the bulk. Alumina is made with densities that range from 85 to almost 100% of its bulk density; most ceramics have a density of about 90% of the bulk. The important physical properties of ceramics are their compression and tensile strength and thermal expansion coefficient. High-density alumina, for example, has a tensile strength between four and five times greater than glass and a compression strength 10 times greater than its tensile strength. Because of its high compression strength, it is not necessary to have a match as close as that of glass to a metal to which it is being sealed. This results in a more rugged seal than is possible between glass and metal. The properties of some ceramics are listed in Table 6.5. Alumina (Al_2O_3) is the most commonly used ceramic in applications such as high-vacuum feed-throughs and internal electrical and thermal standoffs. Machinable glass ceramic also finds wide application in the vacuum industry for fabricating precise and complicated shapes. It is a recrystallized mica ceramic whose machinability is derived from the easy cleavage of the mica crystallites.

Borides and nitrides have found applications in vacuum technology. Evaporation hearths are made from titanium diboride and titanium nitride, alone or in combination. They are available in machinable and pyrolytically deposited form. Forsterite ceramics ($2MgO:SiO_2$) are used in applications in which low dielectric loss is needed, and beryllia (BeO) is used when high thermal conductivity is necessary. Beryllia must be machined while carefully exhausting the dust, because it is extremely hazardous to breathe. General reviews of ceramics and glasses are attributed to Espe [36], and Kohl [37].

Table 6.5   Physical Properties of Some Ceramics

Ceramic	Main Body Composition	Expansion Coefficient ($\times 10^{-7}$)	Softening Temp. (°C)	Tensile Strength (10^6 kg/m^2)	Specific Gravity
Steatite	MgOSiO$_2$	70-90	1400	6	2.6
Forsterite	2MgOSiO$_2$	90-120	1400	7	2.9
Zircon porcelain	ZnO$_2$SiO$_2$	30-50	1500	8	3.7
85% alumina	Al$_2$O$_3$	50-70	1400	14	3.4%
95% alumina	Al$_2$O$_3$	50-70	1650	18	3.6%
98% alumina	Al$_2$O$_3$	50-70	1700	20	3.8%
Pyroceram 9606[a]	Cordierite ceramic	57	1250	14[b]	2.6
Macor 9658[a]	Fluoro-phlogopite	94	800	10[b]	2.52

Source. Reprinted with permission from Vacuum, 25, p. 469, G. F. Weston. Copyright 1975, Pergamon Press, Ltd.
[a] Reprinted with permission from Corning Glass Works, Corning, NY.
[b] Modulus of rupture.

Permeation of gas through glasses and ceramics occurs without molecular dissociation; the permeation constant, which is given by (6.7), depends on the molecular diameter of the gas and the microstructure or porosity of the glass or ceramic. Norton [38] has shown that the permeation of gases through a glass is a function of their molecular diameter. Figure 6.12 contains data for the temperature dependence of the permeation rate of He, D$_2$, Ne, Ar, and O$_2$ through silicon oxide glasses [2]. With the exception of deuterium and hydrogen, it shows that the permeation rate decreases as the molecular diameter increases. Norton suggests that the measured permeation rate of hydrogen is much larger than predicted because of surface reactions and solubility effects. The permeation of a gas through a glass depends on the size of the pores in relation to the diameter of the diffusing species. Permeation is minor through a crystalline material such as quartz but increases with lattice spacing. The non-network-forming Na$_2$O, which is added to SiO$_2$ to form soda glass, plugs these openings (Fig. 6.13) and causes the permeation rate to decrease. Permeation rates for some glasses are given in Fig. 6.14. Shelby [46] has reviewed the diffusion and solubility of gases in glass, and the thermal outgassing properties and helium permeation of Corning

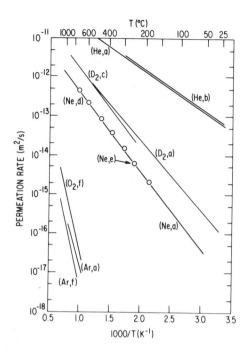

Fig. 6.12 Permeability of He, D_2, Ne, Ar, and O_2 through silicon oxide glasses. Data taken from (a) Perkins and Begeal [39]; (b) Swets, Lee and Frank [40]; (c) Lee [41]; (d) Frank, Swets, and Lee [42]; (e) Shelby [43]; and (f) Norton [38]. Reprinted with permission from *J. Vac. Sci. Technol*, **10**, p. 543, W. G. Perkins. Copyright 1973, The American Vacuum Society.

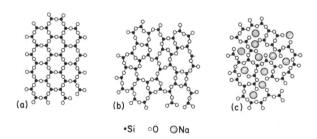

•Si   ∘O   ◯Na

Fig. 6.13 (a) Atomic arrangement in a crystalline material possessing symmetry and periodicity; (b) the atomic arrangement in a glass; (c) the atomic arrangement in a soda glass. Reprinted with permission from the *J. Am. Ceram. Soc.*, **36**, p. 90, F. J. Norton. Copyright 1953, The American Ceramic Society.

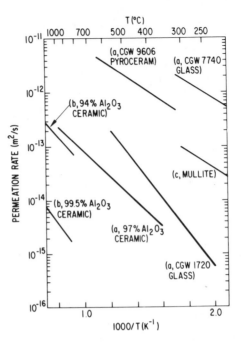

Fig. 6.14 Helium permeability through a number of glasses and ceramics. After Perkins. Data taken from (a) Miller and Shepard [44]; (b) Edwards [45]; (c) Shelby [43]. Reprinted with permission from *J. Vac. Sci. Technol.*, **10**, p. 543, W. G. Perkins. Copyright 1973, The American Vacuum Society.

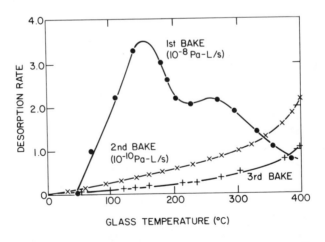

Fig. 6.15 Desorption of water from a Pyrex-glass surface of 180 cm^2 at increasing temperature. Adapted with permission from *Vacuum*, **15**, p. 573, K. Erents and G. Carter. Copyright 1965, Pergamon Press, Ltd.

Macor machinable glass ceramic have been reviewed by Altemose and Kacyon[47].

Gases are physically and chemically soluble in molten glass. The gas on the surface of solid glass is primarily water with some carbon dioxide. Water vapor may exist on glass in layers as thick as 10 to 50 monolayers [48]. Erents and Carter [49] have shown that the first bake of a glass surface releases considerable surface water, whereas the second and succeeding bakes release structural water (Fig. 6.15). Todd [48] has shown that this release of structural water is proportional to $t^{\frac{1}{2}}$ and indicates a diffusion-controlled process. Todd concludes that a high temperature bake should completely eliminate outgassing of water from glass because all the surface water is released in a high temperature bake and the diffusion constant of water vapor is negligible at room temperature. At 25°C gas evolution from glass is dominated by the permeation of helium. Outgassing rates of some unbaked ceramics and glasses are given in Appendix C.3, and Colwell [50] has tabulated the outgassing rates of more than 80 untreated refractory and electrical insulating materials used in the construction of vacuum furnaces.

6.4 POLYMERS

Polymeric materials find applications in several areas of vacuum technology. Elastomers, such as Viton and Buna-N, are used to form O-ring gaskets for static, sliding, or rotary seals. Other polymers such as Vespel, Kapton, Delrin, Teflon, and polyethylene, are used for high-voltage vacuum feed-throughs, and storage boxes for thin-film samples or cleaned metal and glass components. Low vapor pressure epoxies are used for repairing leaks. These materials have certain mechanical and physical qualities that make them suitable for use in vacuum applications.

The important mechanical properties of elastomers are elasticity, plasticity, and hardness. Elastomers are formed by grinding the starting polymer, mixing with plasticizers and stabilizers, and vulcanizing to a state that is largely elastic. These materials are incompressible; that is, any deformation or compression in one direction mut be accompanied by motion in another, so that the total volume remains constant. If the compound is completely elastic, it will return to its exact shape after the applied force has been removed, but if it has some degree of plastic behavior it will flow and not return to its original shape. The measure of its shape change is called compression set. A material like

Teflon with a high compression set will flow and make a poor O-ring seal.

O-ring seals are made by squeezing the elastomer freely between two flat surfaces or in a confined groove of rectangular or triangular cross section; this provides a gas-tight seal between the metal or glass and the polymer.   Many vacuum technology texts and O-ring manufacturers supply tables of groove depths and widths for various O-ring sizes.   These tabulations result in chord compressions ranging from 20 to 38% of the cord diameter and groove filling ratios (chord cross section area/groove cross section area) of 74 to 102%.   In contrast to these many empirical tabulations there is little discussion about the phenomenon of sealing.   Sessink and Verster [51] studied the design of O-ring vacuum seals and found that the general criterion, or principle parameter, for high vacuum sealing is a minimum initial contact pressure of 13 kg/cm^2 for gaskets in the hardness range of 60 to 75 shore. The initial pressure will be slightly reduced with time as the elastomer undergoes some plastic deformation, but it will not affect the integrity of the vacuum seal.   Because sealing is determined by contact pressure, the compression force, deformation, and percentage of groove filling all decrease as the hardness of the O-ring is increased.

Permeation and outgassing are two important physical properties of elastomers.   Gases diffuse through voids in the intertwined polymer chains by a thermally activated process.   The dimensions of the voids are larger than in a glass or metal, and as a result the permeation constants are much larger than for those materials.   Comparison of the permeation rates of gases through polymers, which are given in Appendix C.6, with those for glasses given in Fig. 6.12 shows that the diffusion process in a polymer is not so sensitive to molecular diameter as it is in a glass.   The implication of this is simply that the diffusion of air and other heavy gases through polymers is a serious problem, whereas helium is the only gas of any consequence to diffuse through glass. Elastomers will swell when in contact with certain solvents used for leak detection or cleaning.   This swelling, or increased spacing between molecules, results in an increased permeability [3].

The outgassing of unbaked elastomers has been studied extensively [2, 52–58] and shown to be dominated by the evolution of water vapor.   Baking will reduce this gas load as revealed in the mass scans (Figs. 5.7 and 5.8) taken during the heating of Buna-N and Viton.   As the bakeout temperature was increased, plasticizers added to the polymer before vulcanization were released.   Unreacted polymer is also evolved.   At a higher temperature the elastomer begins to decompose. Sigmond [59], who conducted a detailed study of the evolution of volatile substances from elastomers and plastics, noted that the pertinent outgassing properties of these materials were not dominated by the polymer but by water and the plasticizers and stabilizers of proprietary composition and quantity that were altered from lot to lot without the consumer's knowledge. This led to the observation that one derives from a polymer whatever the manufacturer chooses to put into it, often for reasons unconnected with the user's needs and always at odds with the user's expectations [60].

Cleaning an elastomer by a solvent wash is an ineffective way to reduce outgassing. One effective way is a simple vacuum bake. An unbaked Viton O-ring will have an initial outgassing rate of 10^{-3} W/m^2 (see Appendix C.4). After a 4 h bake at 150°C and 12 h of pumping this value is reduced to 4×10^{-7} W/m^2. The latter value corresponds to approximately 2×10^{-8} Pa-L/s per linear centimeter of gasket material. If the gaskets are not given a vacuum bake before use, their ultimate outgassing rate will be approximately 100 times higher or about 2×10^{-6} Pa-L/s per linear centimeter. Reexposure to atmosphere will result in increased outgassing because some water will be readsorbed. All elastomers are sponges for water vapor and other atmospheric gases; an observation that can be dramatically illustrated by observing a mass scan of a chamber when compressing a gasket in a valve seat.

The most commonly used elastomers for O-ring seals are Buna-N and Viton. A 200°C bake will release adsorbed gases, unreacted polymer, and plasticizers from Viton [51], it cannot, however, be baked at higher temperature because it will decompose. Buna-N is used in less demanding situations in which cost is a factor. Silicone has an unusually high permeation rate and is infrequently used as a gasket material in very high vacuum systems. It is better suited to high-temperature vacuum processes operating at moderate pressures. Silicone compounds are formulated for use at temperatures up to 300°C. Polyimide has a low outgassing rate but absorbs large amounts of water when reexposed. Elastomer gaskets are widely used in systems that pump to the 10^{-6} Pa range, and since the development and widespread application of the copper gasket seal there has been little reason to perfect an elastomer with a lower permeability or to use cooled gaskets [61].

Sigmond [59] found that a volatile organic of mass number 149 was present in ethyl alcohol stored in polyethylene bottles and in samples stored in test tubes stoppered with plastic caps. This organic vapor turned out to be a fragment of di-butyl-phthalate, a commonly used plasticizer. Because this fragment was easily removed by a 100°C bake, it is suggested that it contaminates all materials stored in plastic bags or boxes or touched by plastic gloves.

REFERENCES

1. G. F. Weston, *Vacuum*, **25**, 469 (1975).
2. W. G. Perkins, *J. Vac. Sci. Technol.*, **10**, 543 (1973).
3. R. J. Elsey, *Vacuum*, **25**, 299 (1975).
4. R. J. Elsey, *Vacuum*, **25**, 347 (1975).
5. B. B. Dayton, *Trans. 8th Vac. Symp. and Proc. 2nd Int. Congr. on Vac. Sci. Technol. (1961)*, Vol. 1, Pergamon, New York, 1962, p.42.
6. B. B. Dayton, *Trans. 7th Vac. Symp. (1960)*, Pergamon, New York, 1961, p. 101.
7. D. J. Santeler et al., *Vacuum Technology and Space Simulation*, NASA SP-105, National Aeronautics and Space Administration, Washington, DC, 1966, p. 188.
8. D. G. Bills, *J. Vac. Sci. Technol.*, **6**, 166 (1969).
9. See, for example, P. A. Redhead, J. P. Hobson, and E. V. Kornelsen, *The Physical Basis of Ultrahigh Vacuum*, Chapman and Hall, London, 1968, Chapter 4.
10. M. J. Drinkwine and D. Lichtman, *Prog. Surf. Sci.*, **8**, Pergamon, New York, 1977, p. 123.
11. D. Menzel and R. Gomer, *J. Chem. Phys.*, **41**, 3311 (1964).
12. P. Redhead, *Can. J. Phys.*, **42**, 886 (1964).
13. More recently it has been proposed that the incident electron ejects a core electron of the adsorbate leaving the atom in a nonbonding state after the decay of a valence electron. M. L. Knotek and P. J. Feibelman, *Phys. Rev. Lett.*, **40**, 964 (1978).
14. P. A. Redhead, *Vacuum*, **12**, 267 (1962).
15. T. E. Hartman, *Rev. Sci. Instr.*, **34**, 1190, (1963).
16. P. Marmet and J. D. Morrison, *J. Chem. Phys.*, **36**, 1238 (1962).
17. H. F. Winters and P. Sigmund, *J. Appl. Phys.*, **45**, 4760 (1974).
18. E. Taglauer and W. Heiland, *J. Appl. Phys.*, **9**, 261 (1976).
19. M. J. Drinkwine and D. Lichtman, *J. Vac. Sci. Technol.*, **15**, 74 (1978).
20. F. J. Norton, *Trans. 8th Nat. Vac. Symp. and Proc. 2nd Int. Congr. Vac. Sci. Technol. (1961)*, **1**, Pergamon, New York, 1962, p. 8.
21. D. R. Begeal, *J. Vac. Sci. Technol.*, **15**, 1146 (1978).
22. The Diversey Co. Chicago, Ill.
23. W. A. Rogers, R. S. Buritz, and D. L. Alpert, *J. Appl. Phys.*, **25**, 868 (1954).
24. R. Calder and G. Lewin, *Brit. J. Appl. Phys.*, **18**, 1459 (1967).
25. R. S. Barton and R. P. Govier, *Vacuum*, **20**, 1 (1970).
26. R. P. Govier and G. M. McCracken, *J. Vac. Sci. Technol.*, **7**, 552 (1970).
27. H. J. Halama and J. C. Herrera, *J. Vac. Sci. Technol.*, **13**, 463 (1976).
28. J. R. Young, *J. Vac. Sci. Technol.*, **6**, 398 (1969).
29. R. Dobrozemsky and G. Moraw, *Electron. Fis. Ap.*, **17**, 235 (1974).

30. R. L. Samuel, *Vacuum*, **20**, 195 (1970).
31. Y. E. Strausser, *Proc. 4th Int. Vac. Congr. (1968)*, Institute of Physics and Physical Society, London, 1969, p. 469.
32. R. Nuvolone, *J. Vac. Sci. Technol.*, **14**, 1210 (1977).
33. H. Ishimaru, *J. Vac. Sci. Technol.*, **15**, 1853 (1978).
34. C. Geyari, *Vacuum*, **26**, 287 (1976).
35. T. Takamori, *Treatise on Materials Science and Technology: Glass II*, **17**, M. Tomozawa and R. H. Doremus, Eds., Academic, New York, 1979, p. 117.
36. W. Espe, *Materials of High Vacuum Technology*, **2**, Pergamon Press, New York, 1966.
37. W. H. Kohl, *Handbook of Materials and Techniques for Vacuum Devices*, Reinhold, New York, 1967.
38. F. J. Norton, *J. Am. Ceram. Soc.*, **36**, 90 (1953).
39. W. G. Perkins and D. R. Begeal, *J. Chem. Phys.*, **54**, 1683 (1971).
40. D. E. Swets, R. W. Lee, and R. C. Frank, *J. Chem. Phys.*, **34**, 17 (1961).
41. R. W. Lee, *J. Chem. Phys.*, **38**, 448 (1963).
42. R. C. Frank, D. E. Swets, and R. W. Lee, *J. Chem. Phys.*, (**35**, 1451 (1961).
43. J. E. Shelby, *Phys. Chem. Glasses*, **13**, 167 (1972).
44. C. F. Miller and R. W. Shepard, *Vacuum*, **11**, 58 (1961).
45. R. H. Edwards, M. S. thesis, University of California, Berkeley, 1966.
46. J. E. Shelby, *Molecular Solubility and Diffusion*, in Ref. 35.
47. V. O. Altemose and A. R. Kacyon, *J. Vac. Sci. Technol.*, **16**, 951 (1979)
48. B. J. Todd, *J. Appl. Phys.*, **26**, 1238 (1955).
49. K. Erents and G. Carter, *Vacuum*, **15**, 573 (1965).
50. B. H. Colwell, *Vacuum*, **20**, 481 (1970).
51. B. Sessink and N. Verster, *Vacuum*, **23**, 319 (1973).
52. R. S. Barton and R. P. Govier, *J. Vac. Sci. Technol.*, **2**, 113 (1965).
53. R. R. Addis, Jr., L. Pensak, and N. J. Scott, Trans. *7th Vac. Symp. (1960)*, Pergamon, New York, 1961, p. 39.
54. R. Geller, *Le Vide*, No. 13, **71** (1958).
55. M. M. Fluk, and K. S. Horr, Trans. *9th Vac. Symp. (1962)*, Macmillan, 1963, p. 224.
56. M. Munchhausen and F. J. Schittko, *Vacuum*, **13**, 548 (1963).
57. P. Hait, *Vacuum*, **17**, 547 (1967).
58. J. Blears, E. J. Greer, and J. Nightengale, *Adv. Vac. Sci. Technol.*, **2**, E. Thomas, Ed., Pergamon, New York, 1960, p. 473.
59. T. Sigmond, *Vacuum*, **25**, 239 (1975).
60. Murphy, as quoted by Sigmond.
61. T. M. Miller and K. A. Geiger, Trans. *9th Vac. Symp. (1962)*, Macmillan, 1963, p. 270.

Production

Vacuum pumps are often classified according to the physical or chemical phenomena responsible for their operation. In practice this is a bit awkward because some pumps combine two or more principles to pump a wide range of gases or to pump over a wide pressure range. In this section the discussion of pump operation is divided into three chapters. Chapter 7 discusses mechanical vacuum pumps. Rotary vane, rotary piston, and Roots pumps operate by displacing gas from the work chamber to the pump exhaust. Rotary vane and piston pumps operate in the low vacuum region, whereas the Roots pump operates in the medium vacuum region. The turbomolecular pump is a mechanical high-vacuum pump. It transports gas from regions of low pressure to high pressure by momentum transfer from high-speed blades.

Chapter 8 is devoted to the diffusion pump, which, like the turbomolecular pump, is a momentum transfer pump. Although often maligned, the diffusion pump is the mainstay of the high vacuum industry; its operation should be thoroughly understood by all vacuum-equipment users.

Chapter 9 discusses entrainment pumps. Entrainment pumps bind particles to a surface instead of expelling them to the atmosphere. Getter pumps, like the titanium sublimator, remove gases by chemical reactions that form solid compounds; ion pumps ionize gas molecules and imbed them in a wall. The sputter-ion pump combines gettering and ion pumping. Other entrainment pumps are based on condensation and sorption. Sorption pumps physiorb gas molecules in materials of high surface area. These surfaces are usually cooled to enhance their pumping ability. A modern cryogenic pump uses at least two stages of cooling. The warmer stage pumps by condensation or ad-

sorption on a cooled metallic surface, the colder stage uses in addition
an adsorbate such as charcoal or zeolite. Shown below are the operating pressure ranges of many pumps and pump combinations.

These three chapters do not cover all of the many techniques by
which high vacuum may be achieved. Their purpose is to review the
operation of commonly used pumps in a concise manner and to amplify
the treatment of turbomolecular and cryogenic pumps. The latter
warrant an expanded treatment in view of the rapid changes those
technologies have undergone in recent years.

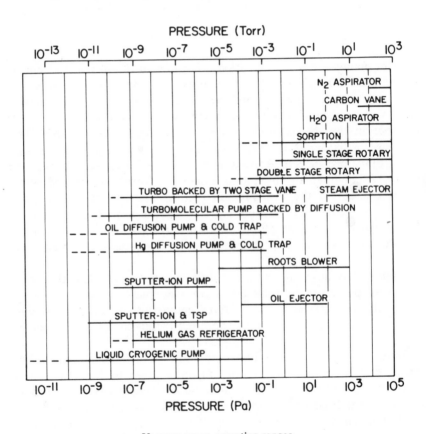

Vacuum pump operating ranges

CHAPTER 7

Mechanical Pumps

In this section we review the operation of two low vacuum pumps (the rotary vane and piston), one medium vacuum pump (the Roots pump), and one high vacuum pump (the turbomolecular pump). In the last two decades the turbomolecular pump has been considerably improved. A detailed review of the turbomolecular pump is in order because of its widespread application in many systems.

7.1 ROTARY VANE PUMPS

The rotary piston pump and the rotary vane pump are two types of oil sealed pumps commercially available for pumping gas in the pressure range of 1 to 10^5 Pa.. Of the two the rotary vane is the most commonly used in small to medium-sized vacuum systems. Rotary vane pumps of 10 to 200 m^3/h displacement are used for rough pumping and for backing diffusion or turbomolecular pumps.

In a rotary vane pump (Fig. 7.1), gas enters the suction chamber (A) and is compressed by the rotor (3) and vane (5) in region B and expelled to the atmosphere through the discharge chamber (C). The exhaust seal is made by the discharge valve (8) and the oil above the valve. An airtight seal is made by two sliding vanes and the closely spaced sealing surfaces (10). Both vanes and surfaces between the rotor and housing are sealed by the low vapor-pressure oil, which also serves to lubricate the pump and fill the volume above the discharge valve. Pumps that use a speed-reduction pulley operate in the 400-to-

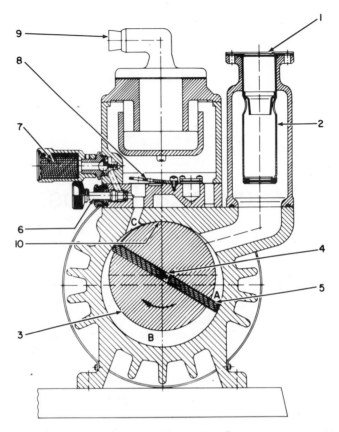

Fig. 7.1 Sectional view of Pfeiffer DUO-35, 35-m^3/hr double-stage, rotary vane pump: (1) intake, (2) filter, (3) rotor, (4) Spring, (5) vane, (6) gas ballast valve, (7) filter, (8) discharge valve, (9) exhaust, (10) sealing surface. Reprinted with permission from A. Pfeiffer Vakuumtechnik, G.m.b.H., Wetzlar, West Germany.

600 rpm range, whereas direct drive pumps operate at speeds of 1500 to 1725 rpm. The oil temperature is considerably higher in the direct drive pumps than in the low speed pumps, typically 80 and 60°C, respectively. These values, however, will vary with the viscosity of the oil and the quantity of air being pumped.

Single stage pumps consist of one rotor and stator block (Fig. 7.1). If a second stage is added, as shown schematically in Fig. 7.2, by connecting the exhaust of the first stage to the intake of the second, lower pressures may be reached. Physically, the second stage is located adjacent to the first and on the same shaft. The pumping speed

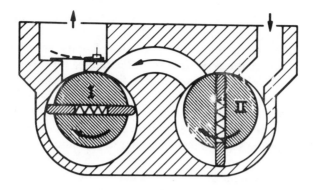

Fig. 7.2   Schematic section through a two-stage rotary vane pump.   Reprinted with permission from *Vacuum Technology*, Leybold-Heraeus, G.m.b.H., Postfach 51 07 60, 5000 Köln, West Germany.

characteristics of single-stage and two-stage rotary vane pumps are shown in Fig. 7.3. The free air displacement and the ultimate pressure are two measures of the performance of roughing pumps. The free air displacement is the volume of air displaced per unit time by the pump at atmospheric pressure with no pressure differential. For the two pumps whose pumping speed curves are shown in Fig. 7.3 this has the value of 30 m³/h (17.7 cfm) at a pressure of 10^5 Pa (1 atm). At the

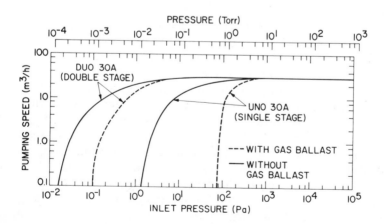

Fig. 7.3   Pumping speed curves for the Pfeiffer UN030A and DUO 30A rotary vane pumps. Reprinted with permission from A. Pfeiffer Vakuumtechnik, G.m.b.H., Wetzlar, West Germany.

ultimate pressure of the blanked-off pump the speed drops to zero because of leakage around seals, dissolved gas in the oil, and trapped gas in the volume below the discharge valve. Rotary vane pumps have ultimate pressures in the 3×10^{-3} to 1 Pa range; the lowest ultimate pressures are achieved with two-stage pumps. The single- and two-stage pumps characterized in Fig. 7.3 have ultimate pressures of 1.4 and 1.5×10^{-2} Pa, respectively. These ultimate pressures are obtained with new pumps and clean, low-vapor pressure oil. As the oil becomes contaminated and the parts wear the ultimate pressures will degrade.

When large amounts of water, acetone, or other condensable vapors are being pumped, condensation will occur on the interior walls. This condensation occurs during the compression stage after the vapor has been isolated from the intake valve. As the vapor is compressed it reaches its condensation pressure, condenses, and contaminates the oil before the exhaust valve opens. Condensation causes a reduction in the number of molecules in the vapor phase and delays or even prevents the opening of the exhaust valve. If condensation is not prevented, the pump will become contaminated, the ultimate pressure will degrade, and gum deposits will form on the moving parts. Some compounds will eventually cause the pump to seize. To avoid condensation and its resulting problems gas is admitted through the ballast valve. The valve is positioned to allow ballast, usually room air, to enter the chamber during the compression stage; the trapped volume is isolated from the intake and exhaust valves. This inflow of gas, which can be as much as 10% of the pump displacement, is controlled by valve 6 (Fig. 7.1). The added gas causes the discharge valve to open before it reaches the condensation partial pressure of the vapor. In this manner the vapor is swept out of the pump and no condensation occurs. The ultimate pressure of the pump is not so low with the gas ballast valve open. Gas ballasting can be used to differentiate contaminated oil from a leak. If the inlet pressure drops when the ballast valve is opened but drifts upward slowly after the valve is closed, the oil is contaminated with a high-vapor pressure impurity. Figure 7.2 shows the effect of full gas ballast on the performance of a single- and double-stage pump. More detailed aspects of gas ballasting are covered in the Leybold-Heraeus reference [1] and Van Atta [2] describes alternative methods of pumping large amounts of water vapor.

There are several good rules for operating rotary mechanical pumps. The exhaust should be vented outside the building. Most pumps are supplied with an oil mist separator, but it does not adequately remove all the oil vapors. In many laboratories and plants safety rules require the use of an outside vent. The vent hose should not run vertically from the exhaust connection because water or other vapors which have condensed on the cooler hose walls will flow into the pump and contaminate the fluid. A satisfactory solution to this problem is the addition of a sump at the exhaust connection to collect the vapors before they can flow into the pump. A roughing pump must also be vented at the time it is stopped to prevent oil from being forced back into the vacuum system by external air pressure. Venting is done automatically in some pumps and can be achieved in others by the addition of a vent valve above the inlet port. The oil level in mechanical pumps should be checked frequently, especially those that are used on systems regularly cycled to atmosphere. Small pumps of capacity less than 30 m^3/h have oil consumption rates (cm^3/h) of about 10^{-6} to $10^{-5}PS$ where P is the inlet pressure in Pa and S is the inlet speed in m^3/h [3]. Larger pumps will use more. The oil should be changed when the pump performance deteriorates or when it becomes discolored or contaminated with particulates. Poor oil maintenance is the major cause of mechanical pump failure. Ninety-five percent of all pump problems can be solved by flushing the pump and changing the oil.

7.1.1   Pump Fluids

The fluid used in a mechanical pump must provide a vacuum seal between the moving surfaces and lubricate the bearings and sliding surfaces. In addition, it assists in transferring heat from the bearings to the pump surface or cooling jacket. To make a vacuum seal the fluid must have low vapor pressure and be viscous enough to form a film that will fill the gap between the moving surfaces. Within a class of fluids vapor pressure is usually inversely proportional to viscosity. Lubricity of a fluid is dependent on its ability to wet the metal surfaces and form a film of shear strength sufficient to support the bearing load. One of the best fluids for this application is a hydrocarbon mineral oil that has been distilled to yield a reasonably narrow range of molecular weights or viscosities. Such a fluid is superior to mechanical pump fluids formulated from a broad spectrum or from a mixture of light and heavy components. The heavy components ultimately form tar deposits, whereas the light fractions increase the vapor pressure. The viscosity required in a particular pump depends on the clearances between the moving parts, the rotational speed, and the operating temperature of the pump. The properties of several fluids are listed in Table 7.1 and Appendixes F.1 and F.3.

Table 7.1   Typical Properties of Mechanical Pump Fluids

Trade Name	Chemical Name	MW (ave)	sp.gr. 25°C	Vapor Pressure (Pa)	Pour Point (°C)	Fire Point (°C)	Oxygen Service
Balzers P-3[4]	Hydrocarbon		0.88	7×10^{-3} (20°)	-16	260	No
Convoil® 20[5][a]	Hydrocarbon	400	0.86	5×10^{-5} (25°)	-8.9	258.9	No
Duo-Seal® 1407K[6]	Hydrocarbon		0.88	2×10^{-3} (50°C)	-6.7	240.6	No
Fomblin® Y-16[7]	Perfluoropoly ether	2600	1.9	4×10^{-3} (25°C)	-40	None	Yes[c]
Fomblin® Y-25[7]	Perfluoropoly ether	3000	1.9	3×10^{-3} (25°C)	-30	None	Yes[c]
Fyrquel® 220[8]	Phosphate ester	410	1.15	(b)	-17.8	365.6	No[e]
Halocarbon 100/100[9]	Polychloro trifluoroethylene	800	1.92	1×10^{-2} (25°C)	-17.8	None	Yes[d]
Halocarbon 25/160[9]	Polychloro trifluoroethylene	850	1.93	1×10^{-2} (25°C)	-9.4	None	Yes[d]
Inland 15[10]	Hydrocarbon	282	0.91	3×10^{-2} (25°C)	-23.3	223.9	No
Inland 19[10]	Hydrocarbon	282	0.88	1×10^{-2} (25°C)	-15	244.5	No
Versilube F-50®[11]	Chlorophenyl-methylpolysiloxane	3000	1.05	(b)	-73	337.8	No[e]

[a] Convoil-20 is a diffusion pump fluid that can be used in rotary vane mechanical pumps.
[b] Vapor pressure is a function of fluid temperature–pressure history; blank-off pressures of order 1 Pa are obtainable,
[c] recommended by the manufacturer for oxygen service;
[d] recommended by the manufacturer and NAEC [12] for oxygen service; not recommended for use in pumps with aluminum threaded connections,
[e] these fluids have a high flash point and are oxidation resistant, but not fireproof.

For some applications a hydrocarbon oil is totally unsuitable because it reacts with certain gases. One application is the pumping of pure oxygen. Special precautions must be taken when pumping pure oxygen because certain mixtures of oxygen and hydrocarbon vapor are explosive in the mechanical pump. These precautions, which include changing the fluid as well as modifying the pump, should also be taken on *any* vacuum system that uses pure oxygen supplied through a regulator and leak valve even if the amount added is only enough to pressurize the high vacuum chamber to the 10^{-2} Pa range. A valve or regulator fault could cause the chamber to be filled with oxygen to atmospheric pressure and result in an explosion. Systems that require oxygen for reactive sputtering, glow discharge cleaning, or oxide evaporation fall in the latter category; the use of a suitable fluid for these applications is mandated by many plant safety departments. A pump, preferably a new one, should be factory-prepared for such service. Modification will include degreasing, replacement of any oil impregnated bearings, replacement of any incompatible elastomer seals, and substitution of the hydrocarbon oil with a nonexplosive fluid. Synthetic fluids for use with oxygen or corrosive gases are listed in Table 7.1. Seal materials that do not react with the process gas and the synthetic fluid will be chosen. The exact procedure will depend on the fluid.

Some of the fluids that are safe for this service have a higher viscosity than the usual hydrocarbon oils and clearances between moving parts in rotary vane pumps may have to be increased. Phosphate ester slowly reacts with water vapor during pump operation. This reaction causes the base pressure of the pump to increase with time with the result that more frequent fluid changes will be needed than when a hydrocarbon oil is used. Phosphate ester and polysiloxane have higher fire points than hydrocarbon oil but are not as safe as a chlorofluorocarbon or fluorocarbon fluid. Although the latter fluids are more expensive than the others, they are safer. See Table 7.2. Silicone fluids are formulated with chlorine or sulfur to form a polar molecule that will bond to the surface and form a lubricating film. They are not so effective as hydrocarbon oils; some users of silicone-based fluids have reported problems with bearing siezures in rotary mechanical pumps.

Chlorofluorocarbon and fluorocarbon fluids react with few chemicals. The chance of either of these fluids creating a hazard when used in a vacuum pump is limited to a few extremely rare instances. Chlorofluorocarbon fluids should not be used in pumps with aluminum threaded connections like those found on the aluminum oil reservoirs of some mechanical vane pumps. The extremely high localized temperatures of minute seizures of aluminum have been reported to cause a chemical reaction with chlorofluorocarbon fluid which has resulted in detonation. They should not be placed in contact with amines, liquid fluorine, liquid boron trifluoride, or sodium or potassium metal. Laboratory experiments have revealed a potential but unlikely problem with perfluoropolyether. This fluid has been shown to decompose when sufficiently heated in the presence of Lewis acids. The trifluorides and trichlorides of aluminum and boron are examples of Lewis acids that may be generated in some high-gas-flow processes. If heated to a high temperature, Lewis acids act as depolymerization catalysts and break the carbon-oxygen bonds in the fluid. This potential reaction has never occurred in practice because the temperature in the pump is too low to allow the reaction to proceed. Neither chlorofluorocarbon nor perfluoropolyether is recommended for applications in which rubbing of magnesium or aluminum could occur because this rubbing action could expose fresh, unoxidized metal surfaces that could catalyze the decomposition of the fluid. This does not present a problem in either fluid at this time, as the moving parts of most pumps are made of cast iron.

Pumps modified for oxygen service should be permanently identified and used only with the specified fluid. Fluids must be compatible with the seal material and oxygen, and should not be mixed. The ultimate pressures of modified pumps will not be so low as others because some of the special fluids have high vapor pressures and, to some extent, because of the large clearances required with the more viscous fluids. Even so, the ultimate pressure attainable by a modified pump is low enough for the pump to back a diffusion or turbomolecular pump. The only exception is perfluoropolyether whose ultimate pressure is nearly equal to that obtained with a high-quality hydrocarbon oil. Some fluids also affect the calibration of thermal conductivity gauges by changing the emissivity of the heated wire. The potential for mechanical pump explosion when pumping pure oxygen is a hazard that the

Table 7.2   Approximate Costs of
Mechanical Pump Fluids[a]

Mechanical Pump Fluid	Cost ($/L)
Perfluoropolyether	260
Chlorofluorocarbon	75
Chlorosiloxane	19
Phosphate Ester	2.10
Hydrocarbon	1.20

[a] 1979 prices based on 55 gallon lots.

vacuum system user must face, a hazard that can be avoided with current technology. Corrosive and toxic gases may also be encountered in certain applications. These applications are treated in Chapter 12.

7.2   ROTARY PISTON PUMPS

Rotary piston pumps are used as roughing pumps on large systems alone or in combination with Roots blowers. They are manufactured in sizes ranging from 30 to 1500 m³/h. A piston pump is a rugged and mechanically simple pump.

Figure 7.4 shows a sectional view of a rotary piston pump. As the keyed shaft rotates the eccentric (1) and piston (2) gas is drawn into the space A. After one revolution that volume of gas has been isolated from the inlet, whereas the piston is closest to the hinge box. During the next revolution the isolated volume of gas (B) is compressed and vented to the exhaust through the poppet valve when its pressure exceeds that of the valve spring. Like the vane pump, the piston pump is manufactured in single and compound or multistage types.

The clearances between the piston and housing are typically 0.1 mm but will be three or four times larger near the hinge box. Because the clearances are greater in a piston than in a vane pump, the piston pump is more tolerant of particulate contamination. Oil is used to seal the spaces between fixed and moving parts and to lubricate and, like a vane pump, it must have low vapor pressure and good lubricity. As a result of the large clearances a rather viscous oil is used in the piston pump. In fact, the piston pump is more tolerant of a higher viscosity oil than a vane pump. There are no spring-loaded vanes to stick in a piston pump and all parts are mechanically coupled to a shaft that can always be powered by a large motor.

The rotational speed of the piston pump is typically 400 to 600 rpm, although some run as slow as 300 rpm and others as fast as 1200 rpm. The maximum rotational speed is limited by vibration from the eccentric. Small piston pumps are air cooled in the same manner as rotary vane pumps, whereas large pumps are water-cooled.

The pumping speed curves for a 51 m³/h single-stage rotary piston pump are shown in Fig. 7.5 with and without gas ballast. The shaft power is also given and it is seen to peak at a pressure of 4 × 10⁴ Pa with or without gas ballast. At lower pressures full gas ballast requires more than twice as much shaft power as without ballast. The ultimate pressure of the single-stage pump shown in Fig. 7.5 is 1 Pa however,

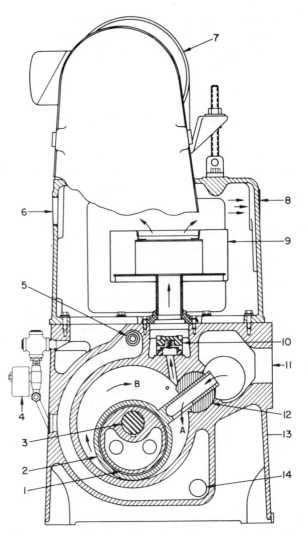

Fig. 7.4 Sectional view of a Stokes 212H, 255-m^3/h rotary piston pump: (1) eccentric, (2) piston, (3) shaft, (4) gas ballast, (5) cooling water inlet, (6) optional exhaust, (7) motor, (8) exhaust, (9) oil mist separator, (10) poppet valve, (11) inlet, (12) hinge bar, (13) casing, (14) cooling water outlet. Reprinted with permission from Stokes Division, Pennwalt Corp., Philadelphia, PA.

compound pumps with ultimate pressures approaching those of rotary vane pumps are available.

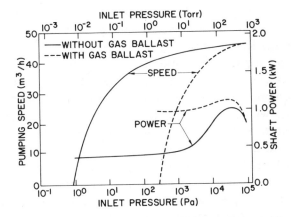

Fig. 7.5 Pumping speed and shaft power for the Stokes 146H, 51-m^3/h rotary piston pump. Reprinted with permission from Stokes Division, Pennwalt Corp., Philadelphia, PA.

7.3 ROOTS PUMPS

Positive displacement blowers are used in series with rotary oil-sealed pumps to achieve higher speeds and lower ultimate pressures in the medium vacuum region than can be obtained with a rotary mechanical pump alone. Roots pumps, or Roots blowers, consist of two lobed rotors mounted on parallel shafts. The rotors have substantial clearances between themselves and the housing—typically about 0.2 mm. They rotate in synchronism in opposite directions at speeds of 3000 to

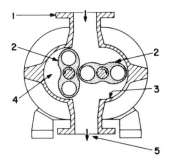

Fig. 7.6 Section through a single-stage roots pump; (1) inlet, (2) rotors, (3) housing, (4) Pump chamber (swept volume), (5) outlet. Reprinted with permission from *Vacuum Technology*, Leybold-Heraeus G.m.b.H., Köln, West Germany.

3500 rpm. These speeds are possible because oil is not used to seal the gaps between the rotors and the pump housing. A sectional view of a single-stage Roots pump is given in Fig. 7.6.

The compression ratio, or ratio of outlet to inlet pressure, is pressure dependent and usually has a maximum near 100 Pa. At higher pressures the compression ratio is lower because the conductance of the gaps increases with pressure. At lower pressures the compression ratio should, theoretically, remain constant, but in practice it decreases. Outgassing and the roughness of the rotor surfaces contribute to this compression loss at low pressures. Each time the rotor surface faces the high pressure side it sorbs gas some of which is released when the rotor faces the low pressure side. The compression ratio K_{omax} for air for a single-stage Roots pump of 500 m^3/h displacement is shown in Fig. 7.7 [6]. It has a maximum compression ratio of 44. Large pumps tend to have a larger compression ratio than small pumps because they have a smaller ratio of gap spacing to pump volume. The compression ratio for a light gas such as helium is about 15 to 20% smaller than the ratio for air. The compression ratio K_{omax} is a static quantity and is measured under conditions of zero flow. The inlet side of the pump is sealed and a pressure gauge is attached. The outlet side is connected to a roughing pump and the system is evacuated. Gas is admitted to the backing line that connects the Roots pump to the rotary pump; the pressure P_b (backing pressure) is measured at the Roots pump outlet

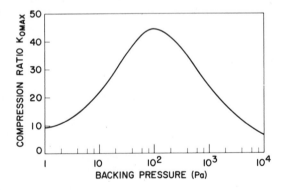

Fig. 7.7 Dependence of the compression ratio K_{omax} of the Leybold WS500 Roots pump on the backing pressure. Valid for air. Values for helium are about 20% smaller. Reprinted with permission from Leybold-Heraeus G.m.b.H., Postfach 51 07 60, 5000 Köln, West Germany.

and the pressure P_i is measured at the inlet. The compression ratio is given by P_b/P_i.

Considerable heat is generated by pumping gas at high pressures with a Roots pump; this causes the rotors to expand, and may harm the pump. To avoid the problems of overheating a maximum pressure difference between the inlet and outlet of a Roots pump is specified. This maximum pressure difference is typically 1000 Pa but that value may be exceeded for a short time without harm to the pump. To avoid heat generation Roots pumps are connected as compression or transport pumps.

In compression pumping, the more common method, a Roots pump is placed in series with a rotary pump whose rated speed is 5 to 10 times smaller than the speed of the Roots pump. When pumping is initiated at atmospheric pressure, a bypass line around the Roots pump is opened or the pump is allowed to free wheel. All the pumping is done by a rotary pump until the backing pressure is below the manufacturer's recommended pressure difference, at which time the Roots pump is activated and the bypass valve is closed. Some Roots pumps have this bypass feature built into the pump housing. The net speed of a Roots pump of 500 m³/h capacity backed by a 100-m³/h rotary piston is shown in Fig. 7.8. The speed curve for the mechanical

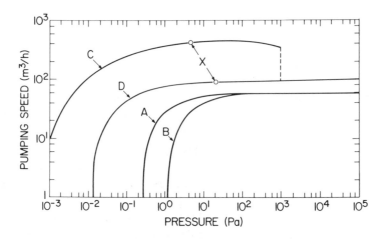

Fig. 7.8 Roots–rotary pump combinations. Transport mode: (a) Leybold RUTA 60 Roots pump and S60 rotary vane; (b) S60 only. Compression mode: (c) Leybold WS500 Roots pump and DK100 rotary piston pump; (d) DK 100 only. Reprinted with permission from *Vacuum Technology*, Leybold-Heraeus G.m.b.H., Köln, West Germany.

pump alone is shown for comparison. Such Roots–rotary pump combinations are often used when speeds of 170 m³/h or greater are required because they cost less than a rotary pump of similar capacity. The second method, transport pumping, uses a Roots pump in series with a rotary pump of the same displacement. Figure 7.8 also shows the pumping speed of a 60-m³/h Roots pump backed by a 60-m³/h rotary vane pump. The pumping speed of the rotary vane pump is also shown. Both pumps are started simultaneously at atmospheric pressure because the critical pressure drop will never be exceeded.

Detailed calculations of the effective pumping speed of the Roots pump have been carried out by Van Atta [2]; however, only approximate formulas for the inlet pressure P_i, and the inlet speed S_i, from reference 1 are given here:

$$P_{inlet} = P_{backing} \left(\frac{1}{K_{omax}} + \frac{S_b}{S_D} \right) \qquad (7.1)$$

$$S_{inlet} = \frac{S_b S_D K_{omax}}{S_D + S_b K_{omax}} \qquad (7.2)$$

where all terms have been defined except S_D, which is the speed of the Roots pump at atmospheric pressure (the pump displacement). By use of these approximate equations the pumping speed curve for the rotary pump, the compression ratio K_{omax}, and the Roots pump displacement, a curve of the speed of the Roots pump versus inlet pressure can be calculated. The line marked X on Fig. 7.8 shows the result of applying (7.1) and (7.2) to calculate one point on this curve. In this example the inlet pressure P_i, and the inlet speed S_i were calculated for a backing pressure of 20 Pa. From Fig. 7.7 we obtain $K_{omax} = 30$ at 20 Pa; Fig. 7.8 gives $S_b = 90$ m³/h. Use of a value of 500 m³/h for S_D, yields $P_i = 4.3$ Pa, and $S_i = 422$ m³/h.

In the high vacuum industry Roots pumps are frequently used to back large diffusion or turbomolecular pumps. For example a diffusion pump with a 35-in. diameter used to evacuate a 2-m³ chamber, is backed by a series connection of a 1300-m³/h Roots pump and a 170-m³/h rotary piston pump.

7.4 TURBOMOLECULAR PUMPS

The axial-flow molecular turbine, or turbomolecular pump as it is known, was introduced in 1958 by Becker [13]. His design originated from a baffling idea with which he had experimented with a few years earlier—a bladed rotating disk mounted above a diffusion pump [14]. When it was introduced commercially, the pump had relatively low speed and high cost, as compared with a diffusion pump. It did not, however, backstream hydrocarbons and did not require a trap of any kind. Since 1958 the turbomolecular pump has undergone rapid development both theoretically and experimentally. The most important theoretical development during this period was the work performed on blading geometry at MIT in the group headed by Shapiro [15, 16]. Many practical advances in lubrication, driving motors, and fabrication techniques have also taken place. Modern turbomolecular pumps have high pumping speeds, large hydrogen compression ratios, and low ultimate pressures. They do not backstream hydrocarbons from the lubricating fluid or mechanical pump and are well suited to pump gas cleanly at high flow rates or low pressures.

This section reviews the mechanism of gas flow in the free molecular pressure range and the relations between pumping speed, compression ratio, backing pump size and gas flow. The differences between vertical and horizontal rotor designs and the problems concerned with bearings and lubrication are also disscused. The operation and performance of pumps in high vacuum, ultrahigh vacuum, and high gas-flow systems are discussed in Chapters 10-12.

7.4.1 Pumping Mechanisms

The turbomolecular pump is a bladed molecular turbine that compresses gas by momentum transfer from the high-speed rotating blades to the gas molecules. The pumps operate at rotor speeds ranging from 24,000 to 60,000 rpm and are driven by solid state power supplies or motor-generator sets. The relative velocity between the alternate slotted rotating blades and slotted stator blades makes it probable that a gas molecule will be transported from the pump inlet to the pump outlet. Each blade is able to support a pressure difference. Because this compression ratio is small for a single stage, a large number of stages is cascaded. For a series of stages the compression ratio for zero flow is approximately the product of the compression ratios for each stage. Figure 7.9 shows a sectional view of a dual-rotor, horizontal-axis, turbomolecular pump. The blades impart momentum to the gas molecules most efficiently in the molecular flow region; therefore this pump, like a diffusion pump, must be backed by a rough mechanical pump.

If the foreline pressure is allowed to increase to a point at which the rear blades are in transition or viscous flow, the rotor will be subjected to an additional torque due to the viscous drag. The power required to rotate a shaft in steady state is proportional to the product of the rotor speed and torque. The power in some pumps is limited by the supply, with the result that an increase to great in foreline pressure causes a sudden reduction in the rotor speed and a loss in gas pumping speed. Another design uses a constant-speed motor that draws more power as the gas load is increased. This design also exhibits a loss in pumping speed after the forechamber enters the viscous flow region. The explanation of the loss of pumping speed in the latter case is not clear. It may be related to a conductance limit. Other effects of backing pump size and some rules for selecting backing pumps are discussed later in this section.

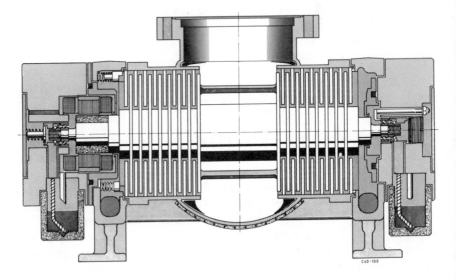

Fig. 7.9 Section view of Pfeiffer TPU-200 turbomolecular pump: (1) inlet, (2) outlet, (3) rotor disk, (4) stator disk, (5) bearing, (6) oil reservoir, (7) motor. Reprinted with permission from A. Pfeiffer Vakuumtechnik, G.m.b.H., Wetzlar, Germany.

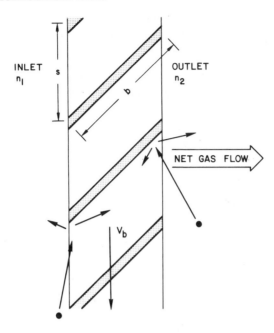

Fig. 7.10   Sectional view of a flat-bladed disk.

7.4.2   Speed–Compression Relationships

Continuum methods that give a reasonable account of pump performance have been developed [14]; however, they are not discussed. A
straightforward way of characterizing the turbomolecular pump has
been worked out by Kruger and Shapiro [15–17] and is presented here.
This method, which uses probabilistic techniques to calculate the
properties of single- and multiple-bladed arrays, is valid for all compression ratios. The model used to analyze a single rotor disk is shown
in Fig. 7.10. This disk, which rotates with a tip velocity near the
thermal velocity of air, imparts a directed momentum to a gas molecule
on collision. The blades are slotted at an angle to make the probability of a gas molecule being transmitted from the inlet to the outlet
greater than in the reverse direction. The stator disks are slotted in
the opposite direction. Γ_1 and Γ_2 are, respectively, the number of
molecules incident upon the disk per unit time at the inlet and the
outlet. a_{12} is the fraction of Γ_1 transmitted from the inlet (1) to the
outlet (2) and a_{21} is the fraction of Γ_2 transmitted from the outlet to

the inlet. Now define the net flux of molecules through the bladed disk in terms of the Ho coefficient W. Recall from Chapter 2 that the Ho coefficient is the ratio of net throughflux to incident flux. In steady state this is

$$\Gamma_1 W = \Gamma_1 a_{12} - \Gamma_2 a_{21} \tag{7.3}$$

or

$$\frac{\Gamma_2}{\Gamma_1} = \frac{a_{12}}{a_{21}} - \frac{W}{a_{21}} \tag{7.4}$$

If the gas temperature and the velocity distributions are the same everywhere, the ratio Γ_1/Γ_2 will be equal to the pressure ratio P_1/P_2. The ratio of outlet to inlet pressure is called the compression ratio K.

$$\frac{P_2}{P_1} = K = \frac{a_{12}}{a_{21}} - \frac{W}{a_{21}} \tag{7.5}$$

If a_{12} and a_{21} are independent of W, the compression ratio will vary in a linear way with the net throughput. Maximum compression occurs at zero flow, whereas unity compression occurs at maximum speed or mass flow. The maximum compression, maximum flow, and the general case for the region between these two extremes are three regions of (7.5), which we now examine in more detail.

Maximum Compression Ratio

For no gas flow $W = 0$ (7.5) reduces to

$$K = K_{\text{max}} = \frac{a_{12}}{a_{21}} \tag{7.6}$$

which states that the maximum compression ratio is that of forward to reverse transmission probabilities. Thus to maximize the compression ratio the ratio a_{12} to a_{21} is maximized. The important problem solved by Kruger and Shapiro was the calculation of a_{12} and a_{21} by Monte Carlo techniques as a function of the blade angle ϕ, the blade spacing-to-chord ratio s/b, and the blade speed ratio $s_r = V_b(M/2kN_oT)^{1/2}$.

Figure 7.11 sketches the results of a calculation for the single-stage compression ratio at zero flow [16]. From this curve we observe that for $s_r \leq 1.5$ the logarithm of the compression ratio is approximately linear with the speed ratio or

$$K_{max} \sim \exp\left[\frac{V_b(M)^{\frac{1}{2}}}{(2kN_oT)^{\frac{1}{2}}}\right]f(\phi) \qquad (7.7)$$

The compression ratio is exponentially dependent upon rotor speed and $(M)^k$. In particular the light gases, such as helium and hydrogen, will have compression ratios much smaller than the heavy gases. For a blade tip velocity of 400 m/s the speed ratio for argon is about unity, whereas for H_2 it is about 0.3. From Fig. 7.11 for $\phi = 30°$ we find that this blade velocity corresponds to compression ratios of $K(H_2) = 1.6$ and $K(Ar) = 4$. If 10 disks (five rotors and five stators) are cascaded, the net compression ratios will be calculated as approximately 100 and 10^6, respectively. A total of 15 disks would raise $K(H_2)$ to 10^3. A stator blade has the same compression ratio and transmission as a rotor; an observer sitting on a stator sees blades moving with the same relative velocity as an observer sitting on a rotor. The linear

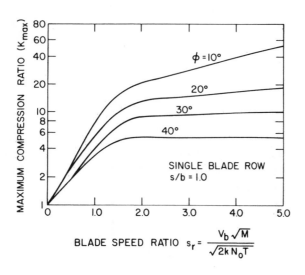

Fig. 7.11 Calculated curve of the compression ratio at zero flow for a single blade row with $s/b = 1$. Reprinted with permission from *Trans. 7th Nat. Vac. Symp. (1960)*, p. 6, C. H. Kruger and A. H. Shapiro. Copyright 1961, Pergamon Press.

blade velocity is proportional to the radius as well as the rotor angular frequency ($V_b = r\omega$). An area closer to the center of the rotor will have a smaller speed ratio and blade spacing-to-chord ratio. The data of Kruger and Shapiro show that the net effect of these changes is a lower compression ratio for the region closest to the rotor axis. For this reason a blade designed to have a high $K(H_2)$ should be slotted to a depth of only about 30% of the radius.

Experimental compression ratios are given for a horizontal-axis, dual-rotor pump in Fig. 7.12. These data were taken in a manner identical to the forepressure tolerance curves for diffusion pumps. Gas is admitted to the foreline of a blanked-off pump and the compression ratio is taken as the ratio of forepressure to inlet pressure. As the foreline or backing pressure is increased, the rear blades go first into transition flow and then into viscous flow and the rotor speed decreases.

Maximum Speed

Maximum speed is achieved when the compression ratio across a blade is unity; that is, when

$$K = 1 = \frac{a_{12}}{a_{21}} - \frac{W}{a_{21}} \qquad (7.8)$$

or

$$W_{max} = a_{12} - a_{21} \qquad (7.9)$$

To maximize W the absolute value of $a_{12} - a_{21}$ is maximized. The Ho coefficient W for a single blade is given in Fig. 7.13 as a function of blade-speed ratio for a spacing-to-chord ratio $s/b = 1$ [16]. For $s_r \leq 1.5$, W is almost linear with s_r,

$$W \sim \left[\frac{V_b(M)^{\frac{1}{2}}}{(2kN_oT)^{\frac{1}{2}}} \right] g(\phi) \qquad (7.10)$$

Because the molecular arrival rate is proportional to thermal velocity $(kT/m)^{\frac{1}{2}}$, the net pumping speed of the blade is independent of the weight of the impinging molecules:

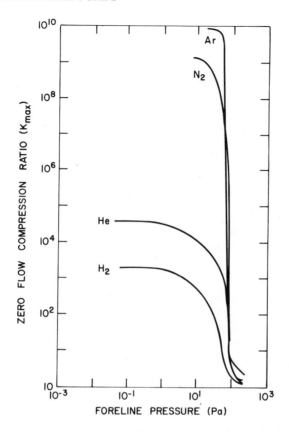

Fig. 7.12 Measured compression ratio for zero flow in a Pfeiffer TPU-400 turbomolecular pump. Reprinted with permission from A. Pfeiffer Vakuumtechnik, G.m.b.H., Wetzlar, West Germany.

$$S \sim V_b g(\phi) \qquad (7.11)$$

Closer examination of Fig. 7.13 and others of different spacing-to-chord ratios [15] reveals that the curves are sublinear; the pumping speed for light gases will then be slightly greater than for heavy gases. Approximation methods have been developed for calculating the net Ho coefficient for a series of blades [15]. The results of this calculation show an increase in Ho coefficient with the number of stages [18].

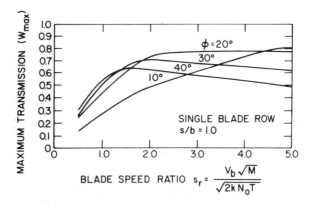

Fig. 7.13 Calculated curve of the Ho coefficient at unity compression ratio for a single blade row with $s/b = 1$. Reprinted with permission from *7th Nat. Vac. Symp. (1960)*, p. 6, C. H. Kruger and A. H. Shapiro. Copyright 1961, Pergamon Press.

General Relation

The maximum compression ratio (7.6) occurs only when the pump is pumping no gas (e.g., at its ultimate pressure), whereas the maximum speed occurs when the pressure drop is zero, a condition that is never reached in actual pump operation. An operating pump works in the region between to these extremes. Equation 7.5 describes the relative speed, or Ho coefficient versus pressure ratio for a series of stages as well as for a single stage, provided that the proper transfer coefficients

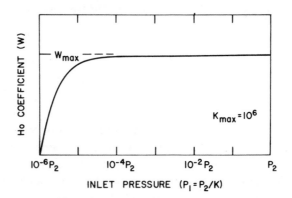

Fig. 7.14 Calculated dependence of Ho coefficient on inlet pressure for a gas with $K_{max} = 10^6$

are used for the series combination of disks. Equation 7.5 states that P_2/P_1 varies linearly between K_{max} and unity because W is varied linearly between zero and W_{max}. If we plot W versus P_2/K on a semi-log scale (Fig. 7.14), the dependence of W on K now looks more like the familiar pumping speed curve. The speed goes to zero at the ultimate pressure (P_1/K_{max}) and approaches a constant value at high inlet pressures. In practice W_{max} is not reached, but only asymptotically approached, because an inlet pressure of 1.0 to 10 Pa is sufficient to cause the rear blades to enter viscous flow with a resultant reduction in speed. Pumping-speed curves for a nominal 400 L/s pump are shown in Fig. 7.15. At a nitrogen inlet pressure of 0.9 Pa the rotor is still running full speed and the gas throughput is 400 Pa-L/s. This is twice the nitrogen throughput of a well-trapped 6-in. diffusion pump operating at a throat pressure of 0.1 Pa.

A different view of (7.5) is obtained if we replace the variables a_{12} and a_{21} with K_{max} and W_{max} from (7.6) and (7.9) respectively. The result is

$$\frac{W}{W_{max}} = \frac{(1 - K/K_{max})}{(1 - 1/K_{max})} \tag{7.12}$$

For any reasonable K_{max} this reduces to

$$\frac{W}{W_{max}} = (1 - \frac{K}{K_{max}}) \tag{7.13}$$

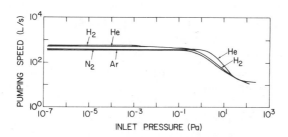

Fig. 7.15 Measured pumping speeds for the Pfeiffer TPU-400 turbomolecular pump. Reprinted with permission from A. Pfeiffer Vakuumtechnik, G.m.b.H., Wetzlar, West Germany.

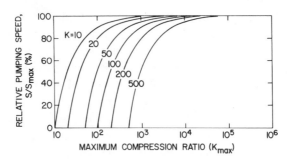

Fig. 7.16 Calculated dependence of relative pumping speed on K and K_{max}.

Equation 7.13 is plotted in Fig. 7.16 for several values of K in the range $10 \le K \le 500$. The compression ratio $K = S_1/S_2 = P_2/P_1$ will be in the range of 50 to 100 for the usual combination of turbomolecular and forepump combinations; for example, a 500 L/s turbomolecular pump will be backed by a 5- to 10-L/s mechanical pump. In Fig. 7.16 we see that if $K_{max} \ge 10^6$ the Ho coefficient and the pumping speed will be at their maximum values. If, however, $K_{max} < 500$, as it is for H_2 in some pumps, the hydrogen pumping speed will be a function of the forepump speed. A large forepump will be required to make $S/S_{max} = 1$ for hydrogen in a pump with a small $K_{max}(H_2)$. A staging ratio (S_2/S_1) of 15:1 to 20:1 is necessary for pumps with $K_{max}(H_2) \le 500$ to pump hydrogen at maximum speed. For large pumps a Roots–rotary combination is the most economical backing-pump system. Pumps with a $K_{max}(H_2) > 1500$ will not exhibit a hydrogen pumping-speed dependence upon backing pump size for staging ratios of the order of 60:1.

Equations 7.12 and 7.13 are not unique to a turbomolecular pump. They are true for any pump that satisfies the conditions of (7.3), that is, in which the reverse leakage is proportional to the pressure difference. These are the equations used to describe the inlet-speed pressure characteristics of a Roots–rotary pump combination in (7.1) and (7.2). For a diffusion pump the situation is even more straightforward. The maximum compression ratio in a modern pump is usually greater than 10^6 for all gases, including H_2, and $S = S_{max}$ for all gases regardless of the size of the backing pump, as long as the backing pump is large enough to maintain the forepressure below the critical value.

Figure 7.17 shows how the various turbomolecular pump and forepump parameters affect the pumping speed of a turbomolecular pump.

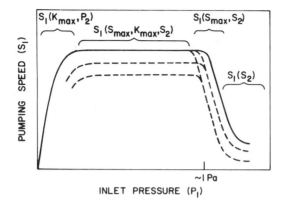

Fig. 7.17 Parameters that control the pumping speed of a turbomolecular pump in its four regions of operation.

The ultimate pressure, assuming no outgassing, is determined by the forepressure P_2 and the maximum compression ratio K_{max}. The pumping speed in the plateau region is determined by S_{max} for heavy gases, whereas for gases with low compression ratios it is also determined by the maximum compression ratio and the size of the forepump. The pressure at which the constant throughput region begins is determined by the ratio S_2/S_{max}.

7.4.3 Ultimate Pressure

The ultimate pressure of a turbomolecular pump is determined by the compression ratio for light gases and by the amount of outgassing of other gases, and discussed in more detail in Chapter 8 [see (8.1) and (8.2)]. The main difference between the turbomolecular pump and the diffusion pump is that the hydrogen compression ratio is low enough in the turbomolecular pump to cause the ultimate hydrogen partial pressure to be determined by $K_{max}(H_2)$ and its pressure in the foreline. In some older pumps, which have a water-vapor compression ratio of less than 10^4, the water-vapor partial pressure may also be compression ratio limited [19]. The partial pressures of all other gases and vapors will be limited by their respective outgassing rates and pumping speeds. During pumpdown of an unbaked turbomolecular pump the slow release of water from the blades closest to the inlet may slightly decrease the rate of H_2O removal; the effective compression ratio for water release from the first few blades is much less than K_{max}.

Henning [20] has shown that the partial pressure of hydrogen found in a turbomolecular pumped system is dominated by the forepump oil. The ultimate pressure varied from 2×10^{-7} to 5×10^{-9} Pa as a function of the kind of oil in the forepump. Because the turbopump oil has a lower vapor pressure than the mechanical pump oil, it will not contribute so much hydrogen to the background as the forepump oil. Henning and Lotz [21] used perfluoropolyether fluids for lubricating the turbomolecular and backing pumps in the presence of corrosive gases. Using a mass spectrometer they observed distinct fluorine peaks as well as hydrogen. This decomposition occurred because local heating of the bearings caused the oil temperature to exceed the range of thermal stability. They concluded from the presence of the hydrogen that the ultimate pressure of the pump was not further improved by the use of hydrogen-free fluids. They postulated that the limiting pressure was caused by hydrogen diffusion through the foreline seals.

Ultimate pressures for baked systems in the range of 2×10^{-8} to 5×10^{-9} Pa are possible with high-compression turbomolecular pumps without assistance in pumping from cryo baffles or titanium sublimation pumps. Hydrogen will constitute more than 99% of the residual gas at the ultimate pressure [20].

7.4.4 Design Considerations

A single blade is inadequate to serve as a high vacuum pump. Multiple-bladed structures that have between 8 and 20 disks will provide adequate compression and speed to make a functional pump. As in the diffusion pump, the stages nearest the high vacuum inlet are designed to serve a purpose different from those near the outlet. The flow through each stage is constant or, stated another way, the product of pressure times pumping speed is a constant. The blades nearest the inlet are designed to have a high pumping speed and a low compression ratio, whereas the blades nearest the foreline entrance are designed to have a high compression ratio and a low pumping speed. For economic reasons it would be impractical to make each blade different from its neighbor. A compromise results in groups of two or three types of blade, in which each is designed for .a particular speed and compression ratio. Each group of blades may be considered analogous to a diffusion pump jet. The pump designer may trade-off pumping speed and light gas compression ratio by the proper choice of a blade-to-chord ratio and blade angle. Pumps exhibiting an overall large compression ratio for H_2 use blades that are optically more opaque

(s/b, ϕ small) than those that are designed to maximize the pumping speed (s/b, ϕ large).

Figure 7.18 shows a view of the three-stage rotor used in a horizontal-axis, dual-rotor pump. In this design the rotors are individually abrasive-machined and balanced. The rotor disks are positioned on a cooled hub that is allowed to equilibrate thermally with the disks and to hold them rigidly in position. Stator disks are manufactured in a similar manner, cut into half-sections, and mounted stage-by-stage as the rotor is moved into the housing. The top portion of the rotor from the vertical-axis, single-rotor pump shown in Fig. 7.19 was machined from a single block of aluminum. Each stage was first machined on the rotor. After all stages were machined, the individual blades were formed by lengthwise sawing. The desired blade angles were then obtained by twisting. The stators were constructed from stampings which were cut and twisted [18].

Modern turbomolecular pumps are constructed in two styles: vertical-axis, single-rotor, and horizontal-axis, double-rotor. In either configuration the designer is free within limits of material stability to choose the number of stages, blade angles, spacings, and blade-to-chord ratios. The horizontal axis pump allows for a somewhat more stable bearing design than the vertical pump. It is possible to optimize both designs for maximum speed or maximum compression. The

Fig. 7.18 Three-stage rotor from a Pfeiffer TPU-200 turbomolecular pump. Reprinted with permission from A. Pfeiffer Vakuumtechnik, G.m.b.H., Wetzlar, West Germany.

Fig. 7.19 Top stages of a rotor from a Leybold-Heraeus TMP-450 turbomolecular pump. Reprinted with permission from Leybold-Heraeus G.m.b.H., Postfach 51 07 60, 5000 Köln, West Germany.

single-rotor, vertical-axis pump has little conductance loss between the inlet flange and the rotor, whereas the horizontal-axis, dual-rotor pump does suffer conductance loss but pumps from two sides. Henning [19] estimates that the pumping speed of a dual-rotor pump is more than 1.6 times that of a vertical pump for the same inlet flange diameter with all other factors constant.

Neither style should be subjected to a steady or transient twisting moment by using the inlet flange to bear the load of a heavy work chamber, especially a cantilever load, or the impulse of a heavy flange closure. Improper loading can cause premature bearing failure. All pumps should be suspended from the system by their inlet flanges; the inlet flange should not be used as a mounting platform for the system. This is usually no problem with a vertical pump because it looks like a diffusion pump and is mounted like one.

The practical upper rotational speed for the rotor is currently ~ 80,000 rpm; commercially, however, the maximum rotational speed is ~ 60,000 rpm or a blade tip velocity of ~ 500 m/s. These limits are due to the bearing tolerances, thermal coefficients of expansion, and material stress limits. Ball bearings are used in most turbomolecular pumps and are the component that is subjected to the greatest wear. Oil, either flowing or in a mist, is used to lubricate and cool the bearings. The oil, in turn, is cooled by mechanical refrigeration or by water. Small diameter bearings are desired to increase the bearing lifetime. Some pumps use grease-packed bearings [22]. Air bearings [23] and magnetic bearings [24] with extremely low wear rates have been reported.

The requirements for a turbomolecular pump oil are somewhat different from those of an ordinary mechanical pump oil. Because the bearing loads are not severe a high shear strength, high viscosity oil is not required. The fluid needs low vapor pressure, and because the bearings rotate at high speed it must have good lubricating properties and moderate viscosity and it must not foam. Although the average oil temperature may be only 70°C, high spot heating on the bearings can cause oil decomposition. A hard cut hydrocarbon oil composed of a narrow range of medium viscosity molecular weights is used. The small amount of heavy fractions that are present are of less concern than the small amount of light fractions that have high vapor pressure and contribute hydrogen to the spectrum. Either a good quality hard-cut hydrocarbon diffusion pump oil or a hydrocarbon mechanical pump oil which has been double distilled to remove light fractions will work well in a turbomolecular pump. It is important that the oil be vacuum degassed to prevent foaming, because oil can absorb up to ~9% gas during its handling and storage.

REFERENCES

1. Leybold-Heraeus Publication HU152, Leybold-Heraeus, G.m.b.H., Köln, West Germany.
2. C. M. Van Atta, *Vacuum Science & Engineering*, McGraw-Hill, New York, 1965, Chapter 5.
3. Reference 1, p. H-B 61.

4. Reprinted with permission from Balzers High Vacuum, Furstentum, Liechtenstein.

5. Reprinted with permission from CVC Products, Inc., 525 Lee Road, Rochester, NY 14603.

6. Reprinted with permission from Sargent-Welch Scientific Co., Vacuum Products Division, 7300 N. Linder Avenue, Skokie, IL 60077.

7. Reprinted with permission from Montedison USA, Inc.,1114 Avenue of the Americas, New York, NY 10036.

8. Reprinted with permission from *Functional Fluids, Synthetic Lubricants and Oil Additives*, Copyright 1975, Stauffer Chemical Company, Specialty Chemical Division, Westport, CN 06880.

9. Reprinted with permission from Halocarbon Products Corp., 82 Burlews Court, Hackensack, NJ 07601.

10. Reprinted with permission from Inland Vacuum Industries Inc., 35 Howard Avenue, Churchville, NY 14428.

11. Reprinted with permission from General Electric Company, Silicone Products Dept., Waterford, NY 12188.

12. T. D. Weikel and H. H. Yuen, *Vacuum Pump Explosion Study*, NAEC-GSED-60, Naval Air Engineering Center, Philadelphia, PA., August 1972.

13. W. Becker, *Vac. Tech.*, 7, 149 (1958).

14. W. Becker, *Vac. Tech.*, 15, 211 (1966).

15. C. H. Kruger and A. H. Shapiro, *Proc. 2nd Int. Symp. Rarified Gas Dynamics*, Berkeley, CA, L. Talbot, Ed., Academic , New York, 1961, pp 117-140.

16. C. H. Kruger and A. H. Shapiro, *Trans. 7th Nat. Vac. Symp. (1960)*, Pergamon, New York, 1961, pp 6-12.

17. C. H. Kruger, *The Axial Flow Compressor In the Free-Molecule Range*, Ph.D. thesis, Department of Mechanical Engineering, M.I.T., Cambridge, MA, 1960.

18. K. H. Mirgel, *J. Vac. Sci. Technol.*, 9, 408 (1972).

19. J. Henning *Proc. 6th Int. Vac. Congr.*, Kyoto, Japan *J. Appl. Phys.* Sup. 2, Pt. 1, 5 (1974).

20. J. Henning, *Vacuum*, 21, (1971).

21. J. Henning and H. Lotz, *Vacuum*, 27, 171 (1977).

22. G. Osterstrom and T. Knecht, *J. Vac. Sci. Technol.*, 16, 746 (1979)

23. L. Maurice, *Proc. 6th Int. Vac. Congr.*, Kyoto, Japan *J. Appl. Phys.* Sup 2, Pt. 1, 21 (1974).

24. Leybold-Heraeus TurboVac 550M turbomolecular pump. Leybold-Heraeus G.m.b.H., Köln, West Germany.

CHAPTER 8

Diffusion Pumps

The diffusion pump has been in existence for more than half a century and in the last 20 years has advanced significantly. The most important developments of this recent period have been the discovery of low vapor pressure pumping fluids and the and control of backstreaming [1]. Today it holds a commanding position as the most widely used high vacuum pump, a position it will continue to maintain for years to come despite the desirability of ion, cryogenic, or turbomolecular pumps for many applications. Because of this long history, the diffusion pump has been the subject of more study and literature than any other high vacuum pump. Its problems are thoroughly understood and its performance is, in some cases, understated. Many excellent reviews of diffusion pumps are available. Examples that summarize the pump's properties for practical applications are those of Hablanian [2, 3] and Singleton [4]. Review articles by Hablanian and Maliakal [1], Florescu [5,6], Tóth [7] and books by Dushman [8], and Power [9] cover its theory of operation and design.

This discussion reviews the basic mechanisms of pump operation, pumping speed and throughput, fluids and heat effects, backstreaming, baffles, and traps. The particular problems associated with the collective operation of a diffusion pump, backing pump, trap or baffle, and work chamber as a complete system are treated in later chapters. The basic high-vacuum diffusion pump system is discussed in Chapter 10. The special requirements of diffusion pumps for ultrahigh vacuum are treated in Chapter 11, and for high gas flow applications in Chapter 12.

8.1 BASIC MECHANISM

The name diffusion pump, first coined by Gaede [10], does not describe the operation of the pump accurately. The diffusion pump is a vapor jet pump which transports gas by momentum transfer on collision with the vapor stream. A motive fluid such as a hydrocarbon oil, an organic liquid, or mercury is heated in the boiler until it vaporizes. The vapors flow up the chimney and out through a series of nozzles. Figure 8.1 [6] sketches a sectional view of a metal-bodied diffusion pump. The nozzles, three in this illustration, direct the vapor stream downward and toward the cooled, outer wall, where it condenses and returns to the boiler. The vapor flow is supersonic and remains so until it hits the wall. Gases that diffuse into this supersonic vapor stream are, on average, given a downward momentum and ejected into a region of higher pressure. Modern pumps have several stages of compression—usually three or four. Each stage compresses the gas to a successively higher pressure than the preceding stage as it transports it toward the outlet.

The boiler pressure in a modern diffusion pump is of an order of 200 Pa. Ideally, the pump cannot sustain a pressure drop any larger than this between its inlet and outlet. In fact, the maximum value of forepressure that the pump can tolerate is less than the boiler pressure. This value, called the critical forepressure, ranges from 25 to 75 Pa; the latter number is typical of modern pumps. Thus the diffusion pump must be "backed" by a mechanical pump of the rotary piston or vane variety large enough to maintain the "fore" pressure or "backing" pressure to a value less than a value known as the critical forepressure. If the forepressure exceeds the critical value, all pumping action will cease. The pumping action ceases at high pressures because the directed supersonic vapor stream no longer extends from the jet to the wall but is ended in a shock front close to the jet [5]. Those vapor molecules beyond the shock front are randomly directed and cannot stop gas molecules from returning to the inlet. As the critical forepressure is exceeded, the inlet pressure will rise sharply and uncontrollably in response to the cessation of pumping. Needless to say, the maximum forepressure should never be exceeded. In newer pumps the inlet pressure and the pumping speed will be unaffected by the value of the forepressure as long as it is below the critical value and the gas throughput is low. At maximum throughput the critical forepressure will be reduced to about ¾ of its normal value [2]. This value is a function of the pump design, heater power, and pump fluid.

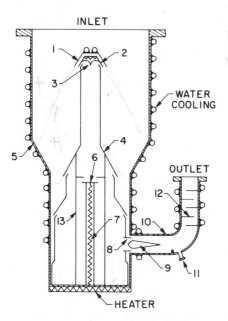

Fig. 8.1 A sectional view of a metal-bodied diffusion pump and some of its innovations: (1) Cooled hood for prevention of vapor backstreaming [11, 12]; (2) heater for the nozzle's cap to compensate for loss of heat [13]; (3) streamlined surface to avoid turbulence [14]; (4) multiple stages to obtain low pressures [15]; (5) enlarged casing to give larger pumping aperture [16]; (6) baffle to impede the access to the jet of liquid splashed up from the boiler [17]; (7) heater for superheating the vapor [18]; (8) lateral ejector stage [19]; (9) conical body allowing operation against higher forepressures [20]; (10) hot maintained diffuser for oil purification [21]; (11) catchment and drain-off of highly volatile oil components [22]; (12) baffle to reduce oil loss [15]; (13) concentric chimneys that allow oil fractionation. Reprinted with permission from *Vacuum*, **13**, p 569, N. A. Florescu. Copyright 1963, Pergamon Press Ltd.

Each stage of the vapor pump has a characteristic speed and pressure drop. The top jet has the largest speed (and the largest aperture) and the lowest pressure drop. The vapor density in the top jet is less than that in the lower jets and cannot sustain a pressure drop as large as the lower jets. Because the gas flow through a series of jets is the same, each successive jet can have a larger pressure drop and a smaller pumping speed. The last jet has the highest pressure drop. Many pumps use a vapor ejector as the last stage because it is efficient at compressing gas in this pressure range. The combination of jets and ejector produces a pump with a higher forepressure tolerance than is possible with vapor jets alone.

8.2   SPEED-THROUGHPUT CHARACTERISTICS

The four operating regions of the diffusion pump are the constant speed, constant throughput, mechanical pump, and compression ratio regions. They are graphically illustrated in Fig. 8.2 [3]. In its normal operating range the diffusion pump is a constant speed device. Its efficiency of pumping gas molecules is about 0.5 for the pump alone but only about 0.3 when the conductance of the traps and valve are included. The usual operating range for constant speed is ~10^{-1} to below 10^{-9} Pa for most gases. The upper pressure limit called the critical inlet pressure corresponds to the point at which the top jet fails. In a 6-in. diffusion pump the top jet becomes unstable at pressures of an order of 0.1 Pa, the middle jet at a pressure of about 3 Pa, and the bottom jet at pressures of about 40 Pa [24].

The gas throughput in the constant speed range is the product of the inlet pressure and the speed of the pump at the inlet flange. It rises linearly with pressure until the critical inlet pressure is reached. Above that pressure the pump thoughput is constant until the jets all cease to function. At higher pressures the throughput again increases in accordance with the speed of the backing pump. The maximum usable throughput of the diffusion pump corresponds to the product of the

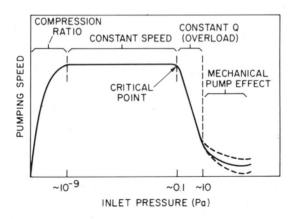

Fig. 8.2 Typical diffusion pump speed curve for a given gas. Four regions are evident: (1) Effect of compression ratio limit; (2) normal operation—constant speed; (3) first jet overloaded—nearly constant throughput; (4) effect of mechanical pump. Reprinted with permission from *Japan. J. Appl. Phys.*, Suppl. 2, Pt. 1, p 25, M. H. Hablanian. Copyright 1974, Japanese Journal of Applied Physics.

inlet speed and the critical inlet pressure. If that pressure is exceeded, the backstreaming may increase substantially and jet instabilities will appear. These instabilities make pressure control difficult. The maximum throughput should not be exceeded in the steady state, although it often happens for short periods of time during crossover from rough pumping to high vacuum pumping.

Exceeding the critical forepressure in a well-designed pump is usually the result of equipment malfunction, whereas the critical inlet pressure is easily exceeded by misoperation. If the pump is equipped with a sufficiently large forepump, the critical forepressure can still be exceeded if a leak occurs in the foreline, the mechanical pump oil level is too low, the mechanical pump belt is loose, or a section of the diffusion pump heater is open. The critical inlet pressure can be exceeded easily by operational error, but otherwise the top jet will continue to pump unless there is a partial heater failure or a large leak.

The speed does not remain constant to absolute zero pressure but rather decreases toward zero as shown in the compression ratio region of Fig. 8.2. This curve decreases at low pressures because of the large but finite compression ratio of the diffusion pump jets. The pump whose hypothetical speed curve for one gas is shown in Fig. 8.2 has an ultimate pressure of 10^{-10} Pa. If, at that point, its forepressure were 1 Pa, its compression ratio would be 10^{10}, a value similar to those quoted in the literature. Figure 8.3 shows the air pumping speed for a 6-in. diffusion pump with and without a liquid nitrogen baffle. All diffusion pumps have some small reverse flow of the gases being pumped, and although this reverse flow is exceedingly small for heavy gases it may be visible for light gases under certain conditions. Because of their

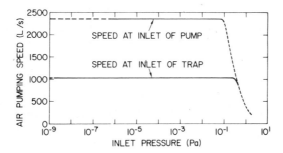

Fig. 8.3 Air pumping speed of the Varian VHS-6, 6-in. diffusion pump with and without a liquid nitrogen trap. Reprinted with permission from Varian Associates, 611 Hansen Way, Palo Alto, CA 94303.

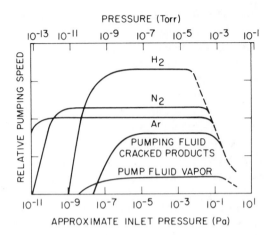

Fig. 8.4 Diffusion pump performance for individual gases. Reprinted with permission from *J. Environ. Sci.*, **5**, p. 7, S. G. Burnett and M. H. Hablanian. Copyright 1964, The Institute of Environmental Sciences.

high thermal velocity and small collision cross section, the compression ratio of light gases such as hydrogen and helium is lower than that of the heavy gases. Figure 8.4 sketches the relative pumping speeds of several gases and vapors as a function of their inlet pressure, and illustrates the effect of a low compression ratio for hydrogen. The compression ratio for heavy gases will be of an order of 10^8 to 10^{10}, whereas for light gases it can be small enough (10^3-10^6) in some pumps to detect a small foreline concentration at the inlet [1, 25]. It is this phenomenon that explains why hydrogen emanating from an ion gauge in the foreline can be detected at the inlet. The operation of one leak detector [26] in which a diffusion pump is operated to produce a low compression ratio for helium is based on this phenomenon. The detector is located at the inlet and the test piece is appended to the foreline. The compression ratio for heavy gases is adequate to produce the required low pressure in the detector while at the same time allowing helium to back diffuse and be counted.

Figure 8.4 is important to an understanding of what determines the ultimate pressure in a diffusion pumped system. For the ideal pump with zero outgassing above the top jet and in the work chamber and using a perfect baffle to collect all oil vapor fragments the ultimate pressure would be the sum of each of the partial pressures in the foreline divided by their respective compression ratios:

$$P_u = \frac{P_{f1}}{k_1} + \frac{P_{f2}}{k_2} + \frac{P_{f3}}{k_3} + \dots \qquad (8.1)$$

For the case in which the base pressure of the system is achieved in the pump's constant speed region the ultimate pressure is the sum of each independent gas flow Q_i, divided by the pumping speed for each gas S_i:

$$P_u = \frac{Q_1}{S_1} + \frac{Q_2}{S_2} + \frac{Q_3}{S_3} + \dots \qquad (8.2)$$

The individual gas flows can originate from outgassing or leaks. In practice the ultimate pressure is determined by (8.2), but in some situations it can be a combination of the two cases; for example, the partial pressure of hydrogen can be determined by (8.1), depending on the pump and the H_2 pressure in the foreline, whereas the partial pressures of the other gases are determined by (8.2). The pump's ultimate pressure will not be limited by the compression ratio for heavy gases but rather by outgassing, the vapor pressure of the lightest pump fluid fractions on the baffle, and the release of gases dissolved in the pump fluid. A single pumping speed curve (Fig. 8.2) representative of all gases cannot be drawn because the pumping speed is not the same for all gases (see Fig. 8.4). The pumping speed is greater for light gases but not in proportion to m^k as predicted by the ideal gas law. Under normal operating conditions helium will be pumped about 20% faster than nitrogen.

8.3 PUMP FLUIDS AND HEAT EFFECTS

The ideal diffusion pump fluid should be stable, have a low vapor pressure, and a low heat of vaporization, and be safe to handle, use, and dispose. It should not decompose, entrap gas, or react with its surroundings. Unfortunately, no fluid exists that meets these criteria. Mercury was the first and only elemental fluid used in diffusion pumps. Today it has been supplanted by distilled hydrocarbons and synthetic fluids for almost all applications. Mercury has a high vapor pressure and must be trapped during operation. It is also toxic. Even so it is still used in some specialized applications like mass spectrometry because it does not decompose or dissolve gas and because it has an easily identifiable mass spectrum. Before leaving the subject of mercury as a pump fluid, it should be noted that it cannot be used interchangeably with oil in a conventional oil diffusion pump. Mercury requires a lower boiler temperature than oil and will react with some materials used in the construction of oil pumps. The sustained operation of a liquid nitrogen trap over a mercury pump will deplete the boiler of its charge after a few days of operation unless an additional baffle, warmer than -40°C, is placed between the liquid nitrogen trap and the pump.

Most diffusion pumps use distilled hydrocarbons, esters, silicones, or ethers. The properties of several diffusion pump fluids are tabulated in Table 8.1. No one fluid has proved superior for all applications. The light oils have a higher pumping speed than the heavy oils and the

Table 8.1   Typical Properties of Some Diffusion Pump Fluids

Trade Name	Chemical Name	MW (ave)	P_v at 25°C (Pa)	Viscosity at 25°C (mm^2/s)	Boiler Temp. at 100 Pa (°C)
Convoil®-20[27]	Hydrocarbon	400	5×10^{-5}	80	210
Balzers-71[28]	Hydrocarbon	425	3×10^{-6}	27	
Octoil-S®[27][a]	Bis (2-ethyl-hexyl) sebacate	427	3×10^{-6}	18.2	220
Invoil®[29][a]	Bis (2-ethyl-hexyl) phthalate	390	3×10^{-5}	51	200
Dow Corning® 704[30][b]	Tetraphenyl-tetra methyl trisiloxane	484	3×10^{-6}	38	220
Dow Corning® 705[30][b]	Pentaphenyl-tri methyl trisiloxane	546	4×10^{-8}	175	250
Santovac 5® [31][a,b]	Mixed 5-ring polyphenylether	447	6×10^{-8}	2400	275
Fomblin® Y VAC 25/9[32][a,b]	Perfluoro-polyether	3400	9×10^{-7}	190	230

[a] Suitable for use in mass spectrometers and other applications where ion or electron beams could cause polymerization;
[b] Excellent oxidation resistance.

heavy fluids have the low vapor pressure necessary to produce the low ultimate pressure needed for more demanding applications. Hydrocarbon oils were the first organic fluids to replace mercury. The ultimate pressure obtainable with them was limited by their decomposition on heating. The light fragments backstreamed into the work chamber, whereas heavy fragments deposited on the pump. Fractionating pumps that allowed the oil to be preferentially directed to the lower jet after condensation so that its light fractions could be removed by the forepump were introduced [23]. Degassing of the oil was accomplished by maintaining a section of the ejector walls at an elevated temperature [21]. Pumps that incorporate these and other design advances and use heavy oils produced by molecular distillation of selected hydrocarbons will reach 5×10^{-5} Pa (untrapped) and the 5×10^{-7} Pa when trapped with liquid nitrogen.

Light hydrocarbon oils are widely used in vacuum metallurgy and other applications in which rapid pumping to the 10^{-3} or 10^{-5} Pa range is desired, low oil cost is important, and trace amounts of hydrocarbon fragment vapors can be tolerated. All hydrocarbon oils have the disadvantage of oxidizing when exposed to air at operating temperatures. The heavy synthetic fluids such as tetra- and pentaphenylsilicones (DC-704 and DC-705), pentaphenylether (Santovac 5, Convalex-10); and other newer fluids were developed to meet these deficiencies. DC-704 is extensively used in quick-cycled, unbaked systems because of its moderate cost, low backstreaming, stability, and oxidation resistance. Hickman [33] first suggested the use of polyphenylether as a diffusion pump fluid because of its exceptional stability and low vapor pressure. DC-705 and pentaphenylether are high quality fluids which are widely available. Solbrig and Jamison [34] were not able to induce explosions when either of these two fluids was used in a system pressurized to ½ atm with pure oxygen.

Regardless of the fluid used, no diffusion pump should be released to air while hot because the fluid may scatter thoroughout the system, and degrade or form varnish deposits on the interior surfaces of the pump. Because of its lower cost (see Table 8.2), DC-705 is used more widely than pentaphenylether, although pentaphenylether replaces silicones or hydrocarbons in mass spectrometer leak detectors, residual gas analyzers, and electron microscopes because it polymerizes on electron impact to form a conducting film rather than an insulating film which can charge up and deflect electrons. Pentaphenylether is also used when silicone contamination cannot be tolerated. The less expensive sebacate ester is suitable for some of these applications. Perfluoropolyether is also stable, has low vapor pressure, and does not

Table 8.2   Approximate Cost of
Diffusion Pump Fluids[a]

Diffusion Pump Fluid	Approx. Cost ($/500 mL)
Perfluoropolyether	258
Polyphenylether	190
Pentaphenyl silicone	85
Tetraphenyl silicone	50
Sebacate ester	45
Phthalate ester	35
Hydrocarbon[b]	5

[a] (1979 $)
[b] Price based on gallon quantity.

polymerize to form an insulator. It is reported, however, to have a lower pumping speed than pentaphenylether [35]. Cost should not be a major factor in the selection of a fluid for critical applications because efficient reclamation is a commercial reality.

The interior of a pump must be thoroughly cleaned before changing from one fluid to another. Hydrocarbon oils are easier to remove than silicones, but severely contaminated pumps that use either fluid may be cleaned successively in decahydronapthalene, acetone, and ethanol. Alternatively, one may use trichloroethelene, acetone, and ethanol. If the pump is relatively clean, the acetone and alcohol are usually adequate. Polyphenylether is soluble in trichloroethelene and in 1,1,1-trichloroethane but the latter is less toxic. Pumps charged with perfluoropolyether must be cleaned in a fluorinated solvent such as trichlorotrifluoroethane or perfluorooctane [36]. Gas bursting may be observed for several days after cleaning and charging with a new fluid. Have patience while waiting for a newly charged pump to reach its ultimate pressure.

The effect of heat input variation is summarized concisely in Fig. 8.5 [3]. The general trends are that the oil temperature, forepressure tolerance, and throughput increase with boiler power, whereas pumping speeds decrease at high heat inputs because of the increased density of oil vapor molecules in the vapor stream [7]. It is not possible to optimize the pumping speed for all gases at the same heater power

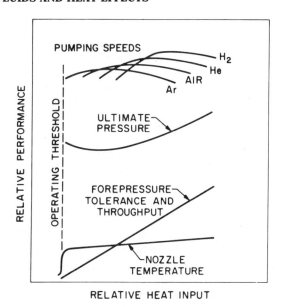

Fig. 8.5   Effect of heat input variations on various diffusion pump parameters. Reprinted with permission from *Japan. J. Appl. Phys.*, Suppl. 2, Pt. 1, p. 25, M. H. Hablanian. Copyright 1974, Japanese Journal of Applied Physics.

because of the differences in mass and thermal velocity.   Each gas reaches maximum speed at a different input power.   Excessively increasing the boiler temperature also hastens oil degradation [3].   Modern pumps will accept most good quality oils equally well, but some pumps, especially those designed to work with older fluids, do not provide adequate heat input for pentaphenylether.   It is advisable to restrict the cooling water flow to the diffusion pump when using the ether [33].   It should be emphasized that there is a direct relationship between throughput and boiler power because they are dimensionally equivalent; 1000 Pa-L/s = 1W; that is, the maximum throughput of a pump is scaled to the boiler power.   For pumps of efficient design this value can be as high as 150 Pa-L/s for each kilowatt of boiler power [2]. . A straight-sided pump with a 6-in. diameter boiler and throat has the same maximum throughput as a pump with a 6-in. boiler and an expanded top like the one sketched in Fig. 8.1.   The maximum speed of a pump in the high vacuum region is proportional to its inlet area, but its maximum throughput is proportional to its boiler power.

8.4 BACKSTREAMING, BAFFLES AND TRAPS

For the purposes of this discussion backstreaming is defined as the transport of pumping fluid and its fractions from the pump to the chamber. Hablanian [37] properly points out that the discussion of backstreaming must not be limited to the pump but must include the trap, baffle, and ductwork as well because all affect the transfer of pumping fluid vapors from the pump body to the chamber. First, let us consider the contributions from the pump. Power and Crawley [12] have determined that steady state backstreaming results from (1) evaporation of fluid condensed on the upper walls of the pump, (2) premature boiling of the condensate before it enters the boiler, (3) the overdivergence of the oil vapor in the top jet, (4) leaks in the jet cap, and (5) evaporation of fluid from the heated lip of the top jet. The backstreaming from (1) can be reduced by low vapor pressure fluids and added trapping over the pump. Modern pump designs eliminate sources (2) and (4). The use of a water cooled cap [1, 12] directly over the top jet assembly substantially reduces (3) and (5), which were found to be the major causes of fluid backstreaming. With these precautions the backstreaming can be reduced to an order of 10^{-3} $(mg/cm^2)/min$ a short distance above the pump inlet.

Further reduction of the backstreaming is possible by geometrical considerations and by the use of a baffle or a trap. The words *trap* and *baffle* are often misused. Operationally, a trap is a pump for condensable vapors, and a baffle is a device that condenses pump fluid vapors and returns the liquid to the pump. Today the two words are used imprecisely and when the baffle is deep cooled this distinction disappears. Pump-fluid molecules or fragments may find their way through the trap by creeping along walls, by colliding with gas molecules, and by reevaporation from surfaces. Creep can be prevented by the use of traps with a creep barrier—a thin membrane extending from the warm, outer wall to the cooled surface [38]—or by use of oleophobic fluids such as pentaphenylsilicone or pentaphenylether which do not creep. Backstreaming due to oil–gas collisions is a linear function of pressure up to the transition pressure region and a function of the trap and pump design. Rettinghaus and Huber [39] have measured this backstreaming. For one 6-in. diffusion pump and trap combination they found that the peak backstreaming was 3×10^{-6} $(mg/cm^2)/min$ at a pressure of 5×10^{-2} Pa. At higher pressures the backstreaming rate was decreased by the flushing action of the gas. In normal operation the diffusion pump will pump through this region quickly; the maximum integrated backstreaming rate from oil–gas collisions is small enough so that contamination from this source is of no concern in an unbaked system.

The problem of reevaporation is more subtle. The vapor pressures of diffusion pump fluids vary widely, (see Appendix F.2). The two fluids with the lowest vapor pressures (DC-705, and Santovac 5) have vapor pressures so low that evaporation from a surface at $10°C$ proceeds at a rate of 5×10^{-10} (mg/cm^2)/min. Some decomposition of the fluid does, however occur in the boiler and lighter fractions are generated. Gosselin and Bryant [40] have studied the residual gases in a diffusion pumped system in which the pump was charged with DC-705. They observed that the light fractions (methane, ethane, and ethylene) were not effectively trapped even on a liquid-nitrogen-cooled surface because of their high vapor pressures. The very heavy fragments (e.g., C_8H_{10}) were quite effectively trapped with only a water-cooled baffle. The partial pressure of an intermediate weight fragment C_6H_6, was reduced by a factor of 10^3 when the trap was cooled from 25 to -196°C. When using modern low vapor pressure fluids such as DC-705 or Santovac 5, the basic operational difference between a liquid nitrogen trap and a cold water baffle is the ability of the liquid

Table 8.3.   Diffusion Pump Backstreaming[a]

Conditions	Duration of Test (h)	Backstreaming Rate (mg/cm^2)/min
(1) Without baffle	165	1.6×10^{-3}
(2) With liquid nitrogen trap	170	5.3×10^{-6}
(3) Same as (2)	380	6.5×10^{-6}
(4) Item (3) plus water baffle	240	2.8×10^{-7}
(5) Item (4) plus creep barrier	240	8.7×10^{-8}
(6) Same as (5)	337	1.2×10^{-7}

Source. Reprinted with permission from *J. Vac. Sci. Technol.*, **6**, p. 265, M. H. Hablanian. Copyright 1969 The American Vacuum Society.

[a] Measurements made with a 6-in. diffusion pump (NRC HS6-1500), DC-705 pumping fluid, liquid-nitrogen-cooled collectors.

nitrogen trap to pump C_6H_6 and to partially trap some of the lighter weight fractions.

The quantitative effects of various trap, baffle, and creep barrier combinations are summarized in Table 8.3 [37]. It was shown that the usual addition of the chevron water baffle between the liquid nitrogen trap and the pump is not much better than the addition of a piece of straight pipe or elbow of the same length. Rettinghaus [41] has shown for one pump and baffle that the addition of throttling structures below the baffle will further reduce the backstreaming. Figure 8.6 summarizes the results of his measurements on the backstreaming of polyphenylether. The addition of a baffle consisting of three circular half-chevrons was shown to give a net backstreaming rate that was

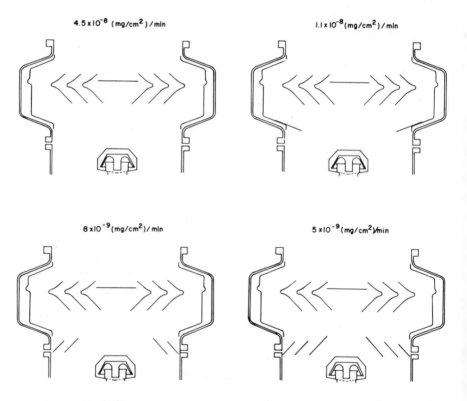

Fig. 8.6 Relationship between backstreaming and added throttling below the baffle for a Balzers 250 diffusion pump stack. Reprinted with permission from G. Rettinghaus, Balzers High Vacuum, Furstentum, Liechtenstein.

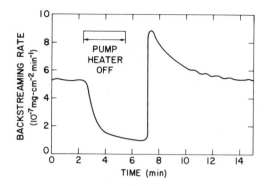

Fig. 8.7 Backstreaming of the parent peak (M/z = 446; Convalex-10) over a liquid nitrogen trap during cool-down and start-up of a diffusion pump. Reprinted with permission from *J. Vac. Sci. Technol.*, **9**, p. 416, G. Rettinghaus and W. K. Huber. Copyright 1972, The American Vacuum Society.

tenfold lower than for the baffle alone but not without further reduction of the pumping speed.

The Herrick effect [42] and the oil burst that resulted from the formation and collapse of the top jet are two transient phenomena that cause backstreaming in addition to the steady-state phenomena already discussed. The Herrick effect (the appearance of oil droplets in the chamber and on fixturing) is the ejection of frozen oil droplets from the surface of an oil covered trap during the initial stages of cooling with liquid nitrogen. The use of a well-designed cold cap and water-cooled partial baffle followed by a continuously operating liquid-nitrogen trap will allow the pump to be run for a year or more without collecting excessive amounts of oil on the trap. The transient back-streaming from the top jet during warm-up and cool-down of the pump is well documented [12, 39, 43]. Figure 8.7 shows an RGA trace of the parent molecule (M/z = 446; Convalex-10 [43]). The back-streaming decreases as the oil is cooled and reaches a peak of about twice the steady-state rate during heating. Power and Crawley [12] also show a peak as the jet is cooling. The total backstreaming was measured as 6 × 10^{-4} mg/cm^2 for a complete start-stop cycle [39]. This kind of backstreaming can be avoided by continuous operation of the diffusion pump or by using the gas-flushing techniques discussed in Chapter 13.

By the use of high-quality, low vapor pressure, anticreep fluid such as DC-705 or Santovac-5 and a continuously operating liquid nitrogen trap contamination due to backstreaming of pump fluid can be made very small. The lowest value of backstreaming shown in Table 8.3 corresponds to a contamination rate of one monolayer per year in a bell jar 500 mm high and 350 mm in diameter. This level of organic contamination is below that produced by O-rings and other sources [2]. Fluid backstreaming in a diffusion pump operating at high vacuum is only one source of organic backstreaming. Additional concerns that relate to specific systems are discussed in Sections 10.1, 11.2, and 12.1.

REFERENCES

1. M. H. Hablanian and J. C. Maliakal, *J. Vac. Soc. Technol.*, **10**, 58 (1973).
2. M. H. Hablanian, *Solid State Technol.*, December 1974, p. 37.
3. M. H. Hablanian, *Proc. 6th Int. Vac. Congr.*, Kyoto, *Japan. J. Appl. Phys.*, Suppl. 2, Pt. 1, 25 (1974).
4. J. H. Singleton, *J. Phys. E.*, **6**, 685 (1973).
5. N. A. Florescu, *Vacuum*, **10**, 250 (1960).
6. N. A. Florescu, *Vacuum*, **13**, 569 (1963).
7. G. Tóth, *Proc. 4th Int. Vac. Congr. (1968)*, Institute of Physics and the Physical Society, London, 300 (1969).
8. S. Dushman, *The Scientific Foundations of Vacuum Technology*, 2nd ed., J. M. Lafferty, Ed., Wiley, New York, 1962, Chapter 3.
9. B. D. Power, *High Vacuum Pumping Equipment*, Reinhold, New York, 1966.
10. W. Gaede, German Pat. 286,404 (filed September 25, 1913).
11. M. Morand, U.S. Pat. 2,508,765 (filed July 27, 1947; priority France, September 25, 1941).
12. B. D. Power and D. J. Crawley, *Vacuum*, **4**, 415 (1954).
13. C. G. Smith, U.S. Pat. 1,674,377 (filed September 4, 1924).
14. W. A. Giepen, U.S. Pat. 2,903,181 (filed June 5, 1956).
15. G. Barrows, Brit. Pat. 475,062 (filed May 12, 1936).
16. J. R. O. Downing, U.S. Pat. 2,386,299 (filed July 3, 1944).
17. B. D. Power, Brit. Pat. 700,978 (filed January 25, 1950).
18. J. R. O. Downing and W. B. Humes, U.S. Pat. 2,386,298 (filed January 30, 1943).
19. R. B. Nelson, U.S. Pat. 2,291,054 (filed August 31, 1939).
20. J. J. Madine, U.S. Pat. 2,366,277 (filed March 18, 1943).
21. N. G. Nöller, G. Reich, and W. Bächler, *Trans. 4th Nat. Symp. Vac. Technol.*, **6** (1957).
22. B. B. Dayton, U.S. Pat. 2,639,086 (filed November 30, 1951).
23. C. R. Burch and F. E. Bancroft, Brit. Pat. 407, 503 (filed January 19, 1933).
24. L. T. Lamont Jr., *J. Vac. Soc. Technol.*, **10**, 251 (1973).

25. S. G. Burnett and M. H. Hablanian, *J. Environ. Sci.*, **5**, 7 (1964).

26. Porta Test,® Varian Associates, 611 Hansen Way, Palo Alto, CA 94303.

27. Reprinted with permission from CVC Products, Inc., 525 Lee Rd., Rochester, NY 14603.

28. Reprinted with permission from Balzers High Vacuum, Furstentum. Liechtenstein.

29. Reprinted with permission from Inland Vacuum Industries, Inc., 35 Howard Avenue, Churchville, NY 14428.

30. Reprinted with permission from Dow Corning Company, Inc., 2030 Dow Center, Midland, MI 48640.

31. Reprinted with permission from Monsanto Company, 800 N. Lindbergh Boulevard, St. Louis, MO 63166.

32. Reprinted with permission from Montedison, USA, Inc., 1114 Avenue of the Americas, New York, NY 10036.

33. K. C. D. Hickman, *Trans. 8th Nat. Vac. Symp. (1961)*, Pergamon, New York, 1962, p. 307.

34. C. W. Solbrig and W. E. Jamison, *J. Vac. Sci. Technol.*, **2**, 228 (1965).

35. G. Caporiccio, R. A. Steenrod Jr., and L. Laurenson, *J. Vac. Sci. Technol.*, **15**, 775 (1978).

36. L. Laurenson, *Ind. Res. Dev.*, November 1977, p. 61.

37. M. H. Hablanian, *J. Vac. Sci. Technol.*, **6**, 265, (1969).

38. N. Milleron, *Trans. 5th Nat. Vac. Symp. (1958)*, Pergamon, New York, 1959, p. 140.

39. G. Rettinghaus and W. K. Huber, *Vacuum*, **24**, 249 (1974).

40. C. M. Gosselin and P. J. Bryant, *J. Vac. Sci. Technol.*, **2**, 293 (1965).

41. G. Rettinghaus, private communication.

42. M. H. Hablanian and R. F. Herrick, *J. Vac. Sci. Technol.*, **8**, 317 (1971).

43. G. Rettinghaus and W. K. Huber, *J. Vac. Sci. Technol.*, **9**, 416 (1972).

CHAPTER 9

Entrainment Pumps

Entrainment pumps operate by capturing gas molecules and retaining them on a surface. The physical or chemical forces that bind molecules to surfaces are sensitive to the gas species. As a result two or more entrainment processes are usually combined to achieve a pump that will effectively entrain the wide range of active and noble gases and vapors encountered in practice. Common to all these processes is the need to remove the entrained material periodically. In some pumps the gas will have reacted with one of the pump materials to form a stable compound and will necessitate cleaning the pump as well as replacing the reacted material. In other pumps the gas is condensed or adsorbed on the surface or held in solid solution, in which case it must be removed by heating. One consequence of entrainment pumping is that each pump has a finite operating time. Further operation is not possible without a regurgitation or replacement procedure.

Entrainment pumps are often referred to as clean pumps. They are clean in the sense that they do not generate the heavy hydrocarbon contamination that is associated with diffusion or oil-sealed mechanical pumps. For many applications cleanliness is not only the freedom from heavy hydrocarbons but from hydrogen, methane, carbon oxides, and inert gases as well. In that sense entrainment pumps can contaminate. Certain gases will displace other adsorbed gases and carbon in some metals can react with water vapor when heated to produce methane or carbon oxides. Hydrogen and other gases may be poorly pumped or released from surfaces on which they were trapped. Each entrainment pump may produce or not pump one or more gas species; the labeling of these gases as contaminants depends on the application.

This chapter reviews evaporable and nonevaporable getter, ion and cryogenic pumps. Cryogenic pumping is not a new technique; cryosorption and cryocondensation, for example, on liquid nitrogen or liquid helium surfaces, are well-established. The use of a gaseous helium refrigerator, however, as the cryo source is a new development and is reviewed here in some depth.

9.1 GETTER PUMPS

Many materials are effective surface and bulk getters for gases. The sublimation pump is primarily a surface getter; gas molecules collide and react with the surface. There is little diffusion into the material. The sorption properties of most gases are highest when the surface is cooled and in that condition diffusion of gases into the material is extremely slow. The pumping speed of a bulk getter material is limited by the diffusion of gas into the bulk and the bulk getter is usually heated to increase the diffusion rate.

9.1.1 Sublimation Pumps

A sublimation pump is a surface getter pump for active gases. Titanium is the most commonly used getter; it is inexpensive, effective, and easily sublimed. It is replenished by depositing a clean film over the

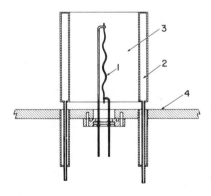

Fig. 9.1 Schematic of a basic titanium sublimation pump. (1) Titanium alloy filament, (2) coolant reservoir, (3) titanium deposit, (4) vacuum wall.

saturated film. Figure 9.1 sketches one form of titanium sublimation pump (TSP). An alternating current is passed through the filament which causes the titanium to sublime and deposit on the adjacent walls. Pump elements are fabricated with three or four separately heated filaments to extend the time between filament replacements. Active gases are captured on the fresh titanium surface which is cooled with water or liquid nitrogen. Because the pumped gases cannot be desorbed by heating, a fresh titanium layer must be deposited periodically to ensure continuous pumping. The pumping characteristics of titanium differ for the active gases, the intermediate gases, and the chemically inactive gases. The active gases (carbon oxides, oxygen, water vapor, and acetylene) are pumped with high sticking coefficients. Water dissociates into oxygen and hydrogen that are then pumped separately. The temperature of the film has no major effect on the pumping speed of these gases because the sticking coefficients are generally near unity in the 77 to 300 K range. The sticking coefficients of the intermediate gases (hydrogen and nitrogen) are low at room temperature but increase at 77 K. After sticking hydrogen may diffuse into the underlying film. The chemically inactive gases such as helium and argon are not pumped at all. Methane has the characteristics of an inactive gas and is only slightly sorbed on titanium at 77 K. Figure 9.2 gives the room temperature sorption characteristics of several active and intermediately active gases [1]. The sticking coefficient is the highest for all gases on a clean film, and for the very active gases remains so until near saturation.

The replacement of one previously sorbed gas by another gas is important. It does create a memory effect, and also results in actual sticking coefficients that depend on the nature of the underlying adsorbed gas. Gupta and Leck [1] observed a definite order of preference in gas replacement. Table 9.1 illustrates the order in which active gases replace less active gases. Oxygen, which is the most active gas, can replace all other gases, whereas methane, which is bound only by van der Waals forces, is displaced by all other active gases.

Gas replacement is a major cause of the large differences in measured sticking coefficients, especially when the films were not deposited under clean conditions. Harra [2] has reviewed the sticking coefficients and sorption of gases on titanium films measured in several independent studies and tabulated their average values in Table 9.2. These coefficients represent the average of the values obtained in different laboratories and under different conditions. They are probably more representative of those in a typical operating sublimation pump whose history is not known than are those measured under clean conditions.

The TSP operates below a pressure of 10^{-1} Pa. Above that pressure surface compound formation inhibits sublimation. A typical pumping speed curve is sketched in Fig. 9.3. At low pressures there is little reaction between the titanium and gas until the titanium atoms reach the surface. This yields a constant pumping speed that is determined by the surface area of the film and the conductance of any interconnecting tubing. At low pressures more titanium is sublimed than is needed when the filament is operated continuously. At high pressures titanium-gas collisions occur before the titanium strikes the surface and the pumping speed is determined by the rate of titanium sublimation. The pumping speed will decrease as $1/P$, as sketched in Fig. 9.3.

The calculation of the pumping speed in the low pressure region is not easy to do precisely because of the sticking coefficient uncertainty and the geometry. In molecular flow the pumping speed S of the geometry shown in Fig. 9.1 is given approximately by

$$\frac{1}{S} = \frac{1}{S_i} + \frac{1}{C_a} \qquad (9.1)$$

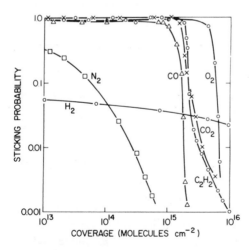

Fig. 9.2 Room-temperature sorption characteristics for pure gases on batch evaporated, clean titanium films. Reprinted with permission from *Vacuum*, **25**, p. 362, A. K. Gupta and J. H. Leck. Copyright 1975, Pergamon Press. Ltd.

where S_i is the intrinsic speed of the surface and C_a is the conductance of the aperture at the end of the cylindrical surface on which the titanium is deposited. This conductance can be ignored if the film is deposited on the walls of the chamber. If a valve or connecting pipe is used, the appropriate series conductance should be added. The intrinsic speed is approximately

$$S_i(\text{L/s}) = 1000A\frac{v}{4}s'$$ 　　　　　　(9.2)

where A is the area of the film, v is the gas velocity, and s' is the sticking coefficient of the gas. Cooling to 77 K provides little additional pumping speed in pumps whose speed is conductance limited by the geometry.

Table 9.1　Order of Preference of Gas Displacement on Titanium Films[a]

Gas Being Pumped	Displaced Gas				
	CH_4	N_2	H_2	CO	O_2
CH_4		N	N	N	N
N_2	Y		N	N	N
H_2	Y	Y		N	N
CO	Y	Y	Y		N
O_2	Y	Y	Y	Y	

Source. Reprinted with permission from *Vacuum*, **25**, p.362, A. K. Gupta and J. H. Leck. Copyright 1975, Pergamon Press, Ltd.

[a] Y = Yes, N = No.

Table 9.2   Initial Sticking Coefficient and Quantity Sorbed for Various Gases on Titanium

Gas	Initial Sticking Coefficient (300 K)	(78 K)	Quantity Sorbed[a] ($\times 10^{15}$ molecules/cm^2) (300 K)	(78K)
H_2	0.06	0.4	8-230[b]	7-70
D_2	0.1	0.2	6-11[b]	-
H_2O	0.5	-	30	-
CO	0.7	0.95	5-23	50-160
N_2	0.3	0.7	0.3-12	3-60
O_2	0.8	1.0	24	-
CO_2	0.5	-	4-24	-
He	0	0		
Ar	0	0		
CH_4	0	0.05		

Source. Reprinted with permission from *J. Vac. Sci. Technol.*, **13**, p. 471, D. J. Harra. Copyright 1976, The American Vacuum Society.
[a] For fresh film thicknesses of 10^{15} Ti atoms/cm^2.
[b] The quantity of hydrogen or deuterium sorbed at saturation may exceed the number of Ti atoms/cm^2 in the fresh film through diffusion into the underlying films at 300 K.

At high pressures the pumping speed is determined by the rate of titanium sublimation. This theoretical maximum throughput is related to the titanium sublimation rate (TSR) by the relation [3]

$$Q(\text{Pa} - \text{L/s}) = \frac{10^8 V_o \text{TSR}(\text{atoms/s})}{nN_o}$$

or

$$Q(\text{Pa} - \text{L/s}) = \frac{10^{-18}}{n} \text{TSR} \ (\text{atoms/s}) \qquad (9.3)$$

V_o is the normal specific volume of the gas, N_o is Avogadro's number, and n is the number of titanium atoms that react with each molecule of

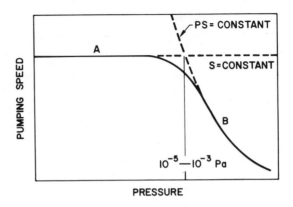

Fig. 9.3 Characteristic pumping speed versus pressure for a TSP: (a) Speed determined by the getter area, sticking coefficient, gas species, and inlet conductance; (b) speed determined by pressure and rate of sublimation.

gas; $n = 1$ for CO and $n = 2$ for N_2, H_2, O_2, and CO_2 [2]. For titanium (9.3) may be rewritten as

$$Q(\text{Pa} - \text{L/s}) = \frac{1.25 \times 10^{-2}}{n} \text{TSR} \ (\mu\text{g/s}) \qquad (9.4)$$

This theoretical throughput can be reached only when the titanium is fully reacted with the gas. The corresponding pumping speed is obtained by dividing the throughput by the pressure in the pump.

The TSP is used in combination with other pumps that will pump inert gases and methane. Small TSPs are used continuously for short periods to aid in crossover between a sorption pump and an ion pump. Large pumps have been developed for use in conjunction with smaller ion pumps for long-term, high-throughput pumping. The TSP is used intermittently for long periods at low pressures to provide high-speed pumping of reactive gases. At low pressures the film need be replaced only periodically to retain the pumping speed. Titanium is sublimed until a fresh film is deposited. The pump is then turned off until the film saturates. Figure 9.4 [3] sketches the pressure rise with decrease in pumping speed as the titanium film saturates in a typical ion pumped system. P_i is the initial pressure with the ion and sublimator operating and P_f is pressure with only the ion pump operating. Not shown on these curves is the pressure burst due to gas release from the titanium during sublimation. Hydrogen, methane, and ethane are

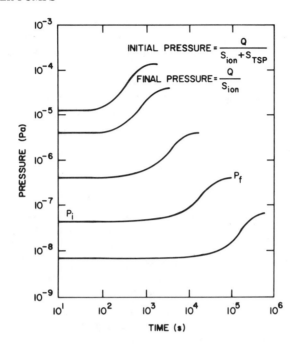

Fig. 9.4 Typical pressure rise due to decrease in pumping speed as a titanium film saturates. Reprinted with permission from *General Characteristics of Titanium Sublimation Pumps*, B. E. Keitzmann, 1965, Varian Associates, 611 Hansen Way, Palo Alto, CA 94303.

released from titanium during heating. The methane and ethane result from a reaction between hydrogen and the carbon impurity in the hot filament [1,4]. After the filament reaches temperature and sublimation begins hydrogen is pumped. Methane and ethane are marginally pumped on surfaces held at 77 K.

Commercially available TSPs use directly heated filaments, radiantly heated sources, or electron-beam heated sources. The most commonly used source is a directly heated filament with a low-voltage ac power supply. Filaments were first made from titanium twisted with tantalum or tungsten and later from titanium wound over niobium and tantalum wire [5]. Because of thermal contact problems, the sublimation rate proved to be unpredictable. Modern pumps use filaments fabricated from an alloy of 85% Ti and 15% Mo [6–8]. This filament has an even sublimation rate and a long life. A typical filament 15-cm long can be operated at sublimation rates of 30 to 90 $\mu g/s$. Large TSPs have been constructed with radiantly heated titanium at sublimation rates as high as 150 μg/s [9,10] and with electron-beam heated, rod fed sources at sublimation rates ranging from 300 μg/s to 0.15 g/s [11]. Electron-beam heated sources do not operate well at pressures higher than 10^{-3} Pa and for most applications are too expensive to operate at pressures below 10^{-5} Pa. They serve best as a high-speed pump in the intermediate region. Radiantly heated sources are best for high speed pumping in the very high vacuum region.

9.1.2 Nonevaporable Getter Pumps

Nonevaporable getters pump by surface adsorption followed by bulk diffusion. Their speed for pumping active gases is determined by the diffusion rate into the bulk. For this reason they are operated at high temperatures. They do not pump inert gases or methane because these gases do not adsorb on the surface. The most effective getter for vacuum use is an alloy of 84% Zr and 16% Al [12] which has been extensively studied [13–15]. This alloy, when heated to 400°C, has a pumping speed of $\sim$ 0.3 L-s^{-1}-cm^{-2} (N$_2$), $\sim$ 1 L-s^{-1}-cm^2 (CO$_2$, CO, O$_2$), and 1.5 L-s^{-1}-cm^{-2} (H$_2$) [12]. At room temperature H$_2$ will still be pumped at about half the speed that it is pumped at 400°C, provided that no oxide or nitride diffusion barriers exist. Other gases are nòt pumped at room temperature because the surface compounds quickly form diffusion barriers. All gases except hydrogen are pumped as stable compounds; hydrogen is pumped as a solid solution and may be released by heating above 400°C. To operate the pump the chamber is first evacuated to a pressure below 1 Pa, after which the pump is activated by heating to 800°C to indiffuse the surface layers. The temperature is then reduced to 400°C. The activation step is repeated each time the pump is cooled and released to atmosphere. Appendage pumps equipped with heaters and getter cartridges fabricated from a plated steel sheet coated with Zr-Al alloy have pumping speeds as high as 10-50 L/s [16]. One getter ion pump package has a combined pumping speed of 1,000 L/s [17]. These pumps have a large capacity for hydrogen[18].

9.2 ION PUMPS

The development of the ion pump has made it possible to pump to the ultrahigh vacuum region without concern for heavy hydrocarbon contamination. This pump resulted from the exploitation of a phenomenon formerly considered detrimental to vacuum gauge operation—pumping gases by ions in Bayard-Alpert and Penning gauges. Ions are pumped easily because they are more reactive with surfaces than neutral molecules and if sufficiently energetic can physically embed themselves in the pump walls. If the ions were generated in a simple parallel-plate glow discharge, for example, the pumping mechanism would be restricted to a rather narrow pressure range. Above about 1 Pa the electrons cannot gain enough energy to make an ionizing collision and below about 10^{-1}–10^{-2} Pa the electron mean free path becomes so long that the electrons collide with a wall before they encounter a gas molecule. Ions can be generated at lower pressures if the energetic electrons can be constrained from hitting a wall before they collide with a gas molecule. This confinement can be realized with certain combinations of electric and magnetic fields. Electrostatic ion pumps use static electric fields to confine the electrons, whereas the sputter-ion pump uses both electric and magnetic fields. In addition to the ion pumping mechanism, these pumps use some form of getter pumping.

9.2.1 Electrostatic Ion Pumps

The Evapor-ion pump and the Orbitron pump are two designs in which the electrons are confined to purely electrostatic fields. The Evapor-ion pump developed by Herb [19] has been described in detail by Swartz [20]. It is for all practical purposes a large ion gauge that contains a hearth for continuously evaporating titanium from a wire source. Active gases are pumped by the continuously renewable getter, whereas inert gases are ionized and implanted in the getter film. This pump, which requires frequent titanium replacement and significant power input, has been commercially displaced by the sputter-ion pump.

The Orbitron pump [21] evolved from attempts by Herb and associates [22] to develop an electrostatic ion gauge. Long mean-free paths are maintained by injecting electrons into a cylindrically symmetric electrostatic field bounded by an outer grounded cylinder and an axial wire held at a high positive voltage (~ 5 kV). Paths as long as tens of meters are possible in a pump 300 mm in diameter [23]. The ionizing electrons, which are generated by a heated cathode located at the end of the pump, orbit the anode many times before colliding with the anode or with chunks of titanium on the ends of the anode wire. Titanium is sublimed onto the pump walls as a result of this electron bombardment heating. The resulting getter film pumps active gases, whereas the inert gases are ion pumped. In this design the clever double use of electrons does not allow independent control of the ion current and the rate of titanium sublimation.

Bills [24] and Denison [25] describe an improved version of the Orbitron, the Electro-Ion pump. In this pump the solid cathode wall is replaced by a grounded cylindrical grid. The grid allows the ions to escape and be accelerated toward the outer wall which is negatively biased. Because the wall is no longer necessary to confine the electrons, several anode and grid stages can be grouped in the center of one pump body. In the Electro-Ion design the titanium is sublimed from a separate heater; the sublimation rate and the ion current can be independently controlled and the negative bias on the wall increases the ion energy and improves the ion pumping.

Electrostatic ion pumps cannot be started at high pressures or the continued formation of titanium nitrides and carbides will effectively prevent Ti sublimation. The pressure should be reduced to 10^{-1} Pa before pumping begins. If the system is rapidly cycled to atmosphere, the pump should be isolated to prevent contamination. In all electrostatic ion pumps titanium getter pumping of active gases is the dominant pumping mechanism; inert gases are pumped at only 1–2% of the getter pumping speed [25]. Because of its low ion pumping speed in relation to the sputter-ion pump, the electrostatic ion pump has found limited application. These applications take advantage of its size, weight, absence of magnetic fields, and sustained pumping speed at low pressures.

9.2.2 Sputter-Ion Pumps

The pumping action of a magnetically confined dc discharge was first observed by Penning [26] in 1937, but it was not until two decades ago that Hall [27] combined several Penning cells and transformed the phenomenon into a functional pump. Some elemental forms of the (diode) sputter-ion pump are shown in Fig. 9.5. Each Penning cell is approximately 12 mm in diameter × 20 mm long with a 4-mm gap between the anode and the cathode. Modern pumps are constructed of modules of cells arranged around the periphery of the vacuum wall with external permanent magnets of 0.1 to 0.2 Tesla strength, and cathode voltages of ~ 5 kV.

The electric fields present in each Penning cell trap the electrons in a potential well between that the two cathodes and the axial magnetic field forces the electrons into circular orbits that prevent their reaching the anode. This combination of electric and magnetic fields causes the electrons to travel long distances in oscillating spiral paths before colliding with the anode and results in a high probability of ionizing collisions with gas molecules. The time from the random entrance of the first electron into the cell until the electron density reaches its steady-state value of ~ 10^{10} electrons/cm^3 is inversely proportional to pressure. The starting time of a cell at 10^{-1} Pa is nanoseconds, whereas at 10^{-9} Pa it is 500 s [31]. The ions produced in these collisions are accelerated toward the cathode, where they collide, sputter away the cathode, and release secondary electrons that in turn are accelerated by the field. Many other processes occur in addition to the processes necessary to sustain the discharge; for example a large number of low-energy neutral atoms are created by molecular dissociation and some high-energy neutrals are created from energetic ions by charge neutralization as they approach the cathode, collide, and recoil elastically.

The actual mechanism of pumping in an ion pump is dependent on the nature of the gas being pumped and is based on one or more of the following mechanisms: (1) precipitation or adsorption following molecular dissociation; (2) gettering by freshly sputtered cathode material; (3) surface burial under sputtered cathode material; (4) ion burial following ionization in the discharge; and (5) fast neutral atom burial. (Ions are neutralized by surface charge transfer and reflected to another surface where they are pumped by burial.) The first four of these

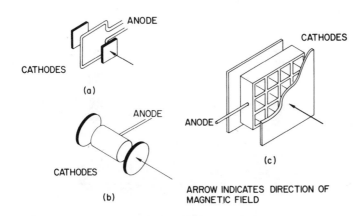

Fig. 9.5 Early forms of the diode sputter-ion pump:. (a) Ring anode cell [26]; (b) long anode cell [28]; same cell with Ti cathodes [29], (c) multicell anode [27]. Reproduced with permission from *Proc. 4th Int. Vac. Cong. (1968)*, p. 325, D. Andrew. Copyright 1969, The Institute of Physics.

mechanisms were elucidated by Rutherford, Mercer and Jepsen [32], and the role of the elastically scattered neutrals was explained by Jepsen [33]. These mechanisms are illustrated in Fig. 9.6.

Organic gases, active gases, hydrogen, and inert gases are pumped in distinctly different ways. There are a few generalities. Initially, gases tend to be pumped rapidly and decay to a steady state [33-35] follows as the layer saturates and reemission rates equal pumping rates. This is more pronounced with noble than with active gases. Pumping speeds cannot be uniquely defined for a gas independent of the composition of other gases being pumped simultaneously. The sputter ion pump is capable of reemitting any pumped gas. This reemission or memory effect complicates the interpretation of some experiments.

Organic gases are easily pumped by adsorption and precipitation after being dissociated by electron bombardment [32].

Active gases such as oxygen, carbon monoxide, and nitrogen are pumped by reaction with titanium, which is sputtered on the anode

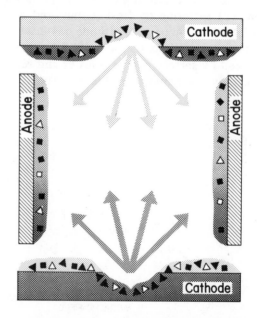

Fig. 9.6 Schematic diagram showing sputter deposition and pumping in a Penning cell: ■ chemically active gases buried as neutral particles; ▲ Chemically active gases ionized before burial; □ inert gases buried as neutral particles; Δ inert gases ionized before burial. Reprinted with permission from *Proc. 4th Int. Vac. Congr. (1968)*, p. 325, D. Andrew. Copyright 1969, The Institute of Physics.

surfaces, and by ion burial in the cathode. These gases are easily pumped because they form stable titanium compounds [32].

Hydrogen behaves differently. Its low mass prevents it from sputtering the cathode significantly. It behaves much like it does in a TSP. It is initially pumped by ion burial and neutral adsorption [33, 36] and diffuses into the bulk of the titanium and forms a hydride. Sustained pumping of hydrogen at high pressures will cause cathodes to warp [32] and release gas as they heat. The hydrogen pumping speed does not rate limit unless cathode surfaces are covered with compounds that prevent indiffusion. The pumping of a small amount of an inert gas, say argon, cleans the surfaces and allow continued hydrogen pumping [37], whereas a trace amount of nitrogen will reduce the speed by contaminating the surface [36].

Noble gases are not pumped so efficiently as active gases in a diode pump. They are pumped by ion burial in the cathodes and by reflected neutral burial in the anodes and cathodes. The noble gas pumping on the cathodes is mostly in the area near the anodes where the sputter build-up occurs. Because most of the neutrals are reflected with low energies in the diode pump, their pumping speed in the anode or other cathode is low; for example, argon is pumped only at 1–2% of the active gas speed.

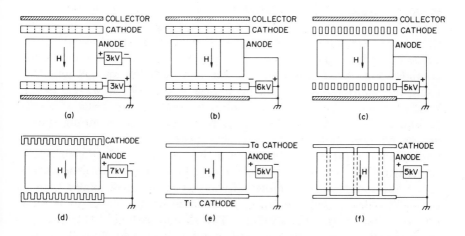

Fig. 9.7 Pump designs for inert gas pumping: (a) The triode pump of Brubaker [38]; (b) triode pump of Hamilton [39]; (c) triode Varian Noble Ion Pump [40]; (d) slotted cathode diode of Jepsen et al. [41]; (e) differential ion pump of Tom and Jones [42]; (f) magnetron pump of Andrew et al. [45]. Reprinted with permission from *Proc. 4th Int. Vac. Congr. (1968)*, p. 325, D. Andrew. Copyright 1969, The Institute of Physics.

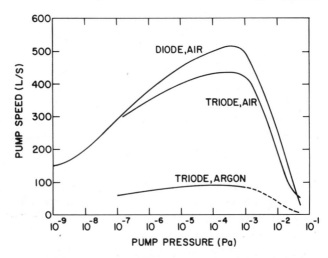

Fig. 9.8 Pumping speeds for air and argon for the 500-L/s Varian diode Vac Ion pump and for the 400-L/s triode Vac Ion pump. Speeds measured at the inlet of the pump. Reprinted with permission from Varian Associates, 611 Hansen Way, Palo Alto, CA 94303.

Argon, in particular, suffers from a pumping instability. Periodically the argon pressure will rise as pumped gas is released from the cathodes. Figure 9.7 [30] illustrates some of the geometries that were devised as a solution to the problem of low argon pumping speed and its periodic reemission. Brubaker [38] devised a triode pump with a collector surface that operated at a potential between the anode and cathode (Fig. 9.7a). Its function was to collect low energy ions that could not sputter. It was shown by Hamilton [39], however, that it worked equally well when the collector surface was held at anode potential (Fig. 9.7b). In the triode pump the argon pumping speeds are as high as 20% of the nitrogen speed. This high speed (high implantation rate) results from the high energy of the neutrals which are scattered at small angles from the cathode walls with little energy loss. Sputtering is much more efficient at these small angles than at normal incidence and sputtering of titanium on the collector is more efficient than in the diode pump. The slotted cathode [41], attempts to accomplish this sputtering with one less electrode than the triode (Fig. 9.7d). This pump has an argon pumping speed of 10% of the speed for air. Tom and Jones [42] devised the differential diode ion pump sketched in Fig. 9.7e. One titanium cathode was replaced by a tantalum cathode. In this manner the recoil energy of the scattered noble gas neutrals, which depends on the relative atomic weight of the cathode material and gas atom, is increased [43, 44]. This gives more effective noble gas pumping than in the diode pump. An argon-stable magnetron structure, (Fig. 9.7*f*), was devised by Andrew et al. [45]. The central cathode rod is bombarded by a high flux of ions at oblique angles of incidence. The sputtering of the rod creates a flux that continually coats the cathode plates and the impinging ions and results in a net argon speed of 12% of the air speed [45]. Among the designs discussed here for increasing the argon pumping speed and reducing or eliminating its instability the triode and differential diode are in most widespread use.

The operating pressure range of the sputter-ion pump extends from 10^{-2} to below 10^{-8} Pa. A characteristic pumping-speed curve of a diode and a triode pump are shown in Fig. 9.8. If starting is attempted at high pressures, say 1 Pa, a glow discharge appears and the elements heat and release hydrogen. As the pressure is reduced, the glow discharge extinguishes and the speed rapidly increases. At low pressures the speed decreases because the sputtering and ionization processes decrease. The exact shape of a pumping speed curve is a function of the magnetic field intensity, cathode voltage, and cell diameter-to-length ratio. As the pressure decreases, the ionization current decreases proportionately and the pressure in the pump may be obtained from the ion current without the need for an ionization gauge tube.

The lifetime of a diode pump is a function of the time necessary to sputter through the cathodes. A typical value is 5000 h at 10^{-3} Pa or 50,000 h at 10^{-4} Pa. The triode, which pumps slightly better than the diode at high pressures, is also easier to start at high pressures and has a lifetime of less than half the diode. In both pumps the life may be shorter due to shorting of the electrodes by loose flakes of titanium.

The sputter ion pump has the advantage of freedom from hydrocarbon contamination and ease of fault protecting but does suffer from the reemission of previously pumped gases, particularly hydrogen, methane and the noble gases.

9.3   CRYOGENIC PUMPS

Cryogenic pumping is the entrainment of molecules on a cooled surface by weak van der Waals or dispersion forces. In principle, any gas can be pumped, provided that the surface temperature is low enough so that arriving molecules will remain on the surface after losing sufficient kinetic energy. Cryogenic entrainment is a clean form of pumping in which the only gas or vapor contaminants are those not pumped by a cryo surface or released from pumped deposits. Unlike the ion pump, the cryopump does not retain the physically adsorbed or condensed gas load after the pumping surfaces have been warmed. Proper precautions must be taken to vent the pumped gas load.

Cryopumps are used in a wide range of applications and in many forms. Liquid nitrogen-cooled molecular sieve pumps are used as roughing pumps. Liquid nitrogen or liquid helium "cold fingers" are used in high vacuum chambers. Primary high vacuum pumps are cooled by liquid cryogens or closed-cycle helium gas refrigerators. The high cost of liquid helium led to the engineering of small, reliable helium gas refrigerators and their incorporation into practical pumps.

The theory and techniques of pumping on a cooled surface have been a part of vacuum technology much longer than the helium gas refrigerator, however, knowledge of both is necessary in order to understand the operation of a modern closed-cycle cryopump. This section reviews the mechanisms of cryocondensation and cryosorption on which all cryogenic pumping is based, discusses pumping speed, ultimate pressure and saturation effects, refrigeration techniques, and pump characteristics. System operation techniques are discussed in Chapters 10 and 11 and the problems of pumping gases at high flow rates are covered in Chapter 12.

9.3.1 Pumping Mechanisms

Low-temperature pumping is based on cryocondensation, cryosorption, and cryotrapping. In Chapter 6 we defined the equilibrium or saturated vapor as that pressure at which the flux of vapor particles to the surface equals the flux of particles leaving the surface and entering the vapor phase, provided that all the particles, solid, liquid, and vapor, are at the same temperature. The arriving particles are attracted to condensation sites on the liquid or solid, where they are held for some residence time after which they vibrate free or desorb into the vapor phase. The vapor pressure and residence time are temperature dependent. As the temperature is reduced the vapor pressure is reduced and the residence time in increased. Tables of vapor pressures of the common gases are given in Appendix B.5. Cryocondensation becomes a useful pumping technique when a surface can be cooled to a temperature at which the vapor pressure is so low and the residence time is so long that the vapor is effectively removed from the system. Liquid nitrogen is a good condenser of water vapor because the water vapor pressure at 77 K is 10^{-19} Pa. The probability c that an atom will condense on collision with a cold surface is called the condensation coefficient. Dawson and Haygood [46], Eisenstadt [47], and Brown and Wang [48] have measured the condensation coefficients of many gases at reduced temperatures and found them all to be in the $0.5 < c < 1.0$ range.

Condensation sites on the solid or liquid state of the vapor are not the only locations at which atoms or molecules can become bound. Any solid surface has a weak attractive force for at least the first few monolayers of gas or vapor. Figure 9.9 describes a typical relationship between the number of molecules adsorbed and the pressure above the adsorbed gas for Xe, Kr, and Ar on porous silver at 77.4 K [49]. These adsorption isotherms tend toward a slope of one at very low pressures, which indicates that the number of adsorbed atoms goes to zero linearly with the pressure. As the sorption sites become increasingly populated the pressure increases. The limiting sorption capacity is reached after a few monolayers have been deposited. A typical monolayer can hold about 10^{15} atoms/cm². The actual number depends on the material. The data shown in Fig. 9.9 saturate at 2×10^{19} atoms

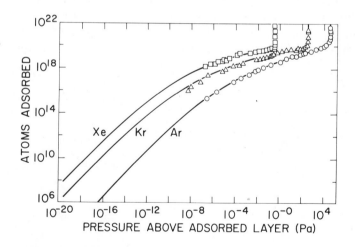

Fig. 9.9 Adsorption isotherms of xenon, krypton, and argon on porous silver adsorbent at 77 K. The lines represent plots of an analytic solution and the points are experimental. Reprinted with permission from *J. Chem., Phys.*, **73**, p. 2720, J. P. Hobson. Copyright 1969, The American Chemical Society.

because the surface area is larger than 1 cm^2. At the vapor pressure condensation begins and the surface layer can increase in thickness. The thickness of the solid deposit is limited only by thermal gradients in the solid and by thermal contact with nearby surfaces of different temperatures.

A curve similar to that in Fig. 9.9 may be measured for each temperature of the sorbate. The effect of temperature on the adsorption isotherm is illustrated in Fig. 9.10 for hydrogen on a bed of activated charcoal. More gas can be adsorbed at a given pressure if the temperature is reduced because the probability of desorption is less than at higher temperatures.

Adsorption is an important phenomenon because it is what allows a vapor to be pumped to a pressure far below its saturated vapor pressure. For gases such as helium, hydrogen, and neon this is the only mechanism by which pumping takes place. The data in Fig. 9.9 show that if the fractional monolayer coverage can be kept small the saturation pressure of the surface can be reduced to 10^{-1}–10^{-12} of the saturated pressure of the vapor over its own condensate. The surface coverage can be minimized by pumping a small quantity of gas or by the generation of a large surface area with porous sorbents such as charcoal or a zeolite. Adsorption isotherms have been measured for many materials and several references are given [49]. The adsorbent properties of charcoal and molecular sieve, which are most interesting for cryopumping, have been the subject of considerable investigation [50–56]. In some cases the isotherms do not approach zero with a slope of one, which indicates that the sorbent was not completely

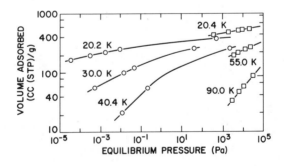

Fig. 9.10 Adsorption of hydrogen on coconut charcoal at low pressures. ○ Gareis and Stern [50], □ Van Dingenan and Van Itterbeek [51]. Reprinted with permission from *J. Vac. Sci. Technol.* **2**, p. 165, S. A. Stern et al. Copyright 1965, The American Vacuum Society.

equilibrated.   Analytical expressions for adsorption isotherms are
discussed elsewhere [49, 57, 58].

The last mechanism of low temperature pumping has been given the
name cryotrapping [58].   Cryotrapping is simply the sorption of one
gas on or in the porous frozen condensate of another [59–61].   Figure
9.11 [59] illustrates the cryotrapping of hydrogen by argon.   In this
experiment a diffusion pump and a cryosurface pumped in parallel on a
known hydrogen gas flow.   This resulted in a steady-state hydrogen
pressure for zero argon flow whose magnitude was determined by the
hydrogen flow rate and diffusion pump speed.   As the argon flow was
increased the cryosurface began to pump the hydrogen and reduce its
partial pressure.   These data show that the efficiency of pumping
hydrogen, or the hydrogen/argon trapping ratio, is much higher at 5
than at 15 K.   At a temperature of more than 23 K cryotrapping of
hydrogen in argon does not occur.   Hengevoss [60] has shown that the
density of the solid argon deposit decreases with its condensation
temperature and that porous argon contains more hydrogen sorption
sites than dense argon.   He also showed that thermal cycling of the
argon to a higher temperature irreversably increased its density.

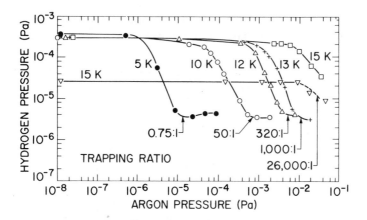

Fig. 9.11   Cryotrapping of hydrogen on solid argon at various temperatures. The drop
in hydrogen pressure corresponds to the onset of cryotrapping at a particular argon
pressure. Reprinted with permission from *Vacuum*, **17**, p. 495, J. Hengevoss and E. A.
Trendelenburg. Copyright 1967, Pergamon Press, Ltd.

9.3.2 Speed, Pressure, and Saturation

In Chapter 2 we outlined kinetic theory and introduced the concepts of
gas flow, conductance, and speed. Whenever the temperature in the
system is the same everywhere, these ideas can be used to predict the
correct performance of a pump or the state of a system. If the temperature varies throughout the system, as it will in a cryogenic-pumped
system, these notions must be applied with care; some are subject to
misinterpretation, whereas others are simply not true. We stated, for
example, that the mean-free path was pressure-dependent. Strictly
speaking, it is only particle-density-dependent. The pressure in a
closed container will increase if the temperature is increased but the
mean-free path will not change because the particle density remains
constant. Such a minunderstanding can easily develop because we
normally associate pressure change with particle density change.

Lewin [62] points out that the definitions of conductance and speed
require that the throughput Q be constant in a series circuit; the
throughput is constant only in an isothermal system. See Section
2.3.1. Particle flow, however, is constant in a nonisothermal system.
It is this concern that directs us to formulate the behavior of a cryogenic pump in terms of particle flow rather than throughput.

The sketch in Fig. 9.12 describes a chamber with gas at pressure P_c
and temperature T_c, connected by area A to a cryogenic pump whose
surfaces are cooled to temperature T_s and in which the gas is in thermal equilibrium with the surface. The temperature of the gas in the
chamber is assumed to be greater than that of the gas in the pump.
The net flux of particles into the pump is $\Gamma_{net} = \Gamma_{in} - \Gamma_{out}$. This may be
written

$$\Gamma_{net} = \frac{A n_c v_c}{4} - \frac{A n_s v_s}{4}$$

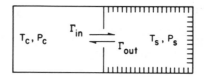

Fig. 9.12 Cryogenic pumping model. The gas in the pump has a temperature T_s equal
to that of the pumping surface and a pressure P_s which is in equilibrium with the gas
condensed or adsorbed on the pumping surface.

$$\Gamma_{net} = \frac{AP_c v_c}{4kT_c} - \frac{AP_s v_s}{4kT_s}$$

$$\Gamma_{net} = \frac{AP_c v_c}{4KT_c}\left[1 - \frac{P_s}{P_c}\left(\frac{T_c}{T_s}\right)^{1/2} \right] \qquad (9.5)$$

In this derivation we have assumed that the condensation coefficient is unity. Equation 9.5 may be simplified by observing that the term outside the brackets is Γ_{in}. The maximum particle flow into the pump corresponds to $\Gamma_{out} = 0$, or $\Gamma_{in} = \Gamma_{max}$. Equation 9.5 may be written as

$$\frac{\Gamma_{net}}{\Gamma_{max}} = \left[1 - \frac{P_s}{P_c}\left(\frac{T_c}{T_s}\right)^{1/2} \right] \qquad (9.6)$$

Now define

$$P_{ult} = P_s\left(\frac{T_c}{T_s}\right)^{1/2} \qquad (9.7)$$

and express Eq. 9.6 as

$$\frac{\Gamma_{net}}{\Gamma_{max}} = c\left(1 - \frac{P_{ult}}{P_c} \right) \qquad (9.8)$$

where the condensation coefficient is now included. Equation 9.7 is the thermal transpiration equation (2.30). It relates the ultimate pressure in the chamber of our model P_{ult} to the pressure over the surface. If the pump is a condensation pump, P_s is the saturated vapor pressure. If the pump is a sorption pump, P_s is the pressure obtained from the adsorption isotherm, knowing the fractional surface coverage and temperature of the sorbent.

The ultimate pressure for the cryosorption or cryocondensation pump modeled in Fig. 9.12 can be determined from (9.7) by use of the proper value of P_s. The ultimate pressure for cryocondensation pumping is equal to the saturated vapor pressure multiplied by the thermal transpiration ratio $(T_c/T_s)^{\frac{1}{2}}$. It is a constant during operation of the pump, provided that the temperature of the cryosurface does not change. The ultimate pressure for cryosorption pumping will increase with time because the saturation pressure over the sorbent is a function of the quantity of previously pumped gas. In either case the ultimate pressure in the chamber will be greater than the saturated vapor pressure or adsorption pressure by the transpiration ratio. For $T_c = 300\ K$ and $T_s = 15\ K$ the ratio is $P_{ult} = 4.47\ P_s$.

Equation 9.8 may also be used to characterize the speed of the pump because $\Gamma_{net}/\Gamma_{max}$ is proportional to S_{net}/S_{max}. The pumping speed of a cryocondensation pump is constant and near its maximum value as long as $P_{ult} \ll P_c$, regardless of the quantity of gas pumped. All gases except H_2, He, and Ne have a saturated vapor pressure of less than 10^{-20} Pa at 10 K.

The pumping speed of a cryosorption pumping surface is affected by its prior use because the saturation pressure of the surface increases as the sites become filled. For high vapor pressure gases such as H_2 and He the pumping speed on a molecular sieve at 10 to 20 K can actually diminish from S_{max} to zero as the clean sorbent gradually becomes saturated with gas during pumping. Figure 9.13 sketches the expected behavior of speed (linear scale) and ultimate pressure (log scale) as a function of the quantity of gas being pumped (log scale) for both cryosorption and cryocondensation pumping. For both cases the net

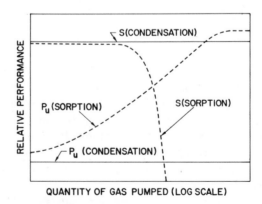

QUANTITY OF GAS PUMPED (LOG SCALE)

Fig. 9.13 Relative variation of pumping speed and ultimate pressure versus quantity of gas pumped for cryosorption pumping and for cryocondensation pumping.

speed goes to zero as the chamber pressure reaches the ultimate pressure.

The simple model presented here is valid for predicting the performance of a cryogenic pump that is connected to the chamber by an aperture or pumping port and in which all of the gas in the pump is cooled to the temperature of the pumping surface.   Unfortunately practical pumps do not meet these criteria.   A chamber that is completely immersed in a liquid cryogen has an ultimate pressure given by P_s, that is, the gas temperature in (9.7) is $T_c = T_s$, whereas Moore [63] has shown that the ultimate pressure in a system consisting of a parallel cryopanel and warm wall is given by

$$P_u = \frac{P_s}{2}\left[1 + \left(\frac{T_1}{T_2}\right)^{\frac{1}{2}} \right] \qquad (9.9)$$

where T_1 is the temperature of the warm wall and T_2 is the panel temperature.   Space chambers are constructed with large cryopanels located inside.   The gas density and temperature in those pumping systems are neither uniform nor in equilibrium and the pressure measured depends upon the orientation of the gauge [63].   The same is true for a cryogenic pumping unit which contains surfaces cooled to several different temperatures.   The model does not account for the heat carried to the cooled surfaces by the gas or by radiation other than to imply that an equal amount of energy must be removed from a cooled surface by the cryogen or refrigerant in order that its temperature remain constant.   The effects of thermal loading on the pumping surfaces are discussed in Section 9.3.4.   This simple model is sufficient to understand conceptually the speed-pressure relationships for the cryocondensation and cryosorption pumping of individual gases.

In the majority of pumping requirements the pump must adsorb or condense a mixed gas load.   Cryotrapping is one instance in which the pumping of one gas aids the pumping of another.   Cryosorption pumping of mixed gases may also cause desorption of a previously pumped gas, or reduced adsorption of one of the components of the mixed gas. Water vapor will inhibit the pumping of nitrogen [64], and CO has been shown to replace N_2 and Kr on Pyrex glass [65].   This is similar to the gas replacement phenomenon for chemisorption that occurs in TSPs; the gas with a small adsorption energy tends to be replaced or be pumped less efficiently than the gas with a large adsorption energy. Hobson [66] has reviewed the basic adsorption processes in cryopumps, and pumping phenomena, while Kidnay and Hiza [67] have summarized the literature on mixture isotherms.

9.3.3 Refrigeration Techniques

Cryogenic pumps are cooled with liquids or gases. Liquid helium and liquid nitrogen are used to cool surfaces to 4.2 and 77 K, respectively. Other cryogens such as liquid hydrogen, oxygen, and argon are used to obtain different temperatures for specific laboratory experiments. In a liquid-cooled pump heat is removed from the cryocooled surfaces to an intermittently filled liquid storage reservoir or to a coil through which the liquid cryogen is continuously circulated. In a two-stage closed-cycle gas refrigerator pump gaseous helium is cooled to two temperature ranges, 10 to 20 K, and 40 to 80 K. Cryopumping surfaces are attached to these locally cooled regions. Both methods of removing heat, liquid cooling and gas cooling, require mechanical refrigeration but in different ways. Liquid cryogens are most economically produced in large refrigerators at a central location and distributed in vacuum-insulated dewars, whereas helium gas refrigerators are economical for locally removing the heat load of a small cryogenic pump. The liquid cryogen requirements of a large cryogenic pumped space chamber are sufficient to warrant the installation of a liquifier at the point of use.

Systems that use liquid cryogens are often referred to as open-loop because the boiling liquid is usually allowed to escape into the atmosphere. This is not necessary; systems have been devised to collect rare gases and return them to the liquifier. Helium gas refrigerators are examples of closed-loop systems. The gas which has adsorbed heat at a low temperature is returned to the compressor.

Many thermodynamic cycles have been developed for the achievement of low temperatures [68–70]. Some are used for the production of liquid helium or other liquid cryogens, some cool semiconducting and superconducting devices, and others produce refrigeration of useful capacity at temperatures ranging from 100 K to a few degrees above liquid helium temperature. It is the latter class of refrigerators that is used to cool modern cryogenic pumps. The refrigerator must be reliable, simple, and easy to manufacture and operate. Cycles embodying these attributes have been developed by Gifford and McMahon [71–74] and by Longsworth [75–77]. These two cycles are variants of a cycle developed by Solvay [68, 70] in 1887. Except for some Stirling cycle machines [78], almost every cryogenic pump in use today is operated on one of these two cycles.

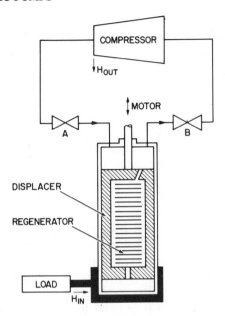

Fig. 9.14 Schematic representation of a single-stage Gifford-McMahon helium gas refrigerator. Adapted with permission from CTI-Cryogenics, Kelvin Park, Waltham, MA 02154.

Figure 9.14 illustrates a basic one-stage Gifford-McMahon refrigerator. The helium compressor is remote from the expander, but is connected with it by two flexible, high-pressure hoses. Within the expander is a cylindrical piston or displacer made from an insulating material. The piston is called a displacer [72] because the regions at each end are connected to give them little pressure difference.

Inside the displacer is a regenerator—a single-channel heat exchanger through which the gas flows at different times in alternate directions. It is tightly packed with a metal of high heat capacity arranged so that the surface-area-to-volume ratio is large. Small spheres, wire mesh, or fine screen will satisfy these criteria. Lead shot and copper or brass screening are the most commonly used packing materials. In the steady state the regenerator will have a temperature gradient. Ambient-temperature helium entering from the warm end will give heat to the metal or cold gas entering from the cooler end will absorb heat from the metal. Even though the regenerator is tightly packed, there is not much flow resistance. The regenerator can transfer thermal energy from the incoming to the outgoing helium quickly and with great efficiency.

The operation of the Gifford-McMahon refrigerator can be understood by following the helium through a complete steady-state cycle. High-pressure gas from the outlet of the compressor is admitted to the regenerator through valve A while the displacer is at the extreme lower end of the cylinder, (see Fig. 9.14). During the time that valve A is open the displacer is raised. The incoming gas passes through the cold regenerator and is cooled as it gives heat to the regenerator. At this point in the cycle the gas temperature is about the same as that of the load. Valve A is then closed before the displacer reaches the top of its stroke. Further movement of the displacer causes the gas to expand slightly. The exhaust valve B is now opened to allow the helium to expand and cool. The expanding helium has performed work. It is this work of expansion which causes the refrigeration effect. No mechanical work is done since expansion did not occur against a piston. Heat flowing from the load, which is intimately coupled to the lower region of the cylinder walls, warms the helium to a temperature somewhat below that at which it entered the lower cylinder area. As the gas flows upward through the regenerator it removes heat from the metal and cools it to the temperature at which it was found at the beginning of the cycle. The displacer is now pushed downward to force the remaining gas from the end of the cylinder out through the regenerator where it is exhausted back to the compressor at ambient temperature. A single stage machine of this design can achieve temperatures in the 30 to 60 K range.

Two-stage machines are constructed to achieve lower temperatures. The first, or warm, stage operates in the region of 30 to 100 K, whereas the second, or cold, stage operates in the region of 10 to 20 K. The exact temperatures depend on the heat load and capacity of each stage. A heat-balance analysis of the refrigeration loss has been performed by Ackermann and Gifford [70]. In the Gifford-McMahon refrigerator the gas is cycled with poppet valves; the valves and the displacer are moved by a motor and all are located on the expander. A Scotch yoke displacer drive is used because it applies no horizontal force to the shaft.

Figure 9.15 illustrates the expander developed by Longsworth [76, 77]. As on the Gifford-McMahon refrigerator, the remotely located compressor is connected to the expander by hoses of high pressure capacity. The helium is cycled in and out of the expansion head through a motor-driven rotary valve. The expander shown here contains a two-stage displacer and two regenerators. The displacer is gas

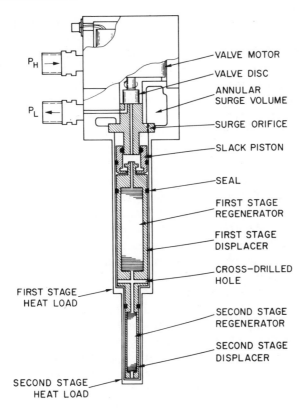

Fig. 9.15 Sectional schematic of the API Model DE-202 expansion head. Reprinted with permission from Air Products and Chemicals, Inc., PA 18105.

driven instead of motor driven, as in the original Gifford-McMahon cycle. A slack piston is incorporated to improve timing. Surrounding the valve stem is an annular surge volume. This volume is maintained at a pressure intermediate to the supply and exhaust pressures by a capillary tube connected to the regenerator inlet line; it provides the reference pressure for pneumatic operation of the displacer.

In the steady state the operating cycle proceeds as follows [80]: the valve is timed to admit high-pressure helium gas through the stem into the volume below the slack piston and in the regenerators while the displacer is in its lower most position. Because the pressure over the slack pistons is less than the inlet pressure, the piston compresses this gas as it moves upward. The gas bleeds into the surge volume through the surge orifice at a constant rate. The surge orifice is like a dashpot;

it controls the speed of the displacer. As the displacer moves upward the high pressure gas flows through the regenerators and is cooled in the process. The inlet valve stops the flow of high pressure gas just before the displacer reaches the top. This slows the displacer and expands the gas in the displacer. The exhaust valve opens and the gas in the displacer expands as it is exhausted to the low pressure side of the compressor. The slack piston moves downward suddenly until it makes contact with the displacer, after which it moves at constant velocity as gas flows from the surge volume into the space over the slack piston. The expansion of gas in the displacer causes it to cool below the temperature of the regenerator. This is the refrigeration effect. Like the Gifford-McMahon cycle, this cycle does no mechanical work on the displacer because both ends are at the same pressure. The exiting gas absorbs heat from the regenerator. Before the end of its stroke the displacer is deaccelerated by closure of the exhaust valve. This completes one cycle of expander operation. Heat is absorbed at two low temperatures and released at a higher temperature.

The compressor used in either of the two Solvay cycle variants is sketched in Fig. 9.16. It uses a reliable oil-lubricated, air-conditioning type of compressor with an inlet pressure of approximately 7×10^5 Pa (100 psig) and an outlet pressure of about 2×10^6 Pa (300 psig). After the gas is compressed the heat of compression is removed by an air or water after cooler. Oil lubrication can be used without fear of contaminating the cold stages because most of it is removed by a two-stage separator and adsorber. Any remaining traces of oil that get into the regenerator are trapped at its warm end and flushed out by the return gas. Helium gas refrigerators operating at these temperatures are more tolerant of many impurities than are helium liquefiers.

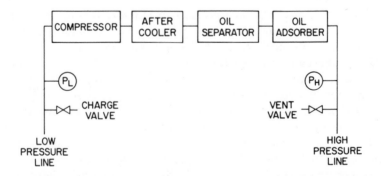

Fig. 9.16 Block diagram of a remotely located helium gas compressor.

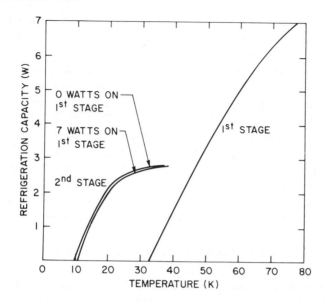

Fig. 9.17 Refrigeration capacity of the API Model CS-202 helium gas refrigerator. Reprinted with permission from Air Products and Chemicals, Inc., Allentown, PA 18105.

The most important attribute of small Solvay-type refrigerators is reliability. The low-pressure differential across the seals in the displacers means light pressure loading and long life. Also contributing to long life is the use of room-temperature valves, a reliable compressor, and oil removal techniques. The result is a compact refrigerator that can be isolated from compressor vibration and located in a vacuum chamber in any attitude.

The coefficient of performance of a refrigerator is defined as the ratio of heat removed to work expended in removing the heat. For an ideal Carnot cycle, the most efficient of all possible cycles, this is [81]

$$\frac{H_{out}}{W_{in}} = \frac{T_1}{T_2 - T_1} \qquad (9.10)$$

where heat is being absorbed at T_1 and released at T_2. For $T_2 = 300$ K a Carnot cycle would require a heat input of 2.9 W to remove 1 W at 77 K and a heat input of 14 W to remove 1 W at 20 K. In practice the efficiency of a refrigerator is defined as the ratio of ideal work to actual work. The refrigerators described here have efficiencies of about 3 to 5% [76, 82].

In a two-stage refrigerator the temperature-dependence of the refrigeration capacity of the two stages is interdependent, as shown in Fig. 9.17 for the API CS-200 refrigerator. The first-stage temperature is in the 33 to 77 K range for heat inputs of 0 to 7 W. The temperature of the second stage, however, is a function of the heat inputs to both stages. The refrigerator shown in this illustration has a rating of 2 W/20 K and 7 W/77 K. Refrigerators are currently available in capacities up to about 10 W/20 K and 35 to 40 W/77 K. The emphasis has been on producing machines with increased capacity in the first stage to isolate radiant heat effectively from the second stage.

9.3.4 Cryogenic Pump Characteristics

At the outset of this discussion we defined cryopumping as the entrainment of molecules on a cooled surface by weak dispersion forces. That definition may be amplified to read entrainment of a gas by cryocondensation, cryosorption, or cryotrapping on a surface that is cooled by a liquid cryogen or a mechanical refrigerator. We also discussed the speed–pressure characteristics of some ideal pumping surfaces. The characteristics of a real cryogenic pump may differ significantly from those ideal cases. A detailed prediction of real pump performance requires a more complete model. The effects of thermal gradients between pumping surfaces and refrigerator, gas and radiant-heat loading, and the geometrical isolation of condensation and sorption stages are three important effects that have not been considered in the ideal model. In the remainder of this section the gas-handling characteristics of rough sorption pumps and refrigerator and liquid-cooled high vacuum pumps are related to the materials, geometry, and heat loading of the pumping surfaces.

Medium Vacuum Sorption Pumps

Seventy-five years ago Dewar first used refrigerated sorption pumping to evacuate an enclosed space. Sorption pumping as we know it today makes use of high-capacity artificial zeolite molecular sieves and liquid nitrogen. A unique feature of cryosorption rough pumping is its ability to pump to 10^{-1} Pa without introducing hydrocarbons into the chamber.

A sorption pump designed for rough pumping is illustrated in Fig. 9.18. It consists of an aluminum body that contains many conducting fins and is filled with an adsorbent. The entire canister is surrounded by a polystyrene foam or metal vacuum dewar filled with liquid nitrogen. Adsorbent pellets are loosely packed in the canister and do not make good thermal contact with the liquid nitrogen. To improve the thermal contact pumps with internal arrays of metal fins are constructed. Even so, the interior of the pump is not in equilibrium with the liquid nitrogen bath, especially during pumping.

The most common adsorbent is a molecular sieve such as Linde 5A. This sieve with an average pore diameter of 0.5 nm exhibits a high capacity for the constituents of air at low pressure. Figure 9.19 illustrates the adsorption isotherms in a pump that contains a charge of 1.35 kg of molecular sieve. The adsorptive capacity for nitrogen is quite high over the pressure range of 10^{-3} to 10^5 Pa, whereas the capacity for helium and neon is quite low. Mixed gas isotherms (e.g., neon in air) will show even less pumping capacity for the least active gas (neon) because it will be displaced by active gases. The preadsorption of water vapor will greatly reduce the sorptive capacity for all gases; as little as 2 wt% water vapor is detrimental to pump operation

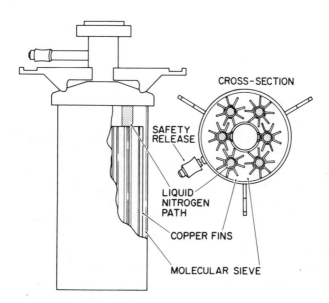

CROSS-SECTION

SAFETY RELEASE

LIQUID NITROGEN PATH

COPPER FINS

MOLECULAR SIEVE

Fig. 9.18 Typical liquid nitrogen cooled sorption pump. Reprinted with permission from Ultek Division, Perkin-Elmer Corp., Palo Alto, CA 94303.

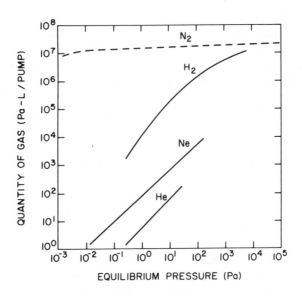

Fig. 9.19 Adsorption isotherms of nitrogen, hydrogen, neon, and helium at 77.3 K in a sorption pump charged with 1.35 kg of molecular sieve. Reprinted with permission from *Cryosorption Pumping*, F. Turner, Varian Report VR-76. Copyright 1973, Varian Associates, 611 Hansen Way, Palo Alto, CA 94303.

[64]. These isotherms are not valid during dynamic pumping because the incoming gas will warm the sieve nonuniformly. They do, however, represent the equilibrium condition of a real pump immersed in a liquid nitrogen bath.

The neon pressure in air is 1.2 Pa and this limits the ultimate pressure. Figure 9.20 sketches the time-dependence of the air pressure in a 100 L chamber for a single stage and two sequential stage pumping with pumps each containing 1.35 kg of molecular sieve and prechilled at least 15 min. Turner and Feinleib [83] have shown that more than 50% of the residual gas present after sorption pumping of air was neon. The ultimate pressure attainable by sorption pumping with a single pump may be reduced by using staged roughing. In staged roughing one pump is used until the chamber pressure reaches approximately 1000 Pa. At that time, the pump is *quickly* valved and a second pump is connected to the chamber. Figure 9.20 shows that the ultimate pressure is lower and the pumping speed higher than when only one pump was used. The improved pumping characteristics are a result of adsorbate saturation and neon removal. The first pump removed 10^7 Pa-L (99%) of air, including 99% of the neon, whereas the second removed only 10^5 Pa-L. The ultimate pressure of the second pump is less than the first stage because it pumped a smaller quantity of gas and because most of the neon was swept into the first pump by the nitrogen stream and trapped there when the valve closed. The valve needs to be closed quickly at the crossover pressure to prevent back diffusion of the neon. Alternatively, a carbon vane pump may be used in place of the first sorption roughing stage.

The ultimate pressure attainable with a sorption pump is a function of its history, in particular the bake treatment and the nature of the gases and vapors. Pressures of an order of 1 Pa are typical. Multistage pumping performance has been described by Turner and Feinleib [83], Cheng and Simpson [84], Vijendran and Nair [85], and Turner [86], whereas Dobrozemsky and Moraw [87] have measured sorption

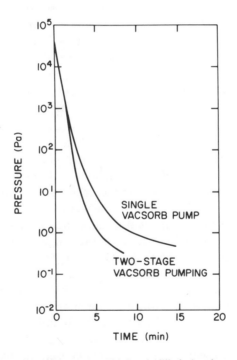

Fig. 9.20 Pumping characteristics of a 100-L, air-filled chamber with (a) one sorption pump, and (b) two sequentially staged sorption pumps. Reprinted with permission from *Cryosorption Pumping*, F. Turner, Varian Report VR-76. Copyright 1973, Varian Associates, 611 Hansen Way, Palo Alto, CA 94303.

pumping speeds for several gases in the pressure range of 10^{-4} to 10^{-1} Pa.

Pumps are normally baked to a temperature of $250°C$ for about 5 h with a heating mantle or reentrant heater element. Miller [88] measured a desorption maxima for water vapor on Linde 5A of 137 to $157°C$. All other gases desorb well at room temperature. The bake is required to release water vapor and obtain the full sorptive capacity of the sieve.

Each sorption pump is equipped with a safety pressure release valve. At no time should the operation of this valve be hindered. Gases will be released when the pump is warmed to atmosphere as well as when it is baked.

A single sorption pump of the size described requires about 5 to 8 L of liquid nitrogen for initial cooling. The amount of liquid nitrogen used to operate the pump, exclusive of boil-off, can be calculated by equating the energy of sorption (number of moles of gas × the heat of sorption) to the energy of vaporization (number of moles of liquid × heat of vaporization). The heat of sorption of nitrogen in 5A molecular sieve is about 16,736 kJ/(kg-mole). This yields a value of 0.46 L of liquid nitrogen to pump each 100 L of nitrogen gas.

High Vacuum Gas Refrigerator Pumps

A typical high-vacuum cryogenic pumping array for a two-stage helium gas refrigerator is sketched in Fig. 9.21. The outer surface is attached to the first (warm) or 80 K stage and the inner pumping surface is attached to the second (cold) or 20 K stage. Indium gaskets are used to make joints of high thermal conductance. In practice the temperature of the stages is a function of the actual heat load and thermal path. The warm stage pumps water vapor and sometimes CO_2. First stage pumping of CO_2 is a function of the temperature and the partial pressure. The vapor pressure of CO_2 on a 77 K surface is 10^{-6} Pa; CO_2 will be pumped only if its concentration is large enough or the surface temperature is adequately low. The second stage contains a cryocondensation and a cryosorption pumping surface. The cryosorption surface is necessary to pump helium, hydrogen, and neon. It is shielded from the inlet aperture as much as possible to increase the probability that all other gases will be condensed on the cryocondensation surface. Charcoal is the most commonly used sorbent because it can be degassed at room temperature. Molecular sieve must be heated to $250°C$ and this is incompatible with the use of indium gaskets.

The temperature of the stages is determined by the total heat flux to the pumping surfaces. Heat is leaked to these surfaces through the expander housing, radiated from high temperature sources, removed from the incoming gas, and conducted from the warm walls by gas molecules at high pressure. Thermal radiation emanates from nearby 300 K surfaces as well as internal sources such as plasmas, electron-beam guns, and baking mantles. Each condensing molecule releases a quantity of heat equal to its heat of condensation plus its heat capacity. Heat can be conducted from a warm to a cold surface by gas-gas collisions if the pressure is high enough so that Kn < 1. In the high vacuum region the radiant flux is much larger than the gas enthalpy or gas conductance. At a pressure of 10^{-4} Pa a 1000 L/s pump is con-

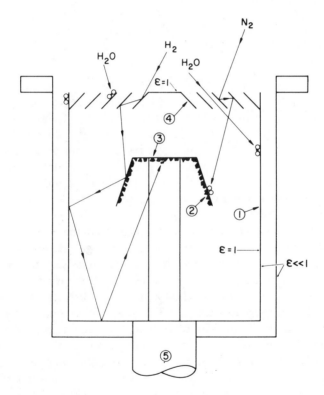

Fig. 9.21 Typical cryogenic pumping array for a two-stage helium gas refrigerator: (1) First stage array, (2) second-stage cryosorption pump, (3) second-stage cryocondensation pump, (4) chevron baffle, (5) refrigerator head.

densing 0.1 Pa-L/s of nitrogen or a heat load of 0.6 mW. At this pressure gas conductance can also be ignored.

The radiant heat flow between two concentric spheres or cylinders [89] is given by

$$\frac{H}{A} = \frac{\sigma \varepsilon_1 \varepsilon_2}{\varepsilon_2 + \frac{A_1}{A_2}(1-\varepsilon_2)\varepsilon_1}(T_2^4 - T_1^4) \qquad (9.11)$$

where the subscripts 1,2 refer to, respectively, the inner and outer surfaces. If $\varepsilon_1 = \varepsilon_2 = 1$, the heat flow will be maximum and

$$\frac{H}{A} = \sigma(T_2^4 - T_1^4) \qquad (9.12)$$

This yields a heat flow of 457 W/m² from the 300 K surface to the 20 K surface. A typical first-stage area of 0.1 m² for a small cryogenic pump would then absorb a heat load of 45.7 W. Clearly, this would overload the refrigerator. The radiant heat load is reduced by reducing the emissivities of the inside surface of the vacuum wall and the outside surface of the first stage. If the walls are plated with nickel whose emissivity is 0.03 [90], this heat load can be reduced to 0.7 W. Even if these surfaces became contaminated so that $\varepsilon_1 = \varepsilon_2 = 0.1$, the heat flow would still be only 2.4 W.

The first stage isolates the second stage from the radiant heat load and pumps water vapor. Some condensable gas deposits, however, have the property of drastically altering the emissivity of the surface on which they condense. Caren et al. [91] have shown that as little as 0.02 mm of water ice on a polished aluminum substrate at 77 K causes the emissivity to increase from 0.03 to 0.8. Their data are in agreement with data taken by Moore [92] at 20 K. Carbon dioxide also absorbs thermal energy [91]. The two purposes of the first stage, pumping water vapor and thermally shielding the second stage, seem contradictory. If the first stage pumps water vapor, its emissivity will rise and it will adsorb excess heat that will overload the refrigerator. This problem is overcome by keeping the outer wall close to the vacuum wall; the water vapor is now pumped along the upper perimeter of the chevron, the chevron itself, and the interior wall of the first stage. It cannot reach the lower portion of the outside wall. Alternatively, for special cases the exterior may be wrapped with multilayer reflective insulation.

The entrance baffle is designed to prevent radiation from illuminating the second stage. It does this by absorbing radiation in the chevron and allowing transmitted radiation to make contact with the blackened inside wall, where it is absorbed. The entrance baffle also impedes the flow of gases to the second stage. An opaque baffle reduces the radiant loading on the second stage but also reduces the pumping speed for all gases. Good baffle designs are a compromise between radiant heat absorption and pumping speed reduction. In some pumps the chevrons are painted black; in others they are highly reflective. It matters little because both will soon be "blackened" with water vapor. Figure 9.22 illustrates a cut-away view of one commercial cryogenic pump. Hands' [93] review of some of the design problems in a cryogenic pump includes the effect of temperature gradients through the deposit and in the arrays on long-term pump performance.

Approximate pumping-speed calculations can be made for each species if the temperatures and geometry are known; for example, gases pumped on the second stage must first pass through the chevron baffle, where they are cooled. If the effective inlet area, inlet gas temperature, second-stage area, and species are known, the approximate speed and ultimate pressure can be estimated. The pumping speed of a gas in a cryogenic pump is not only related to the size of the inlet flange and refrigeration capacity. It is also dependent on the pumping array (relative sizes of the warm and cold stages), gas species, and history (ice, hydrogen, and helium load). All pumping speeds will fall off at about 10^{-1} Pa as the refrigerator becomes overloaded. Because of these effects, it is not possible to draw an illustration analogous to Fig. 8.4, which is representative of all gas refrigerator cooled pumps. The only valid generalization is that helium pumping speeds are usually small and have low saturation values; hydrogen pumping speeds are larger than those of helium but both are nill if the sorbent has saturated or its temperature is greater than 20 K.

When the pump becomes saturated, it is shut down while the cryosurfaces warm and the condensate vaporizes. The vaporizing gases exit through the safety valve or manually operated valve. It is imperative that all cryogenic pumps have a functioning safety release to allow the escape of condensed and adsorbed gases and vapors in the event of pump failure, pump shut-down or utility failure. If there were no safety valve, the pump would become a bomb when warmed.

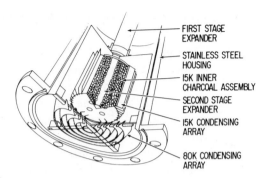

FIRST STAGE
EXPANDER

STAINLESS STEEL
HOUSING

15K INNER
CHARCOAL ASSEMBLY

SECOND STAGE
EXPANDER

15K CONDENSING
ARRAY

80K CONDENSING
ARRAY

Fig. 9.22 Cutaway view of the Cryo-Torr 8 cryogenic pump. Reprinted with permission from CTI Cryogenics, Kelvin Park, Waltham, MA 02154.

High Vacuum Liquid Pumps

Three-stage cryogenic pumps that use liquid helium (4.2 K), gaseous helium boil-off (20 K), and liquid nitrogen are effective in pumping all gases, especially when the molecular sieve is bonded to the third stage [94]. These pumps have been used for pumping large vessels [95–97] and small vacuum chambers [54,98]. The liquid pump has a higher pumping speed for hydrogen and helium than for a gaseous helium refrigerator because that stage is colder and its temperature is more stable. Ultimate pressures of 10^{-11} Pa [95] have been reported.

Few liquid pumps are commercially available. Most are individually constructed or are designed for large projects.

REFERENCES

1. A. K. Gupta and J. H. Leck, *Vacuum*, **25**, 362 (1975).
2. D. J. Harra, *J. Vac. Sci. Technol.*, **13**, 471 (1976).
3. B. E. Keitzmann, *General Characteristics of Titanium Sublimation Pumps*, Varian Associates, Palo Alto, CA 94303.
4. L. Holland, L. Laurenson, and P. Allen, *Trans. 8th Nat. Vac. Symp. (1961)*, Pergamon, New York, 1962, p. 208.
5. R. E. Clausing, *Trans. 8th Nat. Vac. Symp. (1961)*, Pergamon, New York, 1962, p. 345.
6. A. A. Kuzmin, *Prib. Tekh. Eksp.* **3**, 497 (1963).
7. G. M. McCracken and N. A. Pashley, *J. Vac. Sci. Technol.*, **3**, 96 (1966).
8. R. W. Lawson and J. W. Woodward, *Vacuum*, **17**, 205 (1967).

9. D. J. Harra and T. W. Snouse, *J. Vac. Sci. Technol.*, **9**, 552 (1972).
10. D. J. Harra, *J. Vac. Sci. Technol.*, **12**, 539 (1975).
11. H. R. Smith, Jr., *J. Vac. Sci. Technol.*, **8**, 286 (1971).
12. P. della Porta, T. Giorgi, S. Origlio, and F. Ricca, *Trans. 8th Nat. Vac. Symp.. (1961)*, Pergamon, New York, 1962, p. 229.
13. T. A. Giorgi and F. Ricca, *Nuovo Cimento Suppl.*, **1**, 612 (1963).
14. B. Kindl, *Nuovo Cimento Suppl.*, **1**, 646 (1963).
15. B. Kindl and E. Rabusin, *Nuovo Cimento Suppl.*, **5**, 36 (1967).
16. P. della Porta and B. Ferrario, *Proc. 4th Int. Vac. Congr. (1968)*, Institute of Physics and the Physical Society, London, **1**, 369 (1968).
17. S.A.E.S. Getters USA, Buffalo, NY.
18. W. J. Lange, *J. Vac. Sci. Technol.*, **14**, 582, (1977).
19. R. G. Herb, R. H. Davis, A. S. Divatia, and D. Saxon (Abstr.), *Phys. Rev.*, **89**, 897 (1953).
20. J. C. Swartz, *Vacuum Symposium Transactions (1955)*, Committee on Vacuum Techniques, Boston, 1956, p. 83.
21. R. A. Douglas, J. Zabritski, and R. G. Herb, *Rev. Sci. Instrum.*, **36** (1965).
22. W. G. Mourad, T. Pauly, and R. G. Herb, *Bull. Am. Phys. Soc.*, **8**, 336 (1963).
23. J. C. Maliakal, P. J. Limon, E. E. Arden, and R. G. Herb, *J. Vac. Sci. Technol.*, **1**, 54 (1964).
24. D. G. Bills, *J. Vac. Sci. Technol.*, **4**, 149 (1967).
25. D. R. Denison, *J. Vac. Sci. Technol.*, **4**, 156, (1967).
26. F. M. Penning, *Physica*, **4**, 71 (1937).
27. L. D. Hall, *Rev. Sci Instrum.*, **29**, 367 (1958).
28. F. M. Penning and K. Nienhuis, *Philips Tech. Rev.*, **11**, 116 (1949).
29. A. M. Guerswitch and W. F. Westendrop, *Rev. Sci. Instrum.*, **25**, 389 (1954).
30. D. Andrew, *Proc. 4th Int. Vac. Congr. (1968)*, Institute of Physics and the Physical Society, London. 1969, p. 325.
31. R. D. Craig, *Vacuum*, **19**, 70 (1969).
32. S. L. Rutherford, S. L. Mercer, and R. L. Jepsen, *Trans. 7th Nat. Vac. Symp. (1960)*, Pergamon, New York, 1961, p. 380.
33. R. L. Jepsen, *Proc. 4th Int. Vac. Congr. (1968)*, Institute of Physics and the Physical Society, London, 1969, p. 317.
34. A. Dallos and F. Steinrisser, *J. Vac. Sci. Technol.*, **4**, 6 (1967).
35. A. Dallos, *Vacuum*, **19**, 79 (1969).
36. J. H. Singleton, *J. Vac. Sci. Technol.*, **8**, 275 (1971).
37. J. H. Singleton, *J. Vac. Sci. Technol.*, **6**, 316 (1969).
38. W. M. Brubaker, *6th Nat. Vac. Symp. (1959)*, Pergamon, New York, 1960, p. 302.
39. A. R. Hamilton, *8th Nat. Vac. Sump. (1961)*, **1** Pergamon, New York, 1962, p. 338.
40. Varian Associates, 611 Hansen Way, Palo Alto, Ca, 94303.
41. R. L. Jepsen, A. B. Francis, S. L. Rutherford, and B. E. Keitzmann, *7th Nat. Vac. Symp. (1960)*, Pergamon, New York, 1961, p. 45.
42. T. Tom and B. D. Jones, *J. Vac. Sci. Technol.*, **6**, 304 (1969).

43. P. N. Baker and L. Laurenson, *J. Vac. Sci. Technol.*, **9**, 375 (1972).

44. D. R. Denison, *J. Vac. Sci. Technol.*, **14**, 633 (1977). See Reference 1.

45. D. Andrew, D. R. Sethna, and G. F. Weston, *4th Int. Vac. Cong. (1968)*, Institute of Physics and the Physical Society, 1968, p. 337.

46. J. P. Dawson and J. D. Haygood, *Cryogenics*, **5**, 57 (1965).

47. M. M. Eisenstadt, *J. Vac. Sci. Technol.*, **7**, 479 (1970).

48. R. F. Brown and E. S. Wang, *Adv. Cryog. Eng.* **10**, K. D. Timmerhaus, Ed., Plenum, New York, 1965, p. 283.

49. J. P. Hobson, *J. Phys., Chem.*, **73**, 2720 (1969).

50. P. J. Gareis and S. A. Stern, *Bulletin de l'Institut International du Froid*, Annexe 1966-5, p. 429.

51. W. Van Dingenan and A. Van Itterbeek, *Physica*, **6**, 49 (1939).

52. G. E. Grenier and S. A. Stern, *J. Vac. Sci. Technol.*, **3**, 334 (1966).

53. A. J. Kidnay, M. J. Hiza and P. F. Dickenson, *Adv. Cryog. Eng.*, **13** K. D. Timmerhaus, Ed., Plenum, New York, 1968, p. 397.

54. R. J. Powers, and R. M. Chambers, *J. Vac. Sci. Technol.*, **8**, 319 (1971).

55. C. Johannes, *Adv. Cryog. Eng.*, **17**, K. D. Timmerhaus, Ed., Plenum, New York, 1972, p. 307.

56. H. J. Halama and J. R. Aggus, *J. Vac. Sci. Technol.*, **11**, 333, (1974).

57. J. P. Hobson, *J. Vac. Sci. Technol.*, **3**, 281 (1966).

58. P. A. Redhead, J. P. Hobson, and E. V. Kornelsen, *The Physical Basis of Ultrahigh Vacuum*, Chapman and Hall, London, 1968, p. 37.

59. J. Hengevoss and E. A. Trendelenburg, *Vacuum*, **17**, 495 (1967).

60. J. Hengevoss, *J. Vac. Sci. Technol.*, **6**, 58 (1969).

61. J. C. Boissin, J. J. Thibault, and A. Richardt, *Le Vide*, Suppl. 157, 103 (1972).

62. G. Lewin, *J. Vac. Sci. Technol.*, **5**, 75 (1968).

63. R. W. Moore, Jr., *8th Nat. Vac. Symp. (1961)*, **1**, Pergamon, New York, 1962, p. 426.

64. S. A. Stern and F. S. DiPaolo, *J. Vac. Sci. Technol.*, **6**, 941 (1969).

65. Y. Tuzi, M. Kobayshi, and K. Asao, *J. Vac. Sci. Technol.*, **9**, 248 (1972).

66. J. P. Hobson, *J. Vac. Sci. Technol.*, **10**, 73 (1973).

67. J. Kidnay and M. J. Hiza, *Cryogenics*, **10**, 271 (1970).

68. S. C. Collins and R. L. Canaday, *Expansion Machines for Low Temperature Processes*, Oxford University Press, Oxford, 1958.

69. R. Barron, *Cryogenic Systems*, McGraw Hill, New York, 1966.

70. R. Radebaugh, *Applications of Closed-Cycle Cryocoolers to Small Superconducting Devices*, NBS Special Publication 508, U. S. Department of Commerce, National Bureau of Standards, Washington, DC, 1978, p. 7.

71. W. E. Gifford, *Refrigeration Method and Apparatus*, U. S. Pat. 2,966,035 (1960).

72. W. E. Gifford, and H. O. McMahon, *Prog. Refrig. Sci. Technol*, **1**, M. Jul and A. Jul, Eds., Pergamon, Oxford, 1960, p. 105.

73. W. E. Gifford, *Prog. Cryog.*, **3**, K. Mendelssohn, Ed., Academic, New York, 1961, p. 49.

74. W. E. Gifford, *Adv. Cryog. Eng.*, **11**, K. D. Timmerhaus, Ed. Plenum, New York, 1966, p. 152.

75. R. C. Longsworth, *Refrigeration Method and Apparatus*, U.S. Pat. 3,620,029, (1971).

76. R. C. Longsworth, *Adv. Cryog. Eng.*, **16**, K. D. Timmerhaus, Ed., Plenum, New York, 1971, p. 195.

77. R. C. Longsworth, *Adv. Cryog. Eng.*, **23**, K. D. Timmerhaus, Ed., Plenum, New York, 1978, p. 658.

78. For example, Type K-20 Series Cryogenerator, N. V. Philips Gloeilampenfabrieken, Eindhoven, Netherlands.

79. R. A. Ackermann, and W. E. Gifford, *Adv. Cryog. Eng*, **16**, K. D. Timmerhaus, Ed. Plenum, New York, 1971, p. 221.

80. R. C. Longsworth, *An Introduction to the Elements of Cryopumping*, K. M. Welch, Ed., American Vacuum Society, p. II-1.

81. F. W. Sears, *An Introduction to Thermodynamics, The Kinetic Theory of Gases and Statistical Mechanics*, Addison-Wesley, Reading, Ma., 1953, p. 84.

82. T. R. Strobridge, NBS Technical Note 655, U. S. Department of Commerce, National Bureau of Standards, Washington, D. C., 1974.

83. F. T. Turner and M. Feinleib, *8th Nat. Vac. Symp. (1961)*, **1**, Pergamon, New York, 1962, p. 300.

84. D. Cheng and J. P. Simpson, *Adv. Cryog. Eng*, **10**, K. D. Timmerhaus, Ed., Plenum, New York, 1965, p. 292.

85. P. Vijendran and C. V. Nair, *Vacuum*, **21**, 159 (1971).

86. F. Turner, *Cryosorption Pumping*, Varian Report VR-76, Varian Associates, Palo Alto, CA 1973.

87. R. Dobrozemsky and G. Moraw, *Vacuum*, **21**, 587 (1971).

88. H. C. Miller, *J. Vac. Sci. Technol.*, **10**, 859 (1973).

89. R. B. Scott, *Cryogenic Engineering*, Van Nostrand, New York, 1959, p. 147.

90. *Ibid*, p. 348.

91. R. P. Caren, A. S. Gilcrest, and C. A. Zierman, *Adv. Cryog. Eng.*, **9**, K. D. Timmerhaus, Ed., Plenum, New York, 1964, p. 457.

92. B. C. Moore, *9th Nat. Vac. Symp. (1962)*, Macmillan, New York, 1962, p. 212.

93. B. A. Hands, *Vacuum*, **26**, 11, (1976).

94. H. J. Halama and J. R. Aggus, *J. Vac. Sci. Technol.*, **12**, 532 (1975).

95. C. Benvenuti, *J. Vac. Sci. Technol.*, **11**, 591 (1974).

96. C. Benvenuti and D. Blechschmidt, *Japan. J. Appl. Phys.* Suppl. 2, Pt. 1, 77 (1974).

97. H. J. Halama, C. K. Lam and J. A. Bamberger, *J. Vac. Sci. Technol.*, **14**, 1201 (1977).

98. G. Schafer, *Vacuum*, **28**, 399 (1978).

Systems

This group of four chapters focuses on three types of vacuum system used in the thin-film and semiconductor industries: systems that will pump to the high vacuum and ultrahigh vacuum range and systems that will pump a large gas flow in the medium and low vacuum ranges. One way of viewing them is to observe that the end use places emphasis on different facets of the technology. In the high vacuum systems discussed in Chapter 10 the large process gas load and the need for line-of-sight motion of molecules from the hearth to the substrate are paramount in determining pump size and type and chamber design.

The ultrahigh vacuum systems described in Chapter 11 are used to keep surfaces free of monomolecular layers of gas contaminants for long periods. The achievement of ultrahigh vacuum pressures is dependent on the careful selection of materials, joining and cleaning techniques, and the choice of pump.

The systems discussed in Chapter 12 are used when high gas flows are needed in the medium or low vacuum range. Because the gas flow–pressure range is outside that of high vacuum systems, they are throttled or replaced by mechanical pumping systems.

Chapter 13 is a brief review of the economics of purchasing, operating, and owning typical systems. It also summarizes methods of conserving energy that will not impair the system or the product.

CHAPTER 10

High Vacuum Systems

Systems capable of producing a high vacuum environment are used for a variety of tasks; for example, the evaporation and condensation of materials to form thin-film layers. To accomplish its objectives, the system must be capable of producing a pressure low enough to meet two criteria: the mean-free path must be long enough for the evaporant to reach the substrate without making collisions with the gas molecules and the background gas pressure must be low enough to keep the ratio of incident vapor from the source to that of the background very small. At a pressure of 10^{-4} Pa (7.5×10^{-7} Torr) the background gas collides with the substrate at the rate of $3 \times 10^{+14}$ (molecules/cm^2)/s. This is equivalent to the formation of a monolayer every 3 s, if every impinging molecule adhered to the substrate. For reasonable deposition rates, say 2 to 10 nm/s, and for typical sticking coefficients a pressure of 10^{-4} to 10^{-5} Pa is adequate to deposit many films. The total pressure is not the only factor that determines film quality, as Caswell [1] has pointed out. One other system-dependent factor is the partial pressure of the gases that could react with or become incorporated into the growing film. The pressure of each gas or vapor must be kept below a value at which it will have a detrimental effect on film properties. The total pressure, however, is a convenient measure by which we can compartmentalize systems for the purpose of discussion.

The class of high vacuum systems reviewed in this chapter encompasses those that will produce a base pressure of an order of 10^{-6} Pa, and a process pressure less than 10^{-4} Pa. The base pressure of such a system is determined by the pumping speed and by the outgassing of the walls and sealing materials. The work chambers are usually constructed of glass or stainless steel. Elastomers are traditionally used to seal valves and flanges and to make demountable glass-to-metal connections. Clean work chambers on properly designed systems will achieve pressures of 2–5 × 10^{-6} Pa without bakeout. The pressure will be higher during the process because of the gas load evolved from the source and other heated surfaces. In addition, those systems that are used for production applications should be able to reach this base pressure quickly. Because the dominant gas load is water vapor, these systems must minimize its accumulation and make provisions for its rapid pumping. A liquid nitrogen surface will be used over the high vacuum pump and possibly in the work chamber.

This chapter discusses the pumping systems and the chamber with the above objectives in mind. The first four sections discuss the operation of diffusion, turbomolecular, ion, and cryogenic pumped systems, respectively. Included is a procedure for starting, stopping, cycling, and fault protecting each kind of pumping system. The last sections discuss the chamber and leak detection.

10.1 DIFFUSION PUMPED SYSTEMS

The layout of a small diffusion pumped system is sketched in Fig. 10.1. Several accessory items are shown as well. Not every system has or needs all of these additional items but we show them here in order to discuss their proper placement and operation. The main components of this system, the diffusion pump, trap, gate valve, and mechanical pump, together form the single most common high vacuum system.

In less critical applications only a water baffle is necessary. Hablanian [2] measured a backstreaming rate of 5 × 10^{-6} (mg/cm^2)/min for DC-705 fluid over a 6-in. diffusion pump and simple liquid nitrogen trap, whereas a value of 8 × 10^{-6} (mg/cm^2)/min over a 4-in. diffusion baffled with a chevron cooled to 15° C may be calculated from data given by Holland [3]. For many applications the cold-water baffle alone will perform adequately, provided a pentaphenyl ether or pentaphenyl silicone fluid is used in the pump. As Singleton [4] points out, these two fluids are valuable in systems which are trapped with a less than elegant trap. The simplest way to reduce backstreaming in a low-quality system is to use a high-quality fluid.

In the most critical applications a liquid nitrogen trap, a water- or conduction-cooled cap, and possibly a partial water baffle are used.

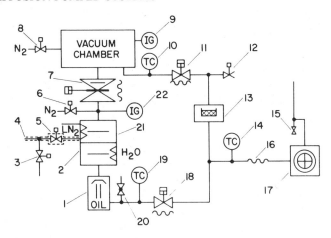

Fig. 10.1 Diffusion pump stack: (1) Diffusion pump, (2) partial water baffle, (3) LN$_2$ vent valve, (4) LN$_2$ inlet, (5) LN$_2$ fill valve, (6) port for gas purging diffusion pump, (7) bellows sealed high vacuum valve, (8) chamber bleed valve, (9) chamber ionization gauge, (10) chamber thermal conductivity gauge, (11) roughing valve, (12) mechanical pump vent, (13) roughing line trap, (14) roughing line thermal conductivity gauge, (15) sump for collecting condensable vapors, (16) bellows for vibration isolation, (17) mechanical pump, (18) foreline valve, (19), foreline thermal conductivity gauge, (20) leak testing port, (21) liquid nitrogen trap, (22) diffusion pump ionization gauge.

The liquid nitrogen reservoir is a trap for water vapor and pump-fluid fragments. A 6-in. diffusion pump stack like the one sketched in Fig. 10.1 will have a net pumping speed of 1000 L/s for air at the baseplate, whereas a matching liquid nitrogen trap can pump water vapor at speeds up to 4000 L/s. Diffusion pumps, fluids and traps of modern design are capable of reducing backstreaming rates to a level such that organics from other sources may be more prevalent in the chamber.

The arrangement shown here for automatically filling the liquid nitrogen trap is convenient when there is a long vacuum-jacketed supply line that must be chilled each time the trap needs filling. This will not eliminate the need for phase separation on large distribution systems. Without this arrangement the incoming liquid boils and the gas quickly warms the trap, thus releasing the trapped vapors. To avoid trap warming a valve is added. When the low level sensor is activated, the vent valve (#3 in Fig. 10.1) is opened and the warm gas is vented to the atmosphere. A second sensor is placed in the supply line near the vent valve. When cooled by the liquid, it closes the vent valve and opens the fill valve. The controller then performs its normal function. A commercial controller with two level sense elements (low level and empty) may be modified to perform this function. If the trap is filled from a dewar by a short length of tubing, this gas bypass operation will not be required.

In smaller systems one mechanical pump is used alternately to rough pump the chamber and back the diffusion pump. During roughing pressure will build up in the valved-off foreline but it is generally not significant if the roughing cycle is shorter than 15 min. Large systems, for example, with 16-in. or larger diffusion pumps, frequently use a Roots blower backed by an oil-sealed piston pump for chamber roughing. A smaller, separate "holding" pump can back the diffusion pump during roughing. At crossover to high vacuum the Roots pump is switched over to back the diffusion pump to provide maximum throughput during crossover.

Thermal conductivity gauges are located in the work chamber and in the foreline for control and fault protection and in the roughing line to check the mechanical pump blank-off pressure. One ionization gauge is located in the chamber, or on the spool piece adjacent to it, and a second is positioned between the cold trap and the gate valve. Although many systems are designed and constructed without a gauge at the latter point, it is extremely useful. The most straightforward diagnostic measurement on a diffusion pump is the blank-off pressure, and that measurement can be made only with a gauge located beneath the main valve. Some larger traps are fabricated with an ion gauge port, whereas gate valves with extra piping or ports on the diffusion pump side facilitate easy mounting of the lower ion gauge tube. In general, it is advisable to install the gate valve with its seal plate facing upward to keep the valve interior under vacuum at all times. This reduces the volume and surface area exposed to the atmosphere in each cycle and thus minimizes the pump-down time. All gauge tubes, both ion and thermal conductivity, should be positioned with their entrances facing downward or to the side to prevent them from becoming traps for particulates.

Some other items shown in Fig. 10.1 are not often needed or found on standard systems. A leak-detection port located in the foreline provides the best sensitivity and speed of response for leak detecting the chamber, trap, and diffusion pump. By use of this port a leak detector may be attached while the system is operating. A flow restrictor placed on the inlet side of the diffusion pump cooling water coil will save water and increase the temperature of the oil ejector stage. Roughing traps are often used in diffusion pump systems and when properly maintained can reduce the transfer of mechanical pump oil to the chamber. In this system the trap is used only as a roughing trap. When placed in the roughing line (Fig. 10.1), it does not become contaminated with diffusion pump fluid. The mechanical pump vent should be located above the trap to permit the oil vapors to be blown away from the chamber each time the mechanical pump is vented.

10.1.1   System Operation

Let us review the operation of a diffusion pumped system by studying the sequence of chamber pumping and venting and system start-up and shut-down. Assume that the system is operating and that the work chamber is at atmospheric pressure. The high vacuum and roughing valves are closed. When the roughing sequence begins, the foreline valve will be closed and the roughing valve opened. In small pumping systems only a single roughing valve is required. In large chambers it is a multistep process. First, the chamber may be rough-pumped through a small diameter orfice or by-pass line that chokes the flow and prevents turbulence from stirring up particulates on the floor of the work chamber. Particulates may become electrostatically attached to insulating substrates. Second, when the pressure has been reduced to a value at which turbulent flow is no longer possible, the main roughing valve is opened. Third, if a Roots blower is used, it will be bypassed or allowed to freewheel until the pressure drop between its inlet and outlet is reduced to a safe operating value, say 1000 Pa, at which time it begins to pump in series with the rotary vane or piston pump. Single set point pressure switches are used to control these functions. Roughing will continue until the pressure reaches 15 Pa. At this pressure the roughing valve will be closed and following a time delay the high vacuum valve will be opened. Never allow the roughing pump to exhaust the system below a pressure of 15 Pa or gross back-streaming of mechanical pump oil will occur.

The pressure at which crossover from rough to diffusion pumping occurs is sensed by a thermal conductivity gauge that operates a relay. The time constant of a 0 to 150 Pa thermocouple gauge is typically 2 s [5]. In a small system which has a system roughing time constant that is much less than the gauge time constant the pressure will reach a level less than that set on the gauge. This problem is easily corrected by adjusting the set point to a pressure higher than actually desired or by using a pressure switch with a faster time constant. The fact that the gauge reading lags the system pressure can be used effectively by the controller to look for leaks or excessive outgassing in the chamber. When the gauge reaches the 15 Pa set point, the roughing valve closes, and a timer delays the opening of the high vacuum valve. The thermal conductivity gauge will reach a pressure minimum that is below the set point and will drift upward at a rate determined by the real or virtual leak. If the pressure increases beyond the set point pressure before the timer opens the valve, the sequence may be programmed to abort; if the pressure is below the set point at the end of the timed interval, the high vacuum valve will be programmed to open. The diffusion pump and liquid nitrogen trap will then pump the system to its base pressure. The time required to reach the base pressure is a function of the speed of the diffusion pump and the liquid nitrogen trap, the chamber volume, surface area, and cleanliness. The main vapor species being pumped is water vapor and a liquid nitrogen trap pumps water vapor efficiently. Diffusion pumps with expanded tops and oversized liquid nitrogen traps have high pumping speeds for water vapor. Pump-down characteristics are discussed in more detail in Section 10.5.

System shut-down begins by closing the high vacuum valve and warming the cryogenic trap as the diffusion pump removes the evolved gases and vapors. If the trap is allowed to equilibrate with its surroundings naturally, it will take between 4 and 20 h, depending on the trap design. This time can be shortened considerably by flushing dry nitrogen gas through the liquid nitrogen reservoir. When the trap temperature reaches $0°C$, the power to the diffusion pump will be turned off. When the pump will be cooled to $50°C$, the foreline valve is closed. The mechanical pump is then shut down and vented. Venting of the diffusion pump should be done with the valve located above the trap (#6 in Fig. 10.1). Valve 20 in Fig. 10.1 may be used to vent the pump, but the pump should be cooled to $50°C$ or lower and venting should proceed slowly; otherwise the fluid and jet assembly may be forced upward. The diffusion pump should not be vented by opening the foreline valve and allowing air to flow back through the roughing trap, roughing line and into the diffusion pump because this would carry excessive mechanical pump oil vapor into the diffusion pump. Last, cooling water in the diffusion pump should be turned off. If the pump interior is exposed to ambient air when cooled below the dew point, water vapor will condense and contaminate the pump fluid.

Start-up of a diffusion pumped system begins by flowing cooling water through the diffusion pump jacket and starting the mechanical pump. After the gas in the roughing line is exhausted to a pressure of, say, 100 Pa the foreline valve may be opened and the diffusion pump heater activated. On a system operated by an automatic controller, this function is done by a preset gauge. Warm-up time for a diffusion pump varies from 15 min for a 4-in. or 6-in. pump to 45 min for a

35-in. pump.   When starting a diffusion pump, precautions may be taken to prevent the initial backstreaming transient from contaminating the region above the trap before it is cooled.   This can be accomplished by precooling the trap or by gas flushing.   Not all system controllers activate the liquid nitrogen trap at the same time in the start-up cycle.   In fact, some controllers leave this as a step to be manually controlled by the operator because a partly cooled trap is prefered when the pump begins to function.

If economy dictates that the pump is to be started and shut down each day, Santeler's [6] gas-purge technique should be used to prevent transient backstreaming from contaminating the system.   When systems are run continuously, the total transient contamination from an occasional start-up or shut-down is so small that it can be ignored, at least for the kind of system described in this chapter.   Daily shut-down multiplies this effect enormously.   Pump fluid fractions desorbing from the trap will backstream into the region between the trap and the high vacuum valve, even though the pump is operating.   In a similar manner the burst of fluid vapor emitted during the collapse of the top jet [7, 8] will diffuse above the trap and travel to the work chamber the next time the high vacuum valve is opened.   These contaminants can be flushed from the system by use of the gas purge described in Section 13.3.

10.1.2   Operating Concerns

The operations sketched in the preceding section did not deal in depth with some of the real and potential problems in diffusion pumped systems—problems such as reduction of backstreaming, operation of the trap, and fault protection.

Backstreaming is a subject for which many vacuum system users have a limited appreciation.   Unfortunately the definition that states that backstreaming is the transfer of diffusion pump fluid from the pump to the chamber is for many a general definition.   Even if it is restricted to diffusion pumped systems, it is still incorrect because mechanical pump oil is the largest organic contaminant in the work chamber.   It need not be, however, because techniques are available to prevent mechanical pump oil from reaching the chamber.

In a diffusion pumped system backstreaming is controlled by three factors: (1) the diffusion pump, traps, and baffles; (2) the roughing pump, traps and piping; (3) the system operating procedures.   Most important is that the backstreaming from all three sources must be reduced to a level that can be tolerated.   It is a meaningless exercise to

totally eliminate backstreaming from diffusion pump fluid while operating the system in a manner that allows gross contamination from
mechanical pump oil. In this section we review ways of reducing
backstreaming from each of the above three sources to a level that will
not affect most processes. Procedures for further reduction of backstreaming to make diffusion pumps suitable for ultrahigh vacuum work
are discussed in Chapter 11.

Methods of reducing fluid contamination from diffusion pumps were
discussed in Chapter 8. The diffusion pump should use a high-quality
fluid and be baffled by a water-cooled or conduction-cooled cap and a
liquid nitrogen trap.

Contamination from the roughing pump may be reduced by use of a
liquid nitrogen or ambient temperature trap and a low vapor-pressure
oil. The liquid nitrogen trap is the most effective but the most difficult
to maintain and consequently is not frequently used. Ambient temperature traps do not require constant refrigeration but do saturate at
some point. Zeolite [9, 10], alumina [3, 10, 11], and bronze or copper
wool have been used for this purpose. Water vapor will soon saturate
a zeolite trap [4, 12] and slow the roughing cycle. Even though zeolite
traps can remove more than 99% of the contamination [9], they suffer
from the fact that particulates can drift into valve seats and into the
mechanical pump, thus hastening wear of the valve seat and pump
interior. All traps except those that are liquid nitrogen-cooled saturate
in at least three months under normal use. If they are not rigorously
maintained, they will loose their value. Experience has shown that
they are almost never maintained and so give a sense of protection that
is false.

Oil backstreaming from the mechanical pump can be reduced effectively by changing to an oil with a lower vapor pressure [6]. The use
of an oil with good lubricity (e.g., Convoil-20 or Invoil-20) can reduce
the backstreaming. Friction will eventually degrade the oil [3] and
periodic changing will be required. When the rotary pump is backing
the diffusion pump, the foreline is in free-molecular flow and mechanical pump oil can freely flow toward the diffusion pump. This is no
problem if the diffusion pump is the fractionating variety because it is
able to reject the high vapor fractions and eject them at the foreline
[4]. Attempts have been made to eliminate the effect of mechanical
pump fluid backstreaming to the diffusion pump by use of a pure
silicone diffusion pump fluid in the mechanical pump. The nonpolar
silicone molecules do not wet and therefore do not form a lubricating
or sealing film.

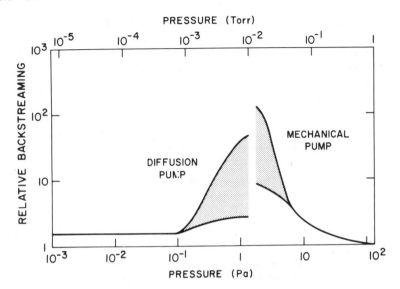

Fig. 10.2 Pumping and lubrication fluid backstreaming in the transition zone between the diffusion and mechanical pump operating regions. Reprinted with permission from *Japan. J. Appl Phys.*, Suppl. 2, Pt. 1, p.25, M. H. Hablanian. Copyright 1974, The Japanese Journal of Applied Physics.

System operation, the third component in the recipe for low backstreaming, requires more emphasis than it is usually given. Improper operation can cause gross contamination of the work chamber, even in a well-designed system. Figure 10.2 from Hablanian [13] qualitatively outlines the problem of backstreaming from a diffusion and a mechanical pump. The pressure region between 10^{-2} and 15 Pa is of most concern. Hablanian and Steinherz [14] and Rettinghaus and Huber [8] discuss the rise in backstreamining of diffusion pump fluids at high pressures due to oil–gas scattering; Baker, Holland, and Stanton [12] measured the pressure-dependent backstreaming of oil vapor from a rotary mechanical pump. The sum and substance of Fig. 10.2 is that both diffusion and roughing pumps have maximum backstreaming in adjacent regions. These effects are of serious concern when roughing the chamber and crossing over to the diffusion pump.

The usual technique for roughing the system is to pump to 15 Pa, close the roughing valve, open the high vacuum valve, and pump to below 10^{-1} or 10^{-2} Pa as quickly as possible. When the high vacuum valve is initially opened at 15 Pa, the viscous gas flow will reduce the backstreaming [8, 15], even though the top jet may be momentarily overloaded. As the pressure is reduced, the backstreaming will peak in the transition flow region, below which the backstreaming is small and linearly proportional to pressure. Rettinghaus and Huber [8] found that the maximum backstreaming rate over a cold trap occurred at a pressure of 5×10^{-2} Pa for one 6-in. diffusion pumped system. The pressure at which peak backstreaming occurs is geometry dependent. Hablanian [13] describes a controlled-opening high-vacuum gate valve which is designed to avoid this kind of backstreaming by keeping the pump out of the overload region. Another scheme accomplishes the same thing by opening the valve in two steps.

The contamination introduced by roughing to the free molecular flow region is illustrated in Fig. 10.3. Baker, Holland, and Stanton [12] measured the backstreaming of oil through a 300-mm-long × 25-mm-dia. roughing line connected to a 4.5-m³/h rotary vane mechanical pump. They found that the backstreaming was small ($\sim 10^{-4}$ mg/min) at high pressures because of the viscous flushing action of the flowing air, but at pressures below approximately 15 Pa the viscous flushing action was diminished until at a pressure of 1.3 Pa the backstreaming was 7×10^{-3} mg/min, or 70 times greater than at high

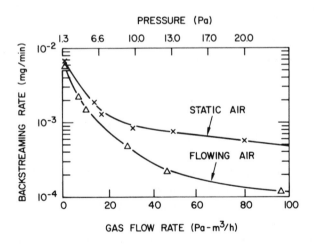

Fig. 10.3 Backstreaming rate as a function of gas flow. The flowing air is traveling the length of the tube. The static air is admitted at the ballast port. Gas pressure is a mean value for flow in the pipe. Edwards 4.5 m³/h rotary vane pump, 300-mm-long 25-mm diameter pipe. Reprinted with permission from *J. Vac. Sci. Technol.*, **9**, p. 412, M. A. Baker, L. Holland, and D. Stanton. Copyright 1972, The American Vacuum Society.

pressures. The data in Fig. 10.3 show that static air, that is gas admitted at the pump ballast port, did not reduce the backstreaming as much
as air that was allowed to flow through the length of the tube as it
would during roughing.   Figure 10.4b illustrates the total integrated
backstreaming calculated for the roughing line described in Fig. 10.2,
assuming the hypothetical pump-down curve given in Fig. 10.4a. At a
backstreaming rate of 10^{-4} mg/min from atmosphere to 15 Pa the total
backstreaming during the 4-min pump down would be 4 x 10^{-4} mg. If
the pump had been permitted to exhaust the chamber to 1 Pa in the
next 11 min, the total backstreaming would have increased from 4 x
10^{-4} mg to 5 x 10^{-2} mg. This illustration should make the danger of
roughing below 15 Pa adequately clear.

   Comparison of the steady-state backstreaming from a diffusion
pump to the backstreaming from the roughing pump cycle reveals that
the roughing pump backstreaming need not dominate the system
contamination. A 6-in. diffusion pumped system has a roughing line
with an inside diameter of approximately 35 mm, or twice the cross
sectional area of the pump used in the example in Fig. 10.3. If we
assume a backstreaming rate of 2 x 10^{-4} mg/min (twice that of the
preceeding example) for the roughing line on the 6-in. system and a
pump-down time of 5 min to a pressure of 15 Pa, the total amount of
oil backstreamed into the system would be 10^{-3} mg per roughing cycle.
The lowest value of vapor backstreaming quoted by Hablanian [2]
would result in a total backstreaming of 3 x 10^{-5} mg/min for a 6-in.
pump. For the worst case the roughing cycle could contribute as much
backstreaming as the diffusion pump does in 30 min of operation at

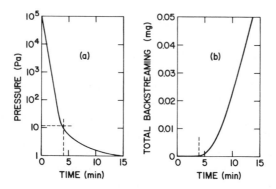

Fig. 10.4   Chamber pressure (a) and total backstreaming (b) as a function of pumping
time for the backstreaming rate (flowing air) given in Fig. 10.3.

Table 10.1   Methods of Reducing Pump Fluid Contamination
in a Diffusion Pumped High Vacuum System

Diffusion Pump	Mechanical Pump	System Operation
Cold cap	Simple trap	Crossover at P>15 Pa
LN_2 trap	Good oil	Minimize time dp is in overload region
Fractionating dp		Mild bake
Pentaphenylether or Pentaphenylsilicone		Gas flush during start/stop[a]

[a] Only if the system is shut down daily.

high vacuum. Even for a diffusion pump operating on a 1 h cycle, contamination from the roughing pump is less than that from the diffusion pump, provided that the roughing cycle is always terminated at approximately 15 Pa.

The two most important system-operation concepts for minimizing backstreaming are to stop roughing at 15 or 20 Pa, and to pump through crossover as quickly as possible or use a conductance-limited high vacuum valve. Table 10.1 summarizes the results of this discussion of methods of reducing contamination in a diffusion-pumped, high vacuum system.

The dominant gas load in most high vacuum systems is water vapor. A liquid nitrogen-cooled surface will provide a high pumping speed (14.5 L/(s-cm^2)) for this gas load. The vapor pressure of water at this temperature is about 10^{-19} Pa; therefore the pumping speed is neither a function of the operating pressure nor sensitive to small trap surface temperature changes. Several gases, however, can reevaporate at low pressures or during trap temperature fluctuations. Carbon dioxide is partly condensed; its vapor pressure is 10^{-6} Pa at 77 K. Methane and CO have high vapor pressures and are not condensed. They are adsorbed to some degree on liquid nitrogen-cooled surfaces.

Hengevoss and Huber [16] have shown how such gases evolve from a liquid nitrogen trap as it slowly warms up: CH_4 is first, followed by CO and CO_2. Santeler [6] has shown that the reevaporation of CO_2 can be eliminated by delaying full cool-down of the trap until the system pressure falls below 10^{-3} Pa or by momentarily heating the trap to 135–150 K after the system pressure falls below 10^{-6} Pa. Because CO_2 is partly condensed on the trap, it can cause problems. Siebert and Omori [17] have shown how certain liquid nitrogen cooling coil designs with temperature variations of only 1 K allow the release of enough CO_2 to cause total pressure variations of 20% in a system operating in the 10^{-4} Pa range.

In any vacuum system faults may occur that could potentially affect equipment performance or product yield. Misoperation of any one of several valves could cause diffusion pump fluid to backstream to the work chamber and contaminate the chamber and product. A leak in the foreline, an inadequate mechanical pump oil level, or a foreline valve failure could cause the forepressure to exceed the critical value. When the forepressure exceeds the critical value, pump fluid will rapidly backstream into the chamber. In addition, many diffusion pump system faults could cause harm to the pumping equipment. Cooling water failure or an inadequate diffusion pump fluid level will result in excessive fluid temperatures, accompanied by decomposition of some fluids. Certain decomposition products are vapors, whereas others form tarlike substances on the jets and pump body.

Modern diffusion pumped systems can be adequately equipped with sensors that will shut the system down or place it in a standby condition in case of loss of utilities, a leak, loss of cryogen, or loss of diffusion pump fluid. Pneumatic and solenoid valves, which leave the system in a safe position during a utility failure, are used on automatically controlled systems. When the compressed air or electrical power fails, all valves except the roughing pump vent should close. The roughing pump vent should open to prevent mechanical pump oil from pushing upward into the roughing line. A flow meter may be installed at the outlet (not the inlet) of the diffusion pump cooling water line. The signal from the flow meter is used by the system controller to remove power to the diffusion pump and to close both high vacuum and foreline valves. The use of a water filter on the inlet to the diffusion pump cooling water line will prevent deposits from clogging the line and causing untimely repairs. A thermostatic switch mounted on the outside of the diffusion pump casing performs these tasks much more simply, and a thermoswitch mounted on the boiler can be used to detect a low fluid level. The set points on the thermocouple or pirani gauge controllers are used to signal an automatic controller to close the high vacuum valve in case of a chamber leak and to isolate and remove the power from a diffusion pump with excessively high forepressure. Many liquid nitrogen controllers have low-level alarm sensors which

Table 10.2   Diffusion Pumped System Operating Concerns

Do	Do Not
Check mp oil level	Exceed dp critical forepressure
Check mp belts	Overload dp in steady state
Measure dp heater power when system is new	Leave cooling water flowing when system is shut down
Clean roughing trap periodically	Vent cold LN_2 trap to air
Pump oxygen with proper fluids	Rough pump below 15 Pa

close the high vacuum valve when the cryogen is exhausted. If not, an insulated thermoswitch can be attached to the trap vent line; the boil-off keeps the vent-line temperature below $0°C$. Table 10.2 lists some of the concerns with which the operator should be familiar.

This section raised several issues in regard to the operation and fault protection of a diffusion pumped system. In many cases fluid backstreaming could result from a utility failure or improper procedure. The easiest way to address these concerns is by fully automatic operation of the system. With automatic control and scheduled maintenance complete fault protection can be provided and backstreaming minimized to a point at which it is only a small source of organic contamination.

10.2   TURBOMOLECULAR PUMPED SYSTEMS

Turbomolecular pumped systems are configured in two ways: they are assembled with a separate roughing line and foreline or as a valveless system in which the chamber is roughed through the turbomolecular pump. Figure 10.5 illustrates the valved system with a separate roughing line. This system is much like a conventional diffusion pumped system. The entire roughing and foreline section apparently is identical to that in a diffusion pumped system. The criteria for sizing the forepump, however, are not the same as for a diffusion pump. Proper diffusion pump operation requires that the forepump speed be large enough to keep the forepressure below the critical forepressure at maximum throughput. In a turbomolecular pumped system a forepump of sufficient capacity should be chosen to keep the blades nearest the foreline in molecular flow or just in transition flow at maximum throughput. For most pumps the maximum steady-state inlet pressure P_{in} is between 0.5 and 1.0 Pa and the maximum forepressure P_f is about 30 Pa. Because the throughput is constant, $P_{in}S_{in} = P_fS_f$, or

$$\frac{P_f}{P_{in}} = \frac{S_{in}}{S_f} \tag{10.1}$$

Equation 10.1 states that the staging ratio, or ratio of turbomolecular pump speed to forepump speed, S_{in}/S_f, should be in the range 30 to 60. In Chapter 7 a more restrictive condition was placed on forepump size for turbomolecular pumps with low hydrogen compression ratios. A staging ratio of 20:1 was found to be necessary for pumps with maximum compression ratios for H_2 less than 500 in order to pump H_2 with the same speed as heavier gases (see Fig. 7.15).

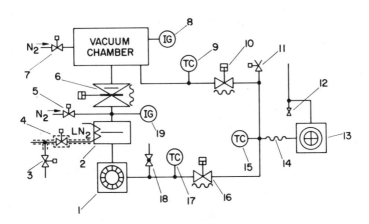

Fig. 10.5 Turbomolecular pump stack with a separate roughing line: (1) Turbomolecular pump, (2) liquid nitrogen trap, (3) LN$_2$ vent valve, (4) LN$_2$ fill valve, (5) turbomolecular pump vent valve, (6) bellows-sealed high vacuum valve, (7) chamber bleed valve, (8) chamber ionization gauge, (9) chamber thermal conductivity gauge, (10) roughing valve, (11) mechanical pump vent valve, (12) sump for collecting condensables, (13) mechanical pump (14) bellows for vibration isolation, (15) mechanical pump thermal conductivity gauge, (16) foreline valve, (17) foreline thermal conductivity gauge, (18) leak testing port, (19) turbomolecular pump ionization gauge.

No conventional liquid nitrogen trap is required on a turbomolecular pump to stop bearing or mechanical pump fluid backstreaming. The compression ratios for all but hydrogen, the lightest gas, are high enough so that none will backstream from the foreline side to the high vacuum side, provided that the pump is rotating at rated angular velocity. A liquid nitrogen-cooled surface will not trap the small amount of hydrogen that does backstream into the work chamber because of its low compression ratio. Liquid nitrogen may be used to increase the system pumping speed for water vapor. Incorrect as it may seem on first consideration, the optimum place to locate this liquid nitrogen-cooled water vapor pump behind the high vacuum valve is directly over the throat of the turbomolecular pump. Alternatively one could conceive of placing a tee behind the high vacuum valve and attaching the turbomolecular pump to one port and a liquid nitrogen-cooled surface to the other port. The conductance between any two ports of the tee is less, however, than the conductance of a liquid nitrogen trap. A 6-in. tee (inside diameter 200 mm) whose length is equal to its inside diameter has an air conductance between any two ports of 1275 L/s, whereas a low-profile liquid nitrogen trap designed for a 6-in. diffusion pump has a conductance of 1700 L/s. Clearly, the optimum arrangement for pumping both water and air, each at their highest speed, is a liquid nitrogen trap located between the turbomolecular pump and the high vacuum valve, even though it may convey the erroneous impression that the trap is present to prevent hydrocarbon backstreaming. In some processes use of a Meissner trap in the chamber is required, just as it would be in a diffusion pumped system.

Neither a roughing trap nor a foreline trap is needed. High crossover pressures prevent oil transfer to the chamber via the roughing line and high compression ratio for hydrocarbons fragments prevents back diffusion from the foreline through the pump to the chamber. Valve 18 in Fig. 10.5 is used for leak testing, not pump venting. Venting the pump at this location would drive oil vapors toward the high vacuum side after the rotor loses about 60% of its speed.

Figure 10.6 shows an unvalved system in which the chamber is roughed directly through the pump. Because there is no roughing line, the problems associated with roughing line contamination and improper crossover are eliminated. Physically the system becomes simpler. There is no roughing valve, piping, or trap. No high vacuum valve is needed and only one ion gauge and thermal conductivity gauge are required. This system must be completely shut down and restarted each time the chamber is opened to the atmosphere. This makes the use of liquid nitrogen-cooled surfaces awkward because they must be warmed each time the system is shut down.

Roughing through the turbomolecular pump places a size restriction on the mechanical pump if the turbomolecular and mechanical pumps are to be started simultaneously. If the mechanical pump is small, it will not exhaust the chamber to the transition region before the turbomolecular pump reaches maximum rotational speed. When this happens, the motor over-current protection circuit will shut down the turbomolecular pump. A properly sized and operated mechanical pump will exhaust the chamber to 20–200 Pa by the time the turbomolecular pump has reached, say, 75% of its rated rotational speed and prevent the backstreaming of the mechanical pump oil. Most turbomolecular pumps are designed with an acceleration time of 5 to 10 min.

10.2.1 System Operation

The operation of the turbomolecular pumped system with a separate roughing line shown in Fig. 10.5 is much like that of the diffusion pumped system described in Fig. 10.1. Before rough pumping is begun, the high vacuum valve and the roughing valve are closed, while the foreline valve is open. Chamber pump-down begins by closing the foreline valve and opening the roughing valve. As in a diffusion or any other high vacuum pump the rough pumping hardware varies in complexity with the chamber size. For 500-L/s or smaller turbomolecular pumps a two-stage rotary vane pump is used; turbomolecular

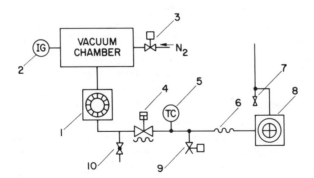

Fig. 10.6 Valveless turbomolecular pump stack: (1) Turbomolecular pump, (2) chamber ionization gauge, (3) chamber pump and vent valve, (4) mechanical pump isolation valve, (5) foreline and mechanical pump thermal conductivity gauge, (6) bellows for vibration isolation, (7) sump for condensables, (8) mechanical pump.

pumps larger than 1000 L/s use a Roots pump backed by a rotary piston pump. At a pressure of 100 to 150 Pa the roughing valve is closed and the chamber is crossed over to the turbomolecular pump. In some pumps this may cause a slight, but delayed, momentary speed reduction which will have no effect on the pump-down. Here, as for a diffusion pumped system, the dominant species is water vapor and the pump-down time will be controlled by the speed of the liquid nitrogen trap. If no liquid nitrogen trap is used, this system will pump water vapor somewhat more slowly than an untrapped diffusion pump of the same speed. The large unbaked internal surface area of the pump adsorbs water during the early stages of the high vacuum pump-down cycle and then reemits it at lower pressures. This effect is more noticeable in a valveless system because it potentially can adsorb more water vapor than a valved system that is not exposed to ambient air at a pressure greater than 150 Pa.

System shut-down begins by closing the high vacuum valve and warming the liquid nitrogen trap, if used, as described in Section 10.1.1 When the trap has equilibrated, the foreline valve is closed and the power to the turbomolecular pump motor is removed. The rotor will now deaccelerate. Typically, it should take 10 min or more for the rotor to come to a complete stop, but if that were to happen hydrocarbons from the foreline would rapidly diffuse to the region above the pump inlet. To prevent backstreaming of mechanical pump oil vapors and turbomolecular pump lubricating-oil vapors the pump is vented with a reverse flow of dry gas. Argon or nitrogen for example, should be admitted at a point above the pump inlet or part way up the rotor stack when the rotor speed has decreased to approximately 50% of maximum rotational speed. The flow should continue until the pump is at atmospheric pressure. This can be properly accomplished by admitting gas through valve 5 (Fig. 10.5). Turbomolecular pumps should not be routinely blasted with atmospheric pressure gas while running at rated speed. It is not good for long-term bearing life. At any time after the foreline valve is closed the mechanical pump system can be shut off and vented by valve 11 in Fig. 10.5 or if so equipped, an internal pump vent. The cooling water should be promptly shut off to prevent internal condensation. Condensation that may form on outer portions of the pump body during normal operation may be eliminated by tempering the water to just above the dew point.

System start-up begins by initiating cooling water flow, opening the foreline valve, and simultaneously starting the mechanical and turbomolecular pumps. After the pump accelerates to rated rotational speed, typically within 5 to 10 min, the liquid nitrogen trap may be filled. At this point the chamber may be pumped as described in the preceding section. Operation of the valveless system illustrated in Fig. 10.6 is considerably simpler than the valved system. Operation begins by opening the cooling water and foreline valves and starting the mechanical and turbomolecular pumps simultaneously. If the roughing pump has been properly chosen to make the chamber roughing cycle equal to the acceleration time, the system will exhaust the chamber to its base pressure without backstreaming pump oil vapors.

The valveless system is vented and shut down by first closing the foreline valve, waiting for the rotor speed to decrease to 50% of maximum rotational speed, and admitting dry gas above the pump throat (valve 3 in Fig. 10.6). The gas vent valve must be shut off when the system reaches atmospheric pressure or it will over-pressurize the work chamber. The mechanical pump is shut down as described and the cooling water flow is stopped.

10.2.2   Operating Concerns

Turbomolecular pumped systems share a few problems with diffusion pumped systems and present some that are unique. All are easily soluble. Backstreaming of mechanical pump oil vapors during roughing or turbomolecular pump shut-down, electrical interference, and damage from utility failure are all potential problems in a turbomolecular pumped system.

Backstreaming of mechanical pump oil to the chamber via the roughing line is even less important than in a diffusion pumped system because the crossover pressure is 100 to 150 Pa and not 15 Pa, as it is in a diffusion pump. If proper procedures are followed, backstreaming by this path is of no concern. The roughing pump in the valveless system (Fig. 10.6) cannot contaminate the chamber because the turbomolecular pump will have reached full rotational speed before the foreline is in free molecular flow.

The compression ratio for hydrocarbon vapors at full rotor speed is high enough in turbomolecular pump to prevent any backstreamed forepump oil vapors from reaching the work chamber. This is not so when the rotor is stopped or operating at reduced speed. Nesseldreher [18] has studied the partial pressures of several gases at the inlet as a function of the rotor speed and found that heavy hydrocarbons were observed at the inlet for rotor speeds of 40% maximum speed and less for the pump under scrutiny (Balzers TPU-400). Stopping the pump with the forepump operating and the work chamber under vacuum will result in rapid backstreaming of oil vapors from the foreline to the clean side of the pump. To prevent this backstreaming the valved and valveless turbomolecular pumped systems are always vented during shut-down. In particular, it is most important that the pump be vented with a dry gas and that the vent gas be admitted in such a way that it will flow to the foreline through at least a portion of the rotor and stator assembly. Oil vapors in the foreline are then flushed away from the high vacuum chamber. The pump must never be vented from the foreline because oil vapors will be forced back toward the pump inlet and the high vacuum chamber.

Proper venting procedures should be followed even after a power failure when the pumping system is unattended. The simplest electronic solution delays the opening of the vent valve until the pump coasts to about 50% of rated speed. A dc solenoid valve is driven by a power supply with a large capacitive output; the capacitor provides sufficient energy to keep the valve closed for a fixed time until the rotor looses speed. Another method uses a battery-operated circuit to delay the opening of the vent until the speed decreases and then closes it after the pump has reached atmospheric pressure. Both circuits allow the pump to coast through momentary power failures without initiating venting. Speed sensing can be used to vent the pump automatically as a result of a catastrophic leak like a broken ion gauge tube.

Nearby amplifiers may pick up mechanical or electrical noise emanating from turbomolecular-pump power supplies. Mechanical vibration may be reduced significantly by use of bellows between the pump and chamber. Improper connection of earth and neutral in three-phase supplies may also generate noise. Other electrical noise can be eliminated by connecting the ground of each piece of equipment, including the pump, to the ground terminal on the most sensitive amplifier stage and then grounding to earth. This is most efficiently done with solid copper strips.

Turbomolecular pumps must be protected against mechanical damage as well as against the loss of cooling water because the pump is a high-speed device with considerable stored energy. If a large, solid particle enters the rotor or a bearing seizes, serious damage may be done to the pump. The expense of repair could easily exceed that of scraping the varnish from a diffusion pump. Such catastrophes need not occur and do not occur if the most elementary precautions are taken. A splinter shield located at the pump throat adequately protects the rotors and stators from physical damage at some loss in pumping speed, and some pumps are available with side entrance ports.

Table 10.3 Turbomolecular Pumped System Operating Concerns

Do	Do Not
Periodically change oils	Use undersized forepump
Vent mp to outside exhaust.	Vent cold LN_2 trap to air
Vent tmp from high vacuum side	Flow cooling water when vented
Keep forechamber in molecular flow	Run tmp without cooling water fault protection
Check mp oil level	Stop tmp while under vacuum
Check mp belts	Run tmp below maximum rotational speed

Water cooling is used in oil-lubricated or grease-packed bearings. Proper cooling is necessary to remove heat from the bearings and to extend bearing life. Most pumps are manufactured with an internal cooling water interlock so that it is generally not necessary to add an external flow sensor. A water-flow restrictor may be added to conserve cooling water. Any device with internal cooling water passages requires a clean water supply. Even though filters are used, it is advisable to reverse-flush the pump water lines once or twice a year to remove material that has passed through the filter. If the water supply is unreliable, a recirculating water cooler is in order. No protection is needed for loss of liquid nitrogen because it does not serve a protective or baffling function; it merely pumps water vapor.

Turbomolecular pumps will give reliable trouble-free operation if they are adequately lubricated and protected against cooling water failure, power failure, mechanical damage, and excessive torque. This protection is easily and routinely provided with the available technology. Table 10.3 lists some of the needs for proper turbomolecular pump operation.

10.3 ION PUMPED SYSTEMS

The sputter-ion pump is the most common of all ion pumps; it is used with a titanium sublimation pump and cryobaffle to form a high vacuum pumping package that is easy to operate and free of heavy hydrocarbon contamination. Invariably it uses a sorption roughing module that is also free of hydrocarbons.

A typical, small, sputter-ion pumped system is shown in Fig. 10.7. The TSP and cryocondensation surface may be in the chamber on which sputter-ion pump modules are peripherally located or they may be in separate units as sketched here. Titanium is most effectively sublimed on a water-cooled rather than a liquid nitrogen-cooled surface in a system that is frequently vented to atmosphere. If liquid nitrogen cooling were used, the film on the cooled surface would be composed of alternate layers of titanium compounds and water vapor. Because water vapor is the dominant condensable vapor in rapid-cycle systems, titanium flaking and pressure bursts would frequently result. In rapid-cycle systems it is best to separate the two functions and condense water vapor on a liquid nitrogen-cooled surface and titanium on a water-cooled surface.

A two- or three-stage sorption pump is used to rough the system. Alternatively, a gas aspirator or carbon vane pump may be used to preexhaust the chamber to about 15,000 Pa before sorption pumping. The gas aspirator requires a high mass flow of nitrogen at high pressure and is noisier and less practical than a carbon vane pump. Neither pump is necessary, but their use does allow more sorption cycles between baking. High-capacity modules equivalent to about 10 to 15 small pumps (10-cm diameter, 25-cm high) are available for roughing large volumes.

If the system is to be cycled frequently, a gate valve should be installed between the high vacuum pumps and the chamber to minimize operation of the ion pump at high pressures. An ion gauge is not needed on the pump side of the gate valve because the ion current is a measure of pressure. A gas-release valve is provided for chamber release to atmosphere. Nitrogen is used for this function because it is easily pumped by the sorption and ion pumps.

10.3.1 System Operation

Evacuation of a small, sputter-ion pumped system begins with sorption pump chilling. Small commercial pumps will equilibrate in 15 min; a chill time of 30 min is not unreasonable. If the sorption pumps have been saturated by prior use, they must be baked at a temperature of 250°C for at least 5 h before they are ready to chill. Pumping on a sorption pump with an oil-sealed mechanical pump is not a good way to speed the outgassing of water vapor, because backstreamed oil vapors contaminate the sieve and reduce its pumping capacity. The

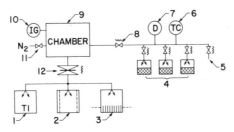

Fig. 10.7 Components of a small sputter-ion pumped system: (1) Titanium sublimation pump, (2) sputter-ion pump, (3) liquid nitrogen-cooled array for pumping condensable gases, (4) sorption roughing pumps. (5) port for the attachment of a gas aspirator or carbon vane roughing pump, (6) thermocouple gauge, (7) diaphragm gauge, (8) roughing valve, (9) work chamber, (10) ionization gauge, (11) chamber release valve, (12) high vacuum valve.

first stage of a two-stage pump manifold is used to rough to 1000 Pa; the pump is then *quickly* valved from the manifold and the second stage is used to pump to 0.4 to 0.2 Pa. If a three-stage manifold is used, a pressure of 3000 to 5000 Pa is obtained in the first stage, 15 Pa in the second, and 0.1 Pa in the third. Staged pumping traps the neon that entered the first stage in viscous flow and also reduces the quantity of gas to be pumped by the last stage. Both effects reduce the ultimate pressure. At a chamber pressure of approximately 0.5 Pa continuous sublimation may begin. The sputter-ion pump can be started when the chamber pressure reaches 0.05 Pa. The roughing line is valved from the system, after which the sputter-ion pumping speed will rapidly increase until it reaches its maximum value at about 10^{-3} Pa. Below 10^{-5} Pa continuous operation of the TSP is not necessary because the sublimation rate of the titanium exceeds the gas flux. As the system pressure decreases, the interval between successive titanium depositions may be increased. When the layer is saturated, the pressure will rise as explained in Fig. 9.4. Timing circuits are available to control the sublimation time and interval between depositions.

A sputter-ion pump which has been exposed to atmosphere before its operation will not pump gas from the chamber as smoothly as if it were clean and under vacuum. Operation of an exposed system begins with chilling the sorption pumps and flowing cooling water to the TSP. The sorption pumps and TSP are used to rough as described for an operating system. When the sputter-ion pump is turned on, however, the system pressure will rise because of outgassing of the pump electrodes. The solution to this dilemma is to continue pumping with the

Table 10.4   Sputter-Ion Pump Operating Concerns

Do	Do Not
Operate s-i pump below 10^{-4} Pa in steady state to extend life	Obstruct the safety vent on the sorption pump
Valve s-i pump when releasing chamber to atmosphere	Pump a large quantity of H_2 at high pressure
Clean Ti flakes when replacing filaments	Operate the pump with the high current switch in the start mode
Adequately bake sorption pumps	Start the pump at high pressure
Sequentially operate sorption pumps	Leave polystyrene dewars on pumps while baking

sorption and TSP until the outgassing load is removed. If the outgassing is not removed in a short time, power to the ion pump should be removed to avoid overheating. The outgassing should be removed after switching the sputter-ion pump on and off a few times for about 5 min each time. When the pump voltage reaches about 2000 V, the roughing line may be valved from the system and operation continued in the normal manner.

The shut-down procedure for an ion pumped system is the simplest of any vacuum system. The high vacuum valve, if any, is closed and the power to the ion pump is removed. The entire system may remain under vacuum until it is needed again. The TSP cooling water should be disconnected if the system is to be vented to atmosphere to prevent condensation on the interior of the system.

10.3.2   Operating Concerns

One advantage of an ion pump is that no fault-protection equipment is needed to prevent damage from a utility failure. Loss of electrical power, cooling water, or liquid nitrogen will not harm the pump. Pumping will simply cease, gases will be desorbed from the walls and cryobaffle, and the pressure will rise. If the pressure does not exceed 10^{-1} Pa, the pump can be restarted by applying power to the ion pump; rough pumping is not required.

Of the severest concern to ion pumps in these applications is that they regurgitate previously pumped gases, in particular hydrogen, and do not pump large gas loads well. Essentially all of the pumping above 10^{-3} Pa is done by the TSP and at that pressure filament life is short. Because of their slow pumping between 10^{-1} and 10^{-3} Pa, ion pumps are not commonly used for routine, unbaked, rapid-cycle systems. Table 10.4 lists some of the needs for proper operation of an ion pumped system.

10.4 CRYOGENIC PUMPED SYSTEMS

The layout of a typical cryogenic pumped system driven by a helium gas refrigerator is sketched in Fig. 10.8. As in Fig. 10.1, more valves and other parts are shown than may be necessary. The system requires no forepump, and mechanical pump operation is required only during roughing. A liquid nitrogen trap is not needed for the prevention of backstreaming from the cryogenic pump, but a Meissner trap, cold-water baffle, or room-temperature baffle, none of which is shown in the sketch, may be necessary in the chamber to baffle the process heat load. These baffles also reduce the overall system pumping speed. Most cryogenic pumps include a hydrogen vapor-pressure gauge for monitoring the temperature of the second stage. The coarsness of the gauge makes it difficult to read extremely low temperatures.

The chamber may be safely crossed over to the high vacuum pump in a range of pressures. The lowest permissible pressure is governed by the desire to prevent oil backstreaming from the mechanical pump. For a 4- to 6-cm-diameter roughing line the minimum pressure is of an order of 10 to 20 Pa. For very large roughing lines it will be less; its value is greater than that which gives a Knudsen number of 0.01. The highest crossover pressure is determined by the cryogenic geometry and the refrigeration capacity of the expander. The cryopumping surfaces have a reasonably large heat capacity and can accept a "burst" of gas without irreversably warming. The maximum quantity of a gas Q_i admissible to the pump in a burst is a constant and is available from the manufacturer. The quantity of gas instantly admitted when the high vacuum valve is opened is given by $Q = P_c V$, where P_c is the crossover pressure and V is the chamber volume. From this we see that the maximum crossover pressure is

$$P_c(\text{max}) = \frac{Q_i}{V} \tag{10.2}$$

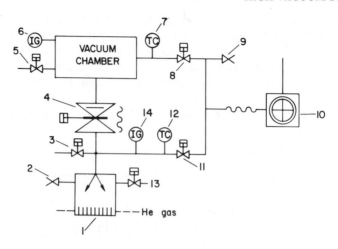

Fig. 10.8 Components of a helium gas refrigerator cryogenic pump stack: (1) Cryo-pumping surfaces, (2) pressure relief valve, (3) flush gas inlet valve, (4) high vacuum valve, (5) chamber vent valve, (6) ion gauge, (7) thermal conductivity gauge, (8) roughing valve, (9) mechanical pump vent, (10) mechanical pump, (11) roughing valve, (12) thermal conductivity gauge, (13) vent valve, (14) ion gauge.

The crossover pressure range for rough pumping air in a typical roughing line of a small pump is

$$15\text{Pa} \leq P_c \leq \frac{Q_i}{V} \tag{10.3}$$

If the value obtained from (10.2) is low enough to allow oil back-streaming before crossover, the cryogenic pump will be too small in relation to the size of the chamber.

Equation 10.2 implies that the crossover pressure could be several orders of magnitude higher than is typical for a diffusion pump. If the gas burst is very large, water vapor in the gas burst could reach the adsorbent stage during viscous or transition flow, coat or saturate it, and hinder the pumping of hydrogen and helium. In most cryogenic pumped systems the crossover pressure will be near 100 Pa. To mini-mize oil backstreaming it is desirable to use a crossover pressure as as large as possible, consistent with sorbent saturation.

10.4.1 System Operation

A cryogenic pumped chamber is evacuated by a roughing pump until the crossover pressure is reached. At that time the roughing valve is closed and the high vacuum valve is opened. The time required to reach the system base pressure is a function of the history of the cryogenic pump and its radiation loading, as well as other characteristics of the chamber which affect all pumps. The chamber is cycled to atmosphere in the usual way by closing the high vacuum valve and venting with dry nitrogen gas.

System shut-down commences by closing the high vacuum valve, removing power to the compressor, and equilibrating any liquid nitrogen trap in the chamber or pump with nitrogen gas. If the pump has been operated for only a short time since the last regeneration and has not accumulated water vapor, it will start without regeneration.

The regeneration procedures that are recommended or are available from automatic controllers make use of various combinations of external heat, rough pumping, and gas flushing. The object is to remove the captured gases from the pump after the power has been removed. In the most straightforward method the pump is flushed with dry gas. Charcoal does not require a high-temperature bake to remove water vapor, but a 50 to 80°C external bake will greatly speed the process. Regeneration is typically an overnight process. It is important that gas flushing be used while the pump is warming. If the pump is simply allowed to equilibrate with its surroundings while venting through the overpressure relief, melted ice from the pump surfaces will form a puddle of water at the base of the pump and seriously impair subsequent pumping. It is also important to regenerate the pump completely. Water vapor will be transferred from the warm to the cold stage if the arrays are only partly warmed before cooling. The time between regenerations, assuming that the pump is continuously run, depends on the application. For many high vacuum applications this may be three months.

The pump is started by roughing to 200 Pa and starting the compressor. Less water vapor will be collected initially on the sorbent stage if the pump body has been purged with gaseous nitrogen. After the pump is cooled, typically 60 to 120 min or longer, depending on the pump size, it is ready for operation.

10.4.2   Operating Concerns

Perhaps the most important concern is for the magnitude of the heat load on the first stage. In addition to the 300 K radiation from near-by chamber walls, the first stage is subject to thermal radiation from any source such as an electron beam hearth, heater lamp, or sputtering discharge. Heat loads up to 100 to 150 W, which are possible in many processes, easily exceed the capacity of the large 35 to 40 W expander stages. To reduce the incident flux on the first stage some form of baffling is necessary. The simplest is a reflective, uncooled baffle. If that is insufficient, a cooled chevron array may be required. Water or liquid nitrogen may be required for baffle cooling. In many instances the manufacturer is unaware of the details of the process and so cannot provide the correct baffling. It is the user's responsibility to ensure that the pump is not thermally overloaded by the process.

Momentary loss of power is a serious concern in continued pump operation, especially if the sorbent is saturated with helium. If power is lost even for a short time, the helium will be released from the sorbent and will conduct a large amount of heat from the chamber walls to the pumping surfaces. It will serve no purpose to rough the pump to 20 Pa because that is not sufficient to prevent continued conductive heat transfer. Stated another way, a pump will wipe out after a burst of helium and require complete regeneration. If the power is off for a longer time, say long enough for water vapor to release from the first stage and deposit on the second stage, the sorbent will saturate and regeneration also will be required. Loss of pumping by power failure or gas overload will not harm the pump. A malfunctioning over pressure relief is about the only way that damage can be done to the operator and pump.

Cryogenic pumps do not handle all gases equally well. The capacities for pumping helium and hydrogen are much less than for other gases. Some gases are easily pumped but they do present safety problems. Because cryosurfaces condense vapors, they can accumulate significant deposits which when warmed are able to react with one another or with the atmosphere. Some combinations such as silane and water vapor can react at 77 K. Even though these vapors can be vented to the atmosphere, the danger of an explosion or reaction is not lessened. Some other concerns for pumping gases at high flow rates are discussed in Chapter 12.

High neutron or gamma radiation will subject internal polymeric parts to degradation. It is not recommended that they be used in such applications [19].

Table 10.5   Cryogenic Pumping Concerns

Do	Do Not
Crossover at high pressure to avoid backstreaming	Obstruct safety release
Periodically change oil adsorption cartridge	Allow oil to collect on sorbent stage
Regenerate completely with gas purging	Vent with helium
Baffle pumping arrays from heated sources	Concentrate dangerous gases

High-purity helium (99.999%) is required to fill the compressors. Neon is the most common impurity in helium and may condense on the low temperature stage and lead to seal wear. Table 10.5 summarizes some of the problems in cryogenic pump operation.

Cryogenic pumps have their own unique features. When they are understood, they can be reliable and safely applied to many pumping situations.

10.5   THE CHAMBER

All-metal high vacuum systems are almost exclusively constructed from TIG welded 300 series stainless steel with elastomer sealed flanges. Viton is the preferred gasket material for high-quality systems because of its low outgassing rate and permeability to atmospheric gases. Buna-N is used when cost is a factor and silicone is used in certain high-temperature applications. Any metal, glass, or ceramic whose outgassing rate and vapor pressure are adequately low can be used in the chamber, assuming that it is compatible with other materials and with the thermal cycle. The use of elastomers in the chamber should be approached with more caution. Extreme heat, as well as excited molecules from glow discharge cleaning, sputtering, or ion etching, will decompose materials like polytetrafluoroethylene. Whenever possible, a ceramic or glass should be used for electrical insulation. Valves with bellows and O-ring stem seals are used on high vacuum chambers.

The bellows stem seal is used in locations such as the high vacuum line, foreline, and roughing line, where a vacuum exists on both sides of the seat. Both sides of this valve are leak tight but the stem side has a larger internal surface area than the seat side. Valves with O-ring stem seals have a high leak rate and are used only for applications like the chamber air release, where the seat side faces the vacuum.

The selection of lubricants for moving components can affect the process. Moving parts may be lubricated with wet lubricants or with dry-lubricant barrier films. Oil or grease lubrication systems can be designed to provide continuous lubrication by migration or evaporation to the moving interface. By the same mechanisms these lubricants can seriously contaminate other parts of the vacuum chamber with organic films. Dry-lubricant barrier films such as sputtered molybdenum disulfide are consumed in the process of lubrication. They have the advantage of having extremely low vapor pressures but they do create particulates and have a finite lifetime. The selection and use of lubricants for vacuum applications has been reviewed in more detail by Friebel and Hinricks [20].

The pumping rate and base pressure of high vacuum systems are limited by the surface (water) desorption rate sketched in Fig. 6.5. A great deal of space is devoted to the presentation of pumping curves in many treatments on vacuum technology, the reasons for which are not clear. The pumping time is not only dependent on the pump size, chamber volume, and internal surface area but also on the state of surface cleanliness. This is especially true when the interior surfaces become covered with deposition residues, some of which are hygroscopic. Consider an electron-beam deposition system capable of coating a 3000 cm^2 substrate area. The 0.4 m^3 chamber is pumped at the rate of 2000 L/s at the high vacuum gate valve. The chamber wall area up to the gate valve is 3.9 m^2. Internal fixturing accounts for 1.5 m^2 of surface area and an additional 6.6 m^2 of stainless steel removable liner plates brings the total internal surface area to 12 m^2. The chamber door, feed throughs, liquid nitrogen trap, and gate valve are sealed with 6 m of Viton gaskets. The equation that describes the time-dependence of the pressure in this problem is

$$SP - Q = -V\frac{dP}{dt} \qquad (10.4)$$

The first term on the left is the quantity of gas exiting to the pump per unit time, the second term is the amount of gas per unit time entering

the chamber from outgassing, permeation, and leaks, and $-VdP/dt$ is
the net rate at which gas is removed from the system.   If Q is constant, or at least changes much more slowly with time than the system
constant (V/S), as it does for outgassing, we may write the approximate solution as

$$P = P_o e^{-St/V} + \frac{Q}{S} \tag{10.5}$$

The first term in the solution represents the time-dependence of the
pressure that is due to the initial gas concentration; the second term
represents the contribution of other gas sources.   If this term represents outgassing, it is a slowly varying function and is the largest of the
two terms after the initial pumping period has passed.   After some
time the pressure is given by

$$P = \frac{Q_1}{S} t^\alpha \tag{10.6}$$

where Q_1 is the value of Q at some initial time, usually 1 h, and α is
often -1.   From this equation the pump-down curve in Fig. 10.9 can be
constructed.   The effect on the base pressure of electropolishing and
chemically cleaning the stainless steel and prebaking the O-rings is

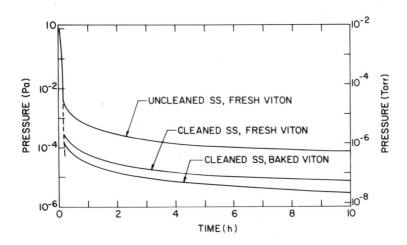

Fig. 10.9 Pumping time for a 0.4 m^3 stainless steel chamber for various steel and
elastomer precleaning conditions.

clearly visible. The variable effects, such as the hygroscopic nature of the film residue and the gas used to release the chamber, cannot be made any more explicit in this example. The amount of water vapor adsorbed by clean stainless steel will rapidly increase with ambient exposure time for the first several hours, after which it will saturate [21]. The saturation time and quantity are a function of the surface cleanliness, nature and temperature. The quantity adsorbed can be reduced by heating the walls and flushing with dry gas while the chamber is open. The gas inlet should be located to allow the gas to flow across the chamber toward the open access port.

Chambers used for routine evaporation are subjected to minimal or no baking. Many chambers are designed with an exterior chamber tubulation that can be heated with (50°C) water to assist in degassing the chamber. Partial outgassing may be achieved by the use of interior heating lamps and glow discharge cleaning, but neither can effectively cover 100% of the interior surface area. A surface cooled by liquid nitrogen or an auxiliary cryogenic pump is often necessary to achieve the necessary base pressure in a time commensurate with an efficient production schedule.

After establishing an operating procedure the serious user may choose to record an initial or "clean and dry" pump-down curve for the system and, if possible, record a residual gas spectrum after the system reaches its base pressure. Preservation of these data in a log book will assist in problem solving and requalifying the equipment

Table 10.6   General System Operating Concerns

Do	Do Not
Keep a system log	Vent the system with air
Record initial pump-down and rate of rise data	Vent system while Meissner trap is cold
Record clean RGA spectrum	Outgas ion gauge at $P > 10^{-3}$ Pa
Operate ion gauge at reduced emission current	Handle fixturing without clean gloves
Vent continuously with dry gas while chamber is open	Apply vacuum grease to static O-ring seals

after routine maintenance. These and some other general operating comments are tabulated in Table 10.6.

10.6 LEAK DETECTION

At some point we will be confronted with a system that does not behave normally—behavior that could be the result of a component malfunction, initial outgassing, or leak. Another occasion may find us checking out a new system, a procedure that includes verification of the integrity of the vacuum envelope. When and how to hunt for leaks are two useful skills discussed in this section. This leak-detection discussion is based largely on the response of a mass spectrometer to a tracer gas flow through a leak in a vacuum wall.

The decision to search for leaks is as important as the method chosen for detecting them. Each new component or subassembly should be routinely leak tested on a helium mass spectrometer leak detector after welding or brazing. It is premature, however, to leak-test a new system a few hours after placing it in operation simply because its performance does not meet the user's expectations. New systems often pump slowly because of outgassing of the fixturing, seals, or fresh pump fluids. The patient operator will usually wait a few days before criticizing the base pressure. At that time it is useful to take an RGA scan of the system background or to leak-test the system.

Searching for a leak in an established system is straightforward when a history of the system is known. A well-documented log book in which the system pressure versus pumping time, base pressure, rate of rise and perhaps a background RGA scan has been recorded will assist the operator in determining the cause of the poor performance. If the system contains an RGA, a quick scan will determine whether the poor performance is due to an atmospheric leak or outgassing, although outgassing of water vapor may be difficult to distinguish from a leak in a water line. One simple way of distinguishing between leaks and outgassing is to examine a plot of the system pressure versus time after closing the high vacuum valve (see Fig. 10.10). A leak will cause a continual increase in pressure; which will be linear in time for a molecular leak. Outgassing will cause the pressure to rise to a steady-state value which is determined by the vapor pressures of the desorbing species.

The sensitive step in the leak-checking procedure is done with a mass spectrometer leak detector (MSLD) or residual gas analyzer. The mass spectrometer leak detector is a mass spectrometer perma-

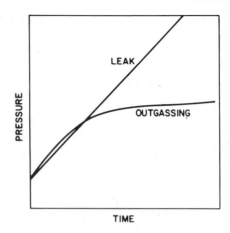

Fig. 10.10 Response of a sealed chamber to a leak and to outgassing from interior walls. Adapted with permission from *Handbook of Vacuum Leak Detection*, N. G. Wilson and L. C. Beavis, Eds. Copyright 1976, The American Vacuum Society.

nently tuned to helium with a separate, self contained pumping system. The operation of the mass filter was described in Chapter 4. The two most important parameters in leak detection with both instruments are sensitivity and response time.

Either the RGA or the MSLD is sensitive to a threshold partial pressure of the tracer gas used to probe the leak. In the best case the minimum detectable partial pressure is the absolute sensitivity of the instrument above the background noise. In a typical operating system a residual background pressure of the tracer gas will exist because it is regurgitated from a pump, back diffuses, or is released from a trap surface. This residual tracer gas pressure may be considerably greater than the ultimate detectability of the instrument and will significantly increase the minimum detectable tracer pressure. The minimum leak flux Q_{min} which can be detected is given by $Q_{min} = P_{min} S$. The leak detector is capable of measuring all leaks that produce a tracer gas partial pressure greater than P_{min} in a system in which the total pumping speed is given by S.

The minimum leak flux Q_{min} is distinct from the size of the leak. A large tracer-gas pressure drop across a small leak conductance can give the same flux as a small tracer-gas pressure drop across a large leak conductance. The minimum detectable leak conductance thus depends on the sensitivity of the instrument, the background tracer gas pressure in the detector, the external pressure of the tracer gas, and the speed of the pumps attached to the chamber. All of these parameters can be optimized to increase leak detection sensitivity. In practice external tracer gas pressures are usually atmospheric, the instrument sensitivity is fixed, and the background tracer gas pressure is dependent on the tracer gas, the type of pump, and the system. Helium is the most frequently used tracer gas. In the molecular flow regime the ratio of helium flow to air flow through a leak (2.46) is $Q(He)/Q(Air) = (M_{air}/M_{He})^{1/2} = 2.69$. For laminar viscous flow this advantage no longer holds true. From (2.37) it is seen that $Q(He)/Q(Air) = (\eta_{air}/\eta_{He}) = 0.92$. Signal strength is not essential in the detection of large viscous leaks, however. If the leak detecting is done with an RGA, a tracer gas other than helium may reduce the background. Increasing the external tracer gas pressure is not an effective technique because the molecular leak flux is only linearly proportional to pressure. With a commercial helium MSLD, the only variable that can be changed easily to improve the sensitivity is the pumping speed.

The maximum detectable sensitivity is reached at zero pumping speed. Closing the valve between the pump and chamber allows the leaking tracer gas to accumulate in the chamber. This is easily accomplished when leak detecting with an RGA and with MSLD units that are equipped with a valve between the ionizer and self-contained pump. For molecular gas flow the partial pressure of the tracer gas will increase linearly with time at zero pumping speed. After time t_1 a detector will measure partial pressure P_1; the leak flux will be given by $Q = P_1V/t_1$. This technique for increasing the basic sensitivity is called the accumulation technique. Small volumes may also be tested effectively by throttling the MSLD to a low but nonzero value. In this technique called foreline sampling S is reduced to increase the sensitivity but some pumping speed is retained to reduce the system background pressure. No increase in sensitivity is realized by this technique for tracer gases such as oxygen which may also desorb from the walls. Desorption increases the background pressure and negates any increase in sensitivity.

The minimum leak flux detectable by an RGA or MSLD is of an order of 10^{-9} Pa-L/s [22]. Under clean conditions the accumulation technique is sensitive to helium leaks as small as 10^{-11} Pa-L/s [23]. Leaks as small as these, which were measured across layers of glass and thin film interconnection metallurgy were soon hydrolized shut by the moisture in the air. In normal vacuum system operation leaks less that 10^{-8} Pa-L/s are rarely found [24].

The maximum sensitivity of the leak detector can be realized only if the tracer gas has had time to reach the steady-state value. For a

system of volume V evacuated by a pump of speed S, the pressure
change due to a sudden application of a tracer gas to a leak is given by

$$P_t = P_{to} + \frac{Q_t}{S}(1-e^{-St/V}) \qquad (10.7)$$

where P_{to} is the background pressure of the tracer gas; 63% of the
steady-state pressure is reached in a time equal to the system time
constant V/S and five time constants are required to reach 99% of the
response. This means that a 100-L system pumped by a leak detector
with a speed of 1 L/s will require application of the tracer gas for at
least 100 s to realize the maximum sensitivity of the instrument. This
is the case when an MSLD is connected to a chamber that has been
valved from the high vacuum pump (see Fig. 10.11a). The time constant can be reduced by placing the pumps in parallel (Fig. 10.11b) but
the sensitivity is also reduced. If the leak detector can handle the gas
load, the fast time constant can be retained in turbomolecular and
diffusion pumps without loss of sensitivity by placing the leak detector
in the foreline and valving the mechanical pump from the system (see
Fig. 10.11c). This technique may still be used if the gas flow is too
large for the leak detector by allowing the leak detector to pump at its
maximum flow rate while pumping the remainder of the gas with the
forepump. In this manner the fast time constant is retained and the
sensitivity is reduced only by the ratio of gas flows to the forepump
and leak detector. Leak detection in an ion or cryogenic pumped

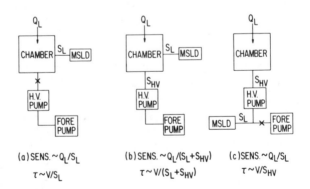

Fig. 10.11 Methods for mass spectrometer leak detecting a vacuum chamber pumped
by a diffusion or turbomolecular pump. Methods a and c have a high sensitivity and
methods b and c have a short response time to the influx of a tracer gas.

system differs from a turbomolecular or diffusion pumped system because the former two pumps have no forepumps. In these systems the MSLD must be appended directly to the chamber, where the high speed of the pump causes a loss of sensitivity. An ion pump may be helium-leak checked by momentarily removing power to the pump. A cryogenic pump cannot be shut down for leak checking because the evolved gas load will overload the pump in the MSLD. If the temperature in the cold stage is increased to 20 K, the helium pumping of the sorbent bed will drop to zero and the leak detector can be operated at its maximum sensitivity. The pump must be equipped with a heating element on the cold stage for this purpose. Moraw and Prasol [25] have discussed the optimum conditions for leak detecting large space chambers.

It is not possible to give a definitive or complete procedure for leak detecting a vacuum system but it is helpful to review some methods. A thorough understanding of the system under test is invaluable. The system volume and pumping speed must be known in order to calculate the system time constant. A system that cannot be pumped below the operating range of the roughing pump has a gross leak or a malfunctioning pump. The search for this leak begins by closing all the valves in the system and observing the pressure in the mechanical pump with a thermal conductivity gauge. If the pump is operating properly, sections of the system can be sequentially pumped until the leaky section is isolated. Helium may be sprayed around suspected seals and welds while listening for a change in the sound of the motor. Alternatively, alcohol sprayed on a leak will cause a large upward deflection of the thermal conductivity gauge. Helium sprayed on a leak in a system roughed by a sorption pump will cause a large increase in pressure.

If the system pumps to the high vacuum range but does not reach the usual base pressure, there may be a leak. There may also be considerable outgassing, a faulty pump, a contaminated gauge, or a leak below the high vacuum valve. A rate of rise measurement will distinguish a leak from outgassing in the chamber. If the pump blank-off pressure is not acceptable, there could be a leak below the high vacuum valve or a faulty pump. Both must be considered. An MSLD may be attached to a system that has a valve on the chamber or foreline without breaking the vacuum or removing power to the system. Leak checking with helium should begin at the top of the chamber; only a small flow rate is necessary. In some cases it may be necessary to wrap plastic around an area and flush it with nitrogen to prevent helium from entering more than one potential leak site. Alcohol will freeze in a small leak and allow adjacent areas to be checked without confusion. The alcohol can be removed with a heat gun. The most obvious places, such as welds and seals should be checked first.

Interior water lines are difficult to check because water may slow the diffusion of helium through the leak. They are best checked by draining, connecting directly to the leak detector, and warming with a heat gun before spraying with helium from the interior of the chamber. The helium background pressure may increase slowly with time when a search is made for small leaks with the MSLD in systems sealed with elastomer O-rings. This pressure rise is brought about by helium permeation [26]. The permeation time is about 1/2 to 1 h for most O-rings. The helium background will not decrease until it has been pumped from the gaskets.

The RGA allows the user to determine the state of the system with little additional data. Air leaks and water outgassing are easily differentiated. Air leaks are discerned by the presence of oxygen at mass 32, except in large TSP systems in which the oxygen pumping speed is so large that the oxygen signal will not be seen. Outgassing and water-line leaks can produce a large mass 18 peak, but they can be distinguished by the rate of rise. The RGA has the added advantage of functioning with tracer gases other than helium. Oxygen is often used for leak checking sputter-ion pumped systems and argon for TSPs. Additional operational hints are given in the AVS leak-detection handbook [25].

The helium mass spectrometer leak detector has been universally accepted for sensitive leak detection of vacuum components and systems. The instrument is most valuable if it has an internal calibration standard and valves for throttling the internal pumping speed (accumulation technique) and the inlet flow (foreline sampling technique). An important consideration in selecting a commercial instrument is ease of operation. Operators are reluctant to use a machine whose operation is difficult. Most instruments are portable to the extent that they can be disconnected for a few minutes while being transported to the test site without going through a complete diffusion pump shutdown procedure. Although it is the most common and sensitive leak detection method, mass spectrometric leak detection is not the only technique. Guthrie [27], Santeler [28], and McKinney [29] discuss leak detection with the ion gauge, the halogen leak detector, and other techniques.

REFERENCES

1. H. L. Caswell, *IBM J. Res. Dev.*, **4**, 130 (1960).

2. M. H. Hablanian, *J. Vac. Sci. Technol.*, **6**, 265 (1969).

3. L. Holland, *Vacuum*, **21**, 45 (1971).

4. J. H. Singleton, *J. Phys. E.*, **6**, 685 (1973).

5. J. M. Benson, *Trans. 8th Nat. Vac. Symp. (1961)*, **1**, Pergamon, New York, 1962, p. 489.

6. D. J. Santeler, *J. Vac. Sci. Technol.*, **8**, 299 (1971).

7. B. D. Power and D. J. Crawley, *Vacuum*, **4**, 415 (1954).

8. G. Rettinghaus and W. K. Huber, *Vacuum*, **24**, 249 (1974).

9. M. J. Fulker, *Vacuum*, **18**, 445 (1968).

10. R. D. Craig, *Vacuum*, **20**, 139 (1970).

11. M. A. Baker and G. H. Staniforth, *Vacuum*, **18**, 17 (1968).

12. M. A. Baker, L. Holland, and D. A. G. Stanton, *J. Vac. Sci. Technol.*, **9**, 412 (1972).

13. M. H. Hablanian, *Proc. 6th Int. Vac. Congr. (1974), Japan. J. App. Phys.* Suppl. 2, 1974, p.25.

14. M. H. Hablanian and H. A. Steinherz, *8th Nat. Vac. Symp. (1961)*, **1**, Pergamon, New York, 1962, p. 333.

15. D. W. Jones and C. A. Tsonis, *J. Vac. Sci. Technol.*, **1**, 19 (1964).

16. J. Hengevoss and W. K. Huber, *Vacuum*, **13**, 1 (1963).

17. J. F. Seibert and M. Omori, *J. Vac. Sci. Technol.*, **14**, 1307 (1977).

18. W. Nesseldreher, *Vacuum*, **26**, 281 (1976).

19. K. M. Welch, *An Introduction to the Elements of Cryopumping*, K. M. Welch, Ed., American Vacuum Society, New York, p. III-20.

20. V. R. Friebel and J. T. Hinricks, *J. Vac. Sci. Technol.*, **12**, 551 (1975).

21. H. Galron, *Vacuum*, **23**, 177 (1973).

22. L. C. Beavis, *Vacuum*, **20**, 233 (1970).

23. J. F. O'Hanlon, K. C. Park, A. Reisman, R. Havreluk, and J. G. Cahill, *IBM J. Res. Dev.*, **22**, 613 (1978).

24. N. G. Wilson and L. C. Beavis, *Handbook of Leak Detection*, W. R. Bottoms, Ed., American Vacuum Society, New York, 1976.

25. M. Moraw and H. Prasol, *Vacuum*, **28**, 63 (1978).

26. J. R. Young, *J. Vac. Sci. Technol.*, **7**, 210 (1970).

27. A. Guthrie, *Vacuum Technology*, Wiley, New York, 1963, p. 456.

28. D. J. Santeler et al., *Vacuum Technology and Space Simulation*, NASA SP-105, National Aeronautics and Space Administration, Washington, DC, p.275.

29. H. F. McKinney, *J. Vac. Sci. Technol.*, **6**, 958 (1969).

CHAPTER 11

Ultrahigh Vacuum Systems

Several fields of science and technology have made remarkable contributions to our understanding of nature because ultrahigh vacuum technology has permitted experimental investigations to be performed in or below the 10^{-7} Pa region. One branch of science that has reaped significant benefits is surface science [1] because it required an environment for the preparation and preservation of atomically clean surfaces. At a pressure of 10^{-9} Pa it takes 50 h to form a monolayer of surface contamination. The realization of an ultrahigh vacuum technology necessitated the development of pumps, gauges, materials, and fabrication and sealing techniques appropriate to this region. The first important development was the invention of a practical gauge by Bayard and Alpert [2] which reduced the x-ray limit below 10^{-6} Pa. Before the development of gauges with a reduced x-ray limit it was not possible to determine directly if ultrahigh vacuum pressures had indeed been reached. This was followed by the development of the ion pump, the cryogenic pump, improved diffusion pump fluids and traps, and a mature components and materials technology.

Today there are commercially available ion, cryogenic, turbomolecular, and diffusion pumped systems that will attain pressures in the ultrahigh vacuum range. There is also a variety of gauges that can be used in this pressure range, some of which were referenced in Chapter 3. Many of the elements of an ultrahigh vacuum technology were included in more general discussions in earlier chapters.

The most important attribute that distinguishes an ultrahigh vacuum system from one not able to reach the pressure is cleanliness, that is,

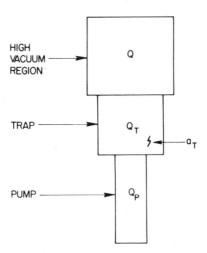

HIGH VACUUM REGION

Q

TRAP

Q_T

a_T

PUMP

Q_p

Fig. 11.1 High-vacuum pumping system (schematic). Reprinted with permission from *J. Vac. Sci. Technol.*, **8**, p.299, D. J. Santeler. Copyright 1971, The American Vacuum Society.

the elimination of contamination from all sources. There are many sources and types of contamination in any vacuum system. Gas can desorb from the chamber walls or evolve from a pump. Gases from either source can be pumped by a cold trap with a probability of being released at a later time. We have chosen to view them after the manner of Santeler [3]. Figure 11.1 shows a generalized vacuum system. The total pressure in the system from vacuum chamber sources is

$$P = \sum_i \frac{Q_i}{S_i} \qquad (11.1)$$

The partial pressure of each gas or vapor desorbing from the chamber is equal to the rate of desorption of that species divided by its pumping speed.

Gases that originate in the pump will flow to the chamber through a trap (if used) in which a fraction of them will be trapped with trapping coefficient a_T. The quantity of gas $Q_p(1-a_T)$ will backstream to the pump where it will be pumped by the pump or the trap, with a resultant pressure contribution to the chamber of

$$P = \sum_i \frac{Q_{pi}\,(1 - a_{Ti})}{S_i} \tag{11.2}$$

If a trap is added between the chamber and the pump, it will be a source of gases that may become partially trapped on other surfaces or flow to the chamber. This contribution to the pressure in the chamber may be expressed as

$$P = \sum_i \frac{K_{ai}\,Q_{Ti}}{S_i} \tag{11.3}$$

K_a depends on the trapping coefficient a_T and the location of the gas molecules in the trap.

The total system pressure results from all of the above sources. It is given by

$$P = \sum_i \frac{Q_i}{S_i} + \frac{Q_{pi}\,(1 - a_{Ti})}{S_i} + \frac{K_{ai}\,Q_{Ti}}{S_i} \tag{11.4}$$

The essence of Santeler's argument is that although the first term behaves in a normal manner (i.e., as the pumping speed increases, the pressure contribution decreases) the second and third terms do not behave in a like manner; for example, if the trap surface and the pump size are increased, the backstreamed gas load and the pumping speed are increased in proportion. The result is an effective base pressure P_o for a particular pump that is affected only by the trapping efficiency. This base pressure will still exist even if the system does not contain a trap between the pump and the chamber; for example, more hydrogen will backstream from a large ion pump than from a small pump. By this argument (11.4) may be simplified to read

$$P = \sum_i \frac{Q_i}{S_i} + P_o \tag{11.5}$$

The production of ultrahigh vacuum is therefore concerned with the pump and chamber. The contamination from the chamber may be reduced by the intelligent choice of materials, fabrication techniques, and operating procedures; the gas load originating in the pump may be reduced by trapping or processing procedures.

This chapter discusses the respective roles of the chamber and diffusion, turbomolecular, ion, and cryogenic pumps in achieving ultrahigh vacuum. The role of the chamber becomes paramount because the clean and reasonable way to reduce the pressure contribution from the chamber is to reduce Q and not to increase S. Recall that increasing S will not decrease P_o. The selection of the pump is also important, however. Each pump type has its characteristic advantages and disadvantages that make it suitable for some applications and not suitable for others.

11.1   THE CHAMBER

The rate limiting steps during the pump-down of a vacuum chamber at ambient temperature were sketched in Fig. 6.5. In a vacuum chamber the volume gas is removed first, followed by surface desorption, outdiffusion from the solid, and last, permeation through the solid wall. All of these processes except volume gas removal are greatly temperature-dependent. At room temperature the pressure decreases so slowly that the permeation limit can never be reached on a practical time scale. High-temperature processing is a necessity if ultrahigh vacuum is to be attained after a reasonable pumping time. The materials, fabrication techniques, and seals used for construction of the chamber walls and internal fixturing must be compatible with this thermal cycling.

In the early days, when ultrahigh vacuum was a laboratory phenomenon systems were constructed from glass because it could be easily baked clean. As the complexity and size of ultrahigh vacuum experimental work grew and fabrication techniques improved stainless steel became the preferred construction material. AISI grades 304, 304L, 316, and 316L or a stabilized grade such as 321 or 347 are the most frequently used; the overwhelming majority of the systems fabricated in this country are made from 304, whereas in England a grade equivalent to 347 is more readily available than 304. In Europe a grade equivalent to 316 is often used. 304L, 321 and 347 steels have the advantage of reducing carbide precipitation which can occur in 304 when welded or subjected to an extended high-temperature bake. All joints are made by TIG welding or metal gasketed flanges; O-rings are not used anywhere in the high vacuum portion of the system, including the high vacuum connection to the pump or trap.

Inside the chamber stainless steel and other metals are used for the fixturing. With the exception of those metals with very high vapor pressures, most metals are suitable as long as they can be outgassed. As discussed in Chapters 3 and 4, some filament materials generate carbon oxides. High density alumina and sapphire are frequently used as insulators because they are stronger than quartz or machinable glass ceramic. The properties of these and other materials are discussed in Chapter 6.

To reduce the outgassing load the components and subassemblies are first chemically cleaned in the same manner as an unbaked high vacuum system and further heat treated to reduce the outgassing to levels acceptable for these low pressure ranges. Several heat treatments are discussed in Chapter 6. For ultrahigh vacuum use the most thorough of these treatments is selected. The two initial treatments that are the most effective on stainless steel are a vacuum bake at 800 to 950°C and a 400°C air bake. The vacuum bake should be performed at a pressure of 10^{-6} Pa to keep the hydrogen concentration in the metal as low as possible. Nuvolone [4] suggests that the 400°C bake be performed in 2,700 Pa of pure oxygen rather than air to avoid any complications that may result from the presence of water vapor during the bake. Metal-to-glass or metal-to-ceramic seals should not be subjected to temperatures higher than 400°C, and sealed copper-gasketed flanges should not be baked at temperatures higher than 450°C. Unsealed flanges, however, will tolerate higher temperatures.

The flange and knife-edge will suffer some loss of temper during an 800 to 950°C bake. The exact amount depends on the grade of stainless steels and the fabrication steps used in flange construction. Most commercially available copper-gasketed flanges are forged from grade 304 and some contain trace additives that will retard grain growth during heat treating. Small grain size is necessary to prevent significant loss of hardness. Not all flanges are forged in the same manner or have the same properties. The extremely large project, such as a particle accelerator installation, can afford to write its own specifications and inspect incoming materials, but the average user must analyze and test a few samples for composition and hardness, before and after heat treatment, or rely on the data provided by the manufacturer.

Subsections that have been prefired by one of the two aforementioned techniques, can be stored until final assembly for periods of months [4]. It is not recommended that parts be stored in boxes or plastic bags and rubber bands because they will quickly contaminate the clean parts with organics [5]. Aluminum foil is frequently used to cover open flange faces or to wrap cleaned parts until final assembly. Aluminum foil that has been specifically degreased for vacuum use will not contaminate the cleaned parts, but ordinary aluminum foil can coat the parts with residues of rolling oils. Clean aluminum foil is commercially available. When the system has been completely assembled, it can be baked under vacuum at 150 to 250°C for 24 h to remove the surface gas [4, 6]. This same bake cycle can be used each time the system is exposed to ambient air. If the system is released with dry nitrogen or argon, each succeeding pump-down cycle will be shorter than if it had been released with air. It is also advantageous to open the smallest possible port and continue the dry gas purge until the flange is closed. Samuel [7] described an alternative procedure for baking a system in which the components were given an 850 to 900°C vacuum bake. The chamber was heated in atmospheric air at a temperature not exceeding 200°C for 2 h, after which the system was rough pumped. When the system reached 10^{-3} Pa, the temperature was reduced to 150°C until the gauges were outgassed; the heat was then removed.

If the system has been properly cleaned and operated, hydrogen will be the dominant residual gas at the ultimate pressure of a stainless steel system and helium will be the dominant residual gas in a glass system. Usually the situation is somewhat more involved than we have described. A system containing any amount of internal fixturing will have gas trapped between flat surfaces and in blind spots (e.g., around screw threads). Slots are usually cut in the screws and in flat mating surfaces to hasten the exit of trapped gas. A system that makes extensive use of stainless steel bellows may have its ultimate pressure limited by hydrogen permeation through the thin walls. Contaminants on the surfaces of copper gaskets can cause oxidation of the copper after repeated baking and open up minute leak paths through an otherwise rugged ultrahigh vacuum seal.

Unbaked areas of the system, which are usually located near the pump entrance, can be responsible for the largest quantity of residual gas in the system. Consider a system in which 99% of the chamber region can be baked and the remainder only chemically cleaned. A typical baked region has a desorption rate of 10^{-11} W/m², whereas an unbaked region has a desorption rate of 10^{-8} W/m². The total gas flux from the unbaked metal surfaces will be 10 times higher than that from the entire baked region.

The area of the system that cannot be baked depends largely on the choice of high vacuum pump. An ion pump and a liquid cryogenic pump can be completely baked. A turbomolecular pump and a gas refrigerator cryogenic pump can be baked at approximately 100°C; a diffusion pump cannot be baked at all. With the exception of the first two pumps, all systems will have some surface that cannot be subjected to a bake at a temperature in excess of 100°C.

Current commercial ultrahigh vacuum systems are limited to the 10^{-9} Pa range by virtue of the baking and construction techniques. Only systems that are completely immersible in liquid helium can be routinely pumped to the 10^{-13} Pa range [8] and, interestingly enough, in these low-temperature systems outgassing from the walls and cleaning procedures are less important because everything sticks to the walls. Bills [9] has provided an interesting discussion of the problems that prevent classical ultrahigh vacuum systems from advancing to the 10^{-14} Pa range.

11.2 PUMPING TECHNIQUES

Selection of the pumps for an ultrahigh vacuum system is important. Equation 11.5 demonstrates that an ultrahigh vacuum pump should act to reduce the pressure by providing high pumping speed (low Q/S) and to reduce the backstreaming from the pump to the chamber (low P_o). Here backstreaming is defined as the flow of *any* gas or vapor from the pump to the system. The goal of reducing contamination from the pump is the same for an ultrahigh vacuum system as it is for any other vacuum system. Only the ultimate pressure and permissible level of contamination are lower in these systems. To reduce the contaminant levels rigorous adherence to exacting pumping procedures is necessary. A liquid nitrogen trap that has been warmed for only a few minutes will backstream condensed vapors to the clean region above the trap. Although rechilling will prevent further contamination, it cannot remove the vapors that have accidentally diffused into the clean region. An accident that would be minor in the high vacuum region can destroy the validity of an experiment taking place at low pressures.

Each pump type has its own characteristic class of gases and vapors that it will not pump and that it will generate as impurities. Consequently, pumps and traps are used in combination to provide adequate pumping speed for all gases while minimizing backstreaming. Diffusion pumps are often combined with titanium sublimation traps, cryocondensation and sublimation pumps are used with ion pumps, and turbomolecular pumps are often assisted by sublimators. In general, no two ultrahigh vacuum systems will use the same combination of pumps. Even though commercial pumping systems are available, the design and selection of pumps for ultrahigh vacuum applications is usually done on an individual basis. Today, ultrahigh vacuum pump selection remains a cottage art with many "best" solutions. The remainder of this section reviews the selection and clean operation of some of the many possible configurations of diffusion, turbomolecular, ion, and cryogenic pumps that are suitable for use in the ultrahigh vacuum region.

11.2.1 Diffusion Pumps

Diffusion pumps are traditionally thought of as sources of hydrocarbon contamination in a vacuum system, but they can pump a chamber to the ultrahigh vacuum region when proper traps and procedures are selected. Diffusion pumps do not show a preference among gases and vapors; the exception is hydrogen or helium at low pressures in some pumps. The light gas compression ratio is not the same in all diffusion pumps. A pump with a hydrogen compression ratio of 10^6 and a hydrogen forepressure of 10^{-4} Pa will have an ultimate hydrogen pressure of 10^{-10} Pa. If that value is too great, the diffusion pump must be replaced, backed by another diffusion pump, or the special trapping techniques discussed in this section must be used. This backstreaming, if present, can easily be observed by admitting helium to the mechanical pump exhaust while watching the output of an RGA tuned to helium.

The major contaminants from a diffusion pumped system operating in the ultrahigh vacuum region result from backstreaming mechanical pump fluid through the roughing line, backstreaming diffusion pump oil, and backstreaming the permanent gases (H_2, CH_4, and C_2H_4) that are formed as a result of pump fluid degradation.

Backstreaming mechanical pump fluid vapors can be prevented by any one of several techniques. A molecular sieve trap will reduce but not totally eliminate mechanical pump fluid back diffusion. It will also transfer particulates to the mechanical pump during roughing or during purge pumping and therefore is not recommended. A liquid nitrogen trap designed for the roughing line is the most effective trap. If a liquid nitrogen trap warms, it will allow oil vapors to backstream when the gas is in the molecular flow region. Purge pumping is even more practical than trapping. Santeler [3] has described the technique of purge gas protection illustrated in Fig. 11.2. A purge gas valve is located on the chamber side of the roughing line trap. Whenever the trap is warm, dry nitrogen gas is admitted through the leak. This flow of gas prevents mechanical pump oil from backstreaming through the trap and cleans the trap. The gas purge can be activated by a thermo-

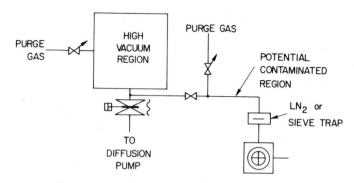

Fig. 11.2 Prevention of hydrocarbon contamination from the roughing pump by use of a purge gas. Adapted with permission from *J. Vac. Sci. Technol.*, **8**. p. 299, D. J. Santeler. Copyright 1971, The American Vacuum Society.

switch in the trap. As we noted in Chapter 10, a reverse purge can contaminate the region above the trap equally well. This will happen if a mechanical pump vent is located between the trap and the mechanical pump. The vent should never be located in that position but always between the trap and the chamber.

Purging can also be used in place of a liquid nitrogen trap by admitting the purge gas flow in the chamber, as shown in Fig. 11.2. The leak valve is set so that the mechanical pump cannot exhaust the chamber below 15 to 30 Pa. At crossover the chamber purge valve is closed, the high vacuum valve is opened, and the roughing line purge is begun. In this manner neither the chamber nor the roughing line is in the molecular flow region when the mechanical pump is operating. The third technique for the prevention of roughing-line backstreaming eliminates the roughing line altogether and roughs through the diffusion pump as it is starting. Here the concerns are the same as those described for the turbomolecular pump in Chapter 10. A gas purge through the chamber, trap, and diffusion pump will prevent oil from backstreaming until the trap and diffusion pump become operative.

Mechanical pumps are traditionally used to rough out diffusion pump systems, but they are in no way a requirement. One of the simplest and most straightforward techniques for roughing out an ultrahigh vacuum chamber uses the two-stage sorption pumping system described in Chapter 10.

Although a fractionating diffusion pump with a side ejector stage will stop mechanical pump oil backstreaming, it will not stop its working fluid from entering the chamber. High-vacuum trapping techniques were discussed in Chapter 8, but more elaborate techniques are necessary for ultrahigh vacuum. A minimum two-bounce array will reduce oil–gas and oil–oil collisional backstreaming. Recall from Chapter 8 that oil–gas backstreaming has a peak at $\sim 5 \times 10^{-2}$ Pa for the dimensions typical of a 6-in. trap. Because this oil transmission is characteristic of the transition flow regime trap dimensions and pressure, it can be reduced by a trap with two totally different sets of dimensions [10]. The two regions of the trap are in series and have backstreaming peaks at different pressures. At lower pressures each oil molecule makes two collisions and the transmission is proportional to $(1 - \alpha)^2$, where α is the accommodation coefficient for oil on the cooled surface. Minimum two-bounce traps can maintain the oil partial pressure in the system at less than that which can be detected [3, 10], which is of an order of 10^{-13} Pa.

Permanent gases that evolve from the pump or trap are also contaminants. In Chapter 10 methods of handling CO_2 reemission from the trap were discussed. Hydrogen, methane, and ethane generated by the decomposition of small amounts of pump fluid in the boiler are permanent gases that are not pumped in the usual sense by a liquid nitrogen trap. Their vapor pressures are so high at 77 K that less than a monolayer of methane, little ethane, and no hydrogen will stick. Some improvement can be obtained by a liquid nitrogen-cooled titanium sublimation trap between the liquid nitrogen trap and the system [11]. To produce an effective trap, the designer should have a clear conception of the difference between a sublimation trap and a sublimation pump. A pump is simply a large surface with a high pumping speed. To trap these gases effectively, this surface should form a two-bounce array and include a creep barrier [3]. Even with these elaborate precautions the elimination of all H_2 and CH_4 is not possible because their accommodation coefficients on titanium at 77 K are low. Table 11.1 summarizes contaminants from an oil diffusion pumped system, how they can be eliminated, and some possible resulting pressures.

Despite the ability of diffusion pumps to achieve ultrahigh vacuum pressures, they are not an overwhelming favorite with users of small systems. Their operation demands perfection to remain free of hydrocarbon contamination. Purge pumping can clean organics from the stainless steel walls of an empty chamber [3] but it will not easily remove them from many of the materials under study or used in the construction of experimental fixturing.

11.2.2 Tubomolecular Pumps

The most significant contaminant from a turbomolecular pump is hydrogen. The amount of hydrogen that backstreams from or is not pumped by these pumps is affected by the design of the pump, the kind of trap, the backing pump, and the operating procedures. The ultimate pressure of a high-compression turbomolecular pump will be 5 × 10⁻⁹ Pa when backed by a two-stage rotary pump, ~10⁻⁹ Pa when backed by a diffusion pump, and <10⁻⁹ Pa when a titanium sublimation trap is installed between the pump and the chamber.

For ultrahigh vacuum pumping the ultimate hydrogen partial pressure is limited by the compression ratio of the pump and the hydrogen partial pressure in the foreline. Modern pumps are available with compression ratios between 1,500 and 5,000. The hydrogen pressure in the foreline can be reduced by the use of a high-quality mechanical pump oil or a diffusion backing pump. Even with a diffusion backing

Table 11.1 Contaminants From an Oil Diffusion Pump, and How They Are Eliminated[a]

Contaminant	Typical Pressure (Pa)	How Eliminated	Resulting Pressure (Pa)
D. P. Oil	10^{-6}-10^{-7}	Good LN_2 trap	$<10^{-13}$
		Sublimation trap	0
M. P. Oil	10^{-4}-10^{-5}	Change oil	10^{-6}-10^{-7}
		Gas purge or valveless system	$<10^{-13}$
		Sorption pump	0
H_2 from D. P.	10^{-7}-10^{-8}	Sublimation trap, pre-bake of pump	$<10^{-11}$
CH_4, C_2H_6 from D.P.	10^{-7}-10^{-8}	Sublimation trap	$<10^{-11}$

Source. Adapted with permission from *J. Vac. Sci. Technol.*, **8**, p. 299, D. J. Santeler. Copyright 1971, The American Vacuum Society.
[a] The pressure values given are representative of a large majority of diffusion pumped systems, but do not cover all possible situations.

pump there will be a finite hydrogen partial pressure at the inlet be-cause hydrocarbon oils lubricate the bearings in the forechamber and elastomers are used for seals between the forechamber and atmos-phere. Lange [12] observed random H_2 pressure bursts in a turbomo-lecular pump when the cooling water temperature was raised above $14°C$. The residual hydrogen can be partially trapped in the titanium sublimation trap described in Section 11.2.3.

Turbomolecular pumps have a large internal surface area and conse-quently must be thoroughly baked for the pump to reach its ultimate pressure. Unfortunately the construction of the pump is such that it is not possible to bake it much over $100°C$. This bake cycle is necessary to remove the water from the interior of the stator and rotor surfaces.

The roughing cycle can contaminate a turbomolecular pumped system just the same as any other system. For ultrahigh vacuum the simplest and safest procedure starts the turbomolecular pump and mechanical pump simultaneously and roughs through the turbopump. Alternatively, gas flushing, a sorption pump, or a second small turbo-mechanical pump set can be used for clean roughing. When properly roughed and baked the turbomolecular pump can pump to the 10^{-9} Pa range without causing organic contamination.

11.2.3 Sputter–Ion Pumps

Sputter-ion pumps are frequently combined with TSPs to produce a compact, bakeable system. Structurally the pumps are identical to those used in high-vacuum systems with the exception of the TSP which must be liquid nitrogen-cooled rather than water-cooled and the pump is provided with a baking mantle or oven.

The main requirement for use of the sputter-ion pump in the ultr-ahigh vacuum region is thorough baking. Assuming that the chamber and components have received an appropriate high temperature pre-bake, a $250°C$ bake of the entire system, including the pump, is ade-quate. The temperature may be increased slowly to $250°C$ after the pressure reaches 10^{-5} or 10^{-6} Pa. A baking time of 10 to 20 h is typi-cal. The ion pump will continue to operate provided that its pressure is below 10^{-4} Pa. If the system contains an additional pump, for example a turbomolecular pump for differentially pumping an RGA, it can be used to reduce the gas load to the sputter-ion pump during baking. After baking is complete, the TSP liquid nitrogen surface may be cooled. The water vapor load in a baked ultrahigh vacuum system is much smaller than it is in an unbaked rapid-cycle system; therefore the sublimed titanium film will not entrap enough water to make titanium flaking a problem.

The main background gas present in a sputter-ion pumped system at low pressures is hydrogen, but other gases may be observed. Previously pumped gases may be reemitted and the presence of impurities in the titanium cathodes and TSP filaments can cause the generation of carbon oxides, methane, and ethane.

Sputter-ion pumped systems will pump routinely to the 10^{-9} Pa range; they are most useful for applications with small gas loads, but are not suitable for applications that require a constant pumping speed over a wide pressure range or in gas sampling systems in which reemitted gases can hopelessly confuse the measurements.

11.2.4 Cryogenic Pumps

Both liquid- and gas refrigerator-cooled pumps are able to reach the ultrahigh vacuum region. Liquid helium- and liquid nitrogen-cooled pumps have attained pressures of 10^{-10} to 10^{-12} Pa [8,13,14], whereas gas refrigerator pumps can reach the mid-to-low 10^{-8} Pa range.

The adsorption isotherms of helium and hydrogen on molecular sieve are strongly temperature dependent in the ultrahigh vacuum region. Because the second stage in a liquid cooled pump operates at 4.2 rather than 10 to 15 K, it has a much lower ultimate pressure than a gas refrigerator pump. A most important consideration in reaching very low pressures is the isolation of thermal and optical radiation from the cold stage. Liquid nitrogen-cooled baffles have been designed to maximize molecular transmission and minimize photon transmission [13,14]. Intermediate baffles located between the 4.2 and 77 K stages, which are cooled to 20 K by the liquid helium boil-off, have also been used [15].

The bonded molecular sieve construction [16,17] used in the 4.2 K stage allows liquid cooled pumps to be baked to temperatures as high as 250°C. Gas refrigerator cooled pumps cannot be baked above 70 to 100°C because their construction makes use of indium gaskets. In both cases there is oil-free rough pumping, and the pumps are exhausted by auxiliary pumps during baking. Sputter-ion or turbomolecular pumps are used for the latter purpose.

Cryogenic pumps easily pump large amounts of hydrogen, have no high voltages, and generate no hydrocarbon, metal film flakes, or other contaminants of their own. Gas refrigerators pump all gases well, with the exception of helium. Helium is pumped well only if the adsorbent is cooled to the 4 to 8 K range, a job best done by a liquid helium pump. Liquid pumps are completely free of magnetic fields and vibration. Gas refrigerator pumps have some vibration that results from the displacer motion. They must be carefully damped for use with sensitive surface-analysis equipment such as ESCA or SIMS. Gas refrigerator pumps have the advantage of not requiring liquid cryogens. Becker [18] discusses the operation of a gas refrigerator cryogenic pump in an ultrahigh vacuum molecular-beam epitaxy system. He has shown that it is a viable alternative to conventional pumping systems.

REFERENCES

1. P. A. Redhead, *J. Vac. Sci. Technol.*, **13**, 5, (1976).
2. R. T. Bayard and D. A. Alpert, *Rev. Sci. Instrum.*, **21**, 571 (1950).
3. D. J. Santeler, *J. Vac. Sci. Technol.*, **8**, 299 (1971).
4. R. Nuvolone, *J. Vac. Sci. Technol.*, **14**, 1210 (1977).
5. T. Sigmond, *Vacuum*, **25**, 239 (1975).
6. J. R. Young, *J. Vac., Sci. Technol.*, **6**, 398 (1969).
7. R. L. Samuel, *Vacuum*, **20**, 295 (1970).
8. W. Thompson and S. Hanrahan, *J. Vac. Sci. Technol.*, **14**, 643 (1977).
9. D. G. Bills, *J. Vac. Sci. Technol.*, **6**, 166 (1969).
10. N. Milleron, *Trans. 5th Nat. Vac. Symp. (1958)*, Pergamon, New York, 1959, p. 140.
11. R. D. Gretz, *J. Vac. Sci. Technol.*, **5**, 49 (1968).
12. W. J. Lange and J. H. Singleton, *J. Vac. Sci. Technol.*, **15**, 1189 (1978).
13. C. Benvenuti and D. Blechschmidt, *Japan. J. Appl. Phys.* Suppl. 2, Pt. 1, 77 (1974).
14. H. J. Halama, and J. R. Aggus, *J. Vac. Sci. Technol.*, **12**, 532 (1975).
15. R. J. Powers and R. M. Chambers, *J. Vac. Sci. Technol.*, **8**. 319 (1971).
16. G. E. Greiner and S. A. Stern, *J. Vac. Sci. Technol.*, **3**, 334 (1966).
17. P. J. Gareis and S. A. Stern, *Bulletin l'Institut International du Froid*, Annexe 1966-5, p. 429.
18. G. Becker, *J. Vac. Sci. Technol.*, **14**, 640 (1977).

CHAPTER 12

High Gas Flow Systems

Not all thin-film deposition processes require high or ultrahigh vacuum environments. In fact, some of the most interesting processes take place in the medium and low vacuum range and in addition require a high gas flow. The pressure–pumping speed ranges for several processes as they are currently commercially practiced are shown in Fig. 12.1.

Sputter deposition is done in the 0.5 to 10 Pa pressure range. For certain materials sputtering is the preferred deposition technique. Various plasma processes are performed in the 5 to 500 Pa range. Plasma deposited films are formed from the reaction of chemical vapors in the glow discharge, and plasma etching and reactive-ion etching are of current interest. Polymer films are formed from the glow discharge polymerization of a monomer such as styrene. Plasma etching is a simple isotropic chemical etching process that uses chemically active neutrals in the discharge; for example, the plasma decomposes CF_4 and creates fluorine atoms that react with a silicon surface to form SiF_4, a volatile product that is pumped away. Reactive ion etching is a directional process useful in fabricating semiconductor microstructures. Its directionality is due to high energy ions that are accelerated through a potential gradient toward the surface on which it is believed that, by some mechanism, they increase the reactivity of the chemically active neutral species with the unmasked portions of a thin film. Because the ion-stimulated neutral reaction proceeds at a rate tenfold faster than simple plasma etching, the thin film etches downward much faster than laterally with little undercutting. This allows the etching of fine lines. Reactive ion etching can be performed over

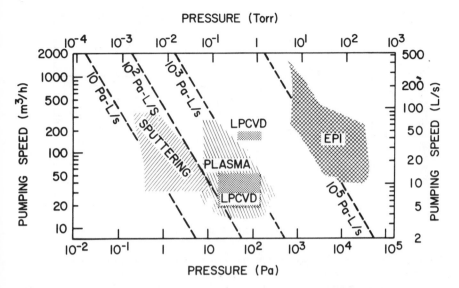

Fig. 12.1 Pressure–speed ranges for some thin-film growth, deposition, and etching processes that require medium to low vacuum and gas flow. Adapted with permission from M. T. Wauk, Applied Materials Inc., 3050 Bowers Avenue, Santa Clara, CA 95051

the entire range of pressures used for sputtering and plasma etching. Any differences attributed to pressure are differences in nomenclature rather than theory. Low pressure chemical vapor deposition (LPCVD) and reduced pressure epitaxy are thermal processes that take place at low pressures. The thermal energy is typically provided by induction heating. LPCVD, which is done in the 50 to 100 Pa range, has attracted wide attention. The high diffusivity of the thermally active species at low pressure improves the transport of the vapor throughout the reactor and allows the growth of more uniform films on a greater number of larger wafers than is possible with atmospheric pressure CVD. Reduced pressure epitaxy takes place in the 500 Pa to atmospheric pressure range. Epitaxial films grown at reduced pressure are higher in quality and have less autodoping than films grown at atmospheric pressure.

Many of the processes done in the medium or low vacuum range also require the use of vapors that are toxic, hazardous, or corrosive. Special precautions must be taken in the design, operation, and mainte-

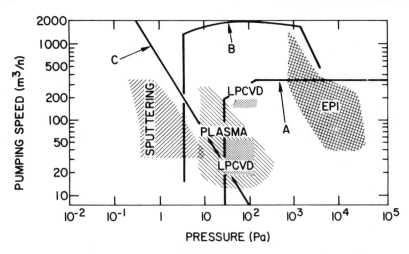

Fig. 12.2 Useful pressure–speed ranges for some pumping systems: (*a*) rotary mechanical pumps; (*b*) Roots pump backed by a rotary pump; (*c*) throttled high-vacuum pump. Adapted with permission from M. T. Wauk, Applied Materials, Inc., 3050 Bowers Avenue, Santa Clara, CA 95051.

nance of these systems to ensure operator safety and equipment protection.

These thin-film deposition and etching processes span a pressure and gas flow range that far exceeds the capability of any one pump. The pressure range of each process is dictated by the physics of the process. Sputtering, for example, cannot commence until the pressure is high enough to initiate a self-sustained glow discharge, but the pressure must be low enough for the sputtered material to reach the anode without suffering a large number of gas collisions. Gas flows ranging from 10^1 to 10^6 Pa-L/s are needed for a different purpose in each of these processes. In some processes the high flow dilutes or replenishes the reactant species and simultaneously flushes away the products of reaction and other impurities; in others it mainly serves to flush away impurities.

The pressure–pumping speed operating ranges of the mechanical pump, Roots pump, and throttled high-vacuum pump are quite different. Rotary vane or piston pumps can pump to the sputtering pressure range and still retain some pumping speed, but below 15 Pa a typical small roughing line is in free molecular flow and backstreaming a large amount of oil vapor is allowed. Rotary vane or piston mechanical pumps are economical up to speeds of 200 to 300 m^3/h and provide effective pumping for all processes in the region deliniated by speeds lower than this value and pressures greater than 15 to 20 Pa. This region is bounded by curve a in Fig. 12.2. For speeds greater than 200 to 300 m^3/h a Roots pump backed by a rotary piston pump is necessary. Again oil backstreaming limits the lowest pressure of operation. At low pressures the roughing line will be in free molecular flow and will allow mechanical pump oil to back diffuse to the Roots pump outlet, creep around the interior surfaces, and enter the process chamber. A molecular sieve trap could be used to reduce the oil backstreaming from the mechanical pump but it generates particulates that contaminate the mechanical pump; it also traps water that will react with certain process gases. Oil contamination may be eliminated by bleeding gas into the roughing line to maintain the Roots pump outlet at a pressure greater than, say, 15 to 20 Pa. This limits the inlet pressure of most Roots–rotary pump combinations to about 3 to 5 Pa [see Fig. 7.8 and the discussion following (7.1) and (7.2)]. The Roots pump has an upper pressure limit of ~1000 Pa which yields the useful operating region outlined by curve b in Fig. 12.2. The only pump systems that will operate at pressures below a few Pa without backstreaming oil vapors are throttled high vacuum pumps. Diffusion, turbomolecular, or cryogenic pumps will maintain chambers at sputtering pressures when the inlet to each is throttled to a pressure below its respective critical inlet pressure. Curve c, in Fig. 12.1 sketches the upper throughput limit of a typical small, throttled high vacuum pump.

This chapter reviews throttled high vacuum systems and unthrottled medium and low vacuum systems. Throttled high vacuum pumps are used mainly for sputtering and to a lesser extent for ion etching. Medium and low vacuum systems are used for ion etching, plasma deposition, LPCVD, and reduced pressure epitaxy. There has been a growing interest in the latter processes and as a result considerable effort has been directed toward the proper pumping of the gases used for these processes. The problem is further complicated by the fact that most of them are explosive, corrosive, or poisonous. An example is used to illustrate proper component placement and safe operation for each pump type.

12.1 THROTTLED HIGH VACUUM SYSTEMS

The pressure range encompassed by sputtering and other plasma processes is above the operating range of all high vacuum pumps. These pumps can be used in the 0.5 to 10 Pa range by placing a throttle valve between the pump and the work chamber. This throttle valve with its low conductance allows gas to flow from the high pressure chamber to the pump while keeping the pressure at the pump entrance below its maximum or critical inlet pressure. A typical sputtering chamber for a 150- to 200-mm-diameter cathode is 500 mm in diameter and 250 mm high. The traditional pumping plant contained a 6-in. diffusion pump, but equivalent cryogenic or turbomolecular pumps have been used. For pumps of this size the maximum throughput will be limited to about 100 to 200 Pa-L/s; larger pumps, although more expensive, are capable of removing gas at a faster rate.

Residual gases pose a greater problem in a sputtering system than in a high vacuum evaporation system because of enhanced plasma desorption of impurities from the walls and because typical sputter deposition rates are much lower than typical evaporation rates. Even when the two processes yield rates of the same order the sputtered films have a greater exposure to residual gas impurities than have films condensed from evaporating sources. Electron and ion impact desorption are efficient at releasing gases from the chamber walls. They are even more effective than mild baking and in the plasma the desorbed gases are likely to exist in the atomic state, where they can easily react with the sputtered film. If hydrogen, for example, is not removed from an argon discharge, the sputtering rate will be reduced [1] and hydrogen will become incorporated in the film [2]. Sputtering discharges with argon or other noble gases can be kept clean by operating the discharge in a static mode with selective pumping or by flowing a large amount of argon through the chamber during sputtering. A static discharge is maintained by exhausting the chamber and refilling it with argon to the operating pressure while other gases are selectively removed by an auxiliary pump located within the chamber. One problem is that the ideal auxiliary pump does not exist. The closest approximation, the titanium sublimator, generates some methane and does not pump noble gas impurities. For these reasons static discharges are not frequently used for sputtering. Here, as for plasma processes with reactive gases, the one reliable technique for impurity removal is viscous flushing.

Viscous flushing will work only if the gas flows through the active sputtering region and chamber and if the arrival rate of impurities from the gas source is much less than the desorption rate from the chamber walls. Lamont [3] has pointed out that high throughput alone does not guarantee adequate flushing; it is necessary that the gas stream velocity be large in the region in which the cleaning action is desired. Contamination originating from within the chamber can be reduced with high gas flow. In the high flow limit the lowest possible level of contamination attainable is that of the source gas. In critical applications the source gas is scrubbed by passing it through a titanium sublimation pump. Because of this, there is little point in flushing a chamber of the size described above at a rate greater than a few hundred Pascal-liters per second with the purest available source gas [4]. Both the source gas cleanliness and the gas flow rate are important to the maintenance of conditions suitable for deposition of pure films.

Gas flushing alone will not adequately clean all the residual gases from the chamber. Contamination-free sputtering requires high vacuum pumping to a suitable base pressure [5] followed by presputter cleaning with the discharge operating and with the shutter covering the samples on which a film is to be deposited. Shirn and Patterson [6] have noted that the flushing time of a typical 6-in. diffusion pumped system is so small (1/7 s) that the system can be cleaned by simply pumping to the process pressure and initiating the glow discharge without pumping to high vacuum. Unfortunately the glow does not clean all surfaces adequately, nor do the surfaces outgas that rapidly. The continued evolution of water vapor from surfaces not exposed to the glow can cause oxygen or hydrogen contamination of deposited films. Most importantly, the only routine way to check for minute leaks in the system without the use of an RGA is to pump to the same low base pressure each time before opening the leak to argon flow. Presputtering cleans by several techniques. If the sputtered material is a getter, it may be allowed to deposit on the chamber walls where it is an effective getter pump [7]. The sputtered material also covers adsorbed gases, whereas the discharge cleans the cathode and other surfaces exposed to the glow.

The value of base pressure and the length of presputtering time are process-, equipment-, and material-dependent. No general observations can be made; for example d'Heurle [8] showed that for aluminum films a 10-minute presputter cleaning was sufficient, whereas Blachman [9] required a minimum of 1 h cleaning time for molybdenum. Some material properties are so critically dependent on film purity that the value of base pressure and minimum presputtering time are of crucial importance in the repeatable fabrication of uniformly high quality films.

The conclusion of this discussion is that high vacuum pumps are needed to establish the initial cleanliness, whereas throttled, high -compression pumps are required for pre-sputter cleaning and contaminant removal from the chamber without permitting backstreaming of hydrogen or a pump fluid. These two requirements can be satisfied by the same pump. Two completely different pumping systems could be used, a high vacuum, low-throughput system for initial cleaning and a medium vacuum system for high gas flow [3], but it would be difficult to design a medium vacuum system with the required hydrogen compression ratio.

The remainder of this section discusses the configuration and operation of diffusion, turbomolecular, and cryogenic pumps for sputtering applications. Ion pumps are not considered because of their inability to handle high gas loads.

12.1.1 Diffusion Pumped

The maximum throughput of a diffusion pump is the product of its critical inlet pressure and its pumping speed at this pressure. The critical inlet pressure is determined by the pressure at which the top jet begins to fail. Typically this happens near 0.1 Pa in a 6-in. diffusion pump. If the pump inlet is maintained at a higher pressure, the pumping action will become unstable and pressure control, difficult. Backstreaming may also increase in the region of jet instability. At 0.1 Pa a 6-in. pump has a speed of about 2000 L/s. Its maximum throughput is therefore 200 Pa-L/s. See for example curve c, Fig. 12.2. A 10-in. diffusion pump with a speed of 5000 L/s at 0.1 Pa has a maximum throughput of 500 Pa-L/s. The maximum throughput is a property of the pump, and is affected by the amount of heat supplied to the boiler by the heater and by the type of fluid. Figure 8.5 illustrates the throughput dependence of the heater power. The maximum throughput will be reduced if the heater power is decreased or if a section of the heater becomes open-circuited. It can be exceeded if the forepump has inadequate capacity. The heater power was chosen by the pump designer to heat a selected pump fluid to a temperature that will produce the desired maximum throughput in the boiler. Use of a fluid other than that for which the pump was designed will cause the critical forepressure to change; for example, a pump that was designed to run with DC-704 but used with DC-705 or Santovac 5 will have a reduced critical forepressure unless the heater power is increased. The pump manufacturer should be consulted to determine the critical forepressure for the fluid and pump in question. This is a concern only in systems that are operated near the forepressure limit. For this reason oversized forepumps are often used to back diffusion pumps used for high gas-flow applications.

The maximum throughput of the pump is not determined by the series conductance of the trap, high vacuum valve, and throttle valve;

it is a property of the pump. The inlet gas leak valve controls the throughput, whereas the throttle valve controls the chamber pressure. Because the throughput at any point in a series flow path is constant, the pumping speed at the chamber under throttled conditions can be calculated as

$$S_c = \frac{P_p S_p}{P_c} = \frac{Q_p}{P_c} \qquad (12.1)$$

A 2000-L/s diffusion pump will have a throttled pumping speed, measured at the base plate of ~ 100 L/s for a chamber pressure of 2.0 Pa. The maximum speed of the pumping system is naturally affected by the series conductances of the trap and opened valves and the pump can operate at its maximum speed only when the inlet pressure is less than the critical inlet pressure. Diffusion pumps with expanded inlets will have greater speed in the high vacuum region but cannot pump any larger gas quantity than a straight-sided pump of the same boiler diameter. The increased speed of the expanded inlet pump will decrease the time required to reach the base pressure only slightly, because the slow surface release of gas molecules limits the removal rate of the outgassing species.

The one question that has not been adequately resolved is the optimum location of the throttle valve in the pump stack. The two possi-

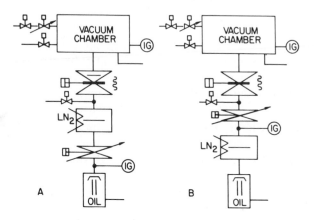

Fig. 12.3 Throttle-valve location in a high-throughput diffusion pump stack: (*a*) throttle located in between the liquid nitrogen trap and the pump, (*b*) throttle located between the liquid nitrogen trap and the chamber.

ble locations of the throttle are illustrated in Fig. 12.3. It may be placed below the liquid nitrogen trap directly over the throat of the diffusion pump (Fig. 12.3a) or it may be located upstream from the liquid nitrogen trap, between the liquid nitrogen trap and the chamber (Fig. 12.3b). The arrangement shown in (a) allows the trap to pump water vapor. At sputtering pressures the trap is in transition, or perhaps viscous flow, and its conductance for pumping water vapor is considerably more than it is in the free molecular flow range, although much of the water vapor does not strike the cooled surface because of vapor-gas collisions. Even though this is not the most efficient location, the trap pumping speed for water vapor is maximized when placed over the throttle.

The trap should capture oil vapors at a greater rate when placed below the throttle because the probability of an oil-trap collision is greater at low pressures where the mean free path is longer than it is at high pressures. The sweeping action of the gas flow from the sputtering chamber is constant and cannot exceed the maximum throughput; therefore as the trap pressure is raised the trap's effectiveness as an oil baffle is reduced by the short mean free path but not enhanced by any increased gas streaming. From a backstreaming consideration the location of the throttle over the liquid nitrogen trap will yield a lower net backstreaming rate provided that the added baffling action of the throttle is the same in both cases. This is not normally so, especially when the throttle is not cooled. If an ambient temperature throttle is located over the trap, any oil vapor condensed on the throttle may escape into the chamber, whereas for a throttle located between the trap and the pump the evaporation may be small compared with the primary backstreaming rate [10]. Rettinghaus and Huber [11] have suggested that the pump may be effectively throttled by a cooled throttle plate located near the cooled cap of the top jet, where it can intercept the primary backstreaming.

No general conclusions in regard to the best arrangement for minimizing backstreaming can be drawn from these considerations. The circumstances of each application must be considered. In many sputtering applications, however, a large water-vapor pumping speed is essential and takes precedence over the differences between two already small backstreaming rates in choosing to locate the throttle below the liquid nitrogen trap.

Occasionally throttled diffusion pumps will be used to pump reactive gases. In those situations a silicone or perfluoropolyether fluid should be used because hydrocarbons will react with many gases. Certain gases will condense on the liquid nitrogen trap and either affect the process or become a safety hazard.

12.1.2 Turbomolecular Pumped

A turbomolecular pumped system suitable for high-flow applications must attain an adequate base pressure as well as exhaust a high gas flow at medium vacuum pressures. The two requirements are different but not conflicting. In Chapter 7 we discussed the requirements of a turbomolecular pump for good high vacuum pumping: high compression ratio for light gases, high pumping speed for all gases, and a deep cooled trap for increasing the water-vapor pumping speed. The selection of a pump for a particular application is then made on the basis of pumping speed and compression ratio data like those in Figs. 7.12 and 7.14. These data describe the speed of the pump and the compression ratio for each gas with zero flow. These data, necessary for characterizing the pump at high vacuum, are inadequate for describing the pump's performance during high gas flow.

High gas flow imposes other constraints on the pump in addition to the obvious one of pumping a considerable quantity of process gas, as in sputtering. The pump must also maintain a reasonably high compression ratio and pumping speed for hydrogen while pumping a high throughput of a heavy gas such as argon. A flow of argon that is large enough to cause viscous flow in the blades or to reduce the rotational velocity will alter the compression and speed characteristics of the pump. Few data are available to show how the design or the operation of the pump are affected by high gas flow.

Visser [4] has measured the compression characteristics of methane in a turbomolecular pump as a function of the quantity of the argon gas flow. His results, shown in Fig. 12.4, demonstrated that a flow of argon from 2 to 8 Pa-L/s increased the compression ratio for methane in a 250 L/s pump. Data at higher argon flow rates have not been obtained.

Figure 12.5 shows the pumping speed for hydrogen and the argon flow rate as a function of argon gas pressure [12]. The argon flow is the product of the argon pumping speed (Fig. 7.15), and the inlet pressure. The pumping speed for hydrogen remained constant for argon inlet pressures up to 0.9 Pa, above which it dropped precipitously. The sudden drop in hydrogen pumping speed corresponded to a similar sharp decrease in the rotational speed of the turbomolecular pump blades in this pump. The exact pressure at which the rotational speed begins to decrease is a function of the pumping speed of the forepump. As the inlet pressure increased, the gas flow in the blades closest to the fore chamber changed from molecular to transition and then to viscous. Near the onset of viscous flow the added frictional drag demanded more torque from the motor; the constant power motor responded by losing speed. The knee of curve (*a*) in Fig. 12.5 will move slightly to the right for a forepump larger than the 35 m²/h pump used here, and to the left for a smaller forepump. When using the pump described here, it is important to keep the pump running at full rotational velocity to maintain adequate hydrogen pumping speed. The blades will run at full velocity as long as the inlet pressure is suitably throttled; for example, the pump used to take the data in Fig. 12.5 should be throttled to an inlet pressure of 0.9 Pa or less when

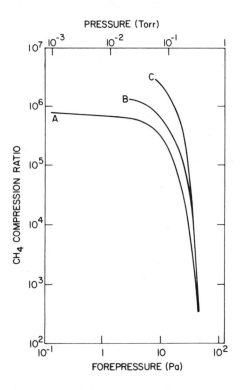

Fig. 12.4 Methane compression ratio as a function of argon gas flow in a Balzers 250 L/s turbomolecular pump. (*a*) $Q(Ar) = 0$, (*b*) $Q(Ar) = 2$ Pa-L/s, (*c*) $Q(Ar) = 8$ Pa-L/s. Adapted with permission from *Trans. Conference and School on Elements and Techniques and Applications of Sputtering*, Brighton, p. 105. Copyright 1971, Materials Research Corp., Orangeburg, NY 10962.

backed by a 35 m³/h mechanical pump. This pump has an inlet argon speed of 225 L/s at an inlet pressure of 0.9 Pa of argon when backed by the 35 m³/h mechanical pump; it yields a maximum argon through-put of 200 Pa-L/s, the same value that can be obtained with a 6-in. diffusion pump. The numerical values recorded for this pump are not directly applicable to any other. Some turbomolecular pumps are driven by constant-speed rather than constant-power motors. Even so, the pressure at which the argon speed decreases is dependent in a similar manner on the size of the forepump. At the time of this writing the dependence of the hydrogen pumping speed on argon flow rate is not available from any pump manufacturer.

The maximum inlet pressure is a function of the size of the fore-pump and the design of the turbomolecular pump. To obtain the maximum suitable throughput from the pump the staging ratio, or ratio of the turbomolecular pump speed to the forepump speed should be small, perhaps as low as 20:1 or 30:1.

Figure 12.6 shows a turbomolecular pump configuration suitable for a sputtering application. A high-pressure ionization gauge of the Schulz-Phelps type is located at the pump inlet, below the throttle valve, to monitor the inlet pressure. The gauge tube should be mount-ed off-axis and its entrance protected by a wire mesh. The gas flow can be increased and the throttle adjusted to maintain the chamber at the desired pressure and the pump inlet pressure should be low enough to allow the pump to run at full rotational velocity. A liquid nitrogen trap for pumping condensables is shown below the high vacuum valve.

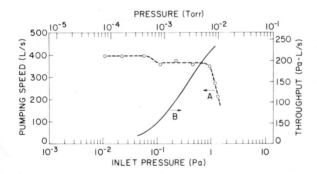

Fig. 12.5 Hydrogen pumping speed (*a*), and argon throughput (*b*) as a function of argon inlet pressure in a Balzers 400 L/s turbomolecular pump backed by a 35-m³/h rotary vane pump. Reprinted with permission from *J. Vac. Sci. Technol.*, **16**, p. 724, J. F. O'Hanlon. Copyright 1979, The American Vacuum Society.

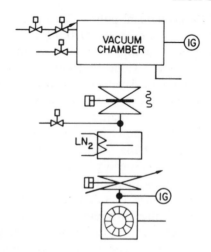

Fig. 12.6 Component location in a turbomolecular pumped sputtering system.

Alternatively, it may be placed in the sputtering chamber as a Meissner trap, where it will have increased pumping speed but will require cyclical venting.

The turbomolecular pump is well suited for use as a high-flow, medium vacuum pump, provided that it is throttled to keep the blades running at full velocity and is backed by an adequately large mechanical pump. For large turbomolecular pumps (1000 L/s and larger) this will usually mean use of a Roots pump and a suitably large foreline. When operated in this manner, it will have a high pumping speed and a high compression ratio for light gases. The turbomolecular pump thus serves as a one-way baffle for light gases and hydrocarbons, whereas most of the pumping is done by the forepump. The turbomolecular pump is not well suited for pumping on high-pressure plasma polymer deposition systems because material may deposit on the rotors, unbalance them, and destroy the pump [13].

12.1.3 Cryogenic Pumped

Like turbomolecular and diffusion pumps, the cryogenic pump should be capable of evacuating the chamber to an adequate base pressure and pumping a large gas flow. The ultimate, or base, pressure is determined by chamber outgassing and the temperature and history of the cold stage. In Chapter 9 it was noted that the temperature was determined by a balance between the refrigeration capacity and the heat loads. The temperature alone was not the only factor that determined the pumping speed for a gas. It was found to be a function of the nature and quantity of gases previously sorbed on the cryo surface.

It was also observed that the heat load carried to the pumping surfaces by the incoming gases under high vacuum conditions (low gas throughput) was insignificant in comparison to the radiant flux. If nitrogen was pumped with a typical cryogenic pump consisting of two cooled stages, one at 80 K, and the other at, say, 20 K, then the time to build up a condensed layer of solid nitrogen 1 mm thick would be about 10^4 h at a pressure of 10^{-5} Pa [14]. Therefore neither the heat load of the incoming gas nor the resulting solid deposit is a major concern in the high vacuum region. This is not so at high gas flows. As the gas flow to a cryogenic pump is increased, the pumping speed changes. Figure 12.7 sketches the pressure dependence of the pumping speed over several flow regions. In the free molecular flow region the pumping speed is constant. At somewhat higher pressures the speed increases due to the increased conductance as the gas enters the transition flow region. Under some circumstances this flow will reach a maximum value (choked or critical flow) that is characteristic of the sonic velocity of the gas. At higher pressures the heat conductivity of the gas becomes large and heat from the walls of the chamber flows to the cooled surfaces by gas collisional energy transfer. As these surfaces warm the sticking coefficients decrease and pumping ceases. This behavior has been observed by Dawson and Haygood [15] for CO_2 and by Bland [16] for water vapor. Loss of pumping speed in a practical pump usually occurs at about 0.2 to 0.4 Pa, where the heat loads exceed the capacity of the refrigerator. The pumping speed of a typical pump in the high pressure region is sketched in Fig. 12.7b.

At high gas throughputs the major heat flow to the cryogenic surface is carried by the gas molecules; for example an argon flow of 180 Pa-L/s corresponds to an incident heat flux of 1 W on a 20 K surface. For gases such as nitrogen which have heats of condensation greater than argon the heat flux will be proportionately larger. Most of this heat flux for nitrogen or argon will be absorbed by the cold stage. As an example consider a two-stage cryogenic pump with surfaces at 20 and 80 K, respectively, in which the argon collides with the 80 K baffle, passes through the baffle, and finally collides with the 20 K surface, where it is pumped. For each mole of argon that flows into the pump a total of 13,862 KJ must be removed by the expander. This value is obtained by taking the difference in enthalpy between 300 and 20 K. See Table 12.1. If all of the argon were to be cooled to 80 K on impact with the 80 K baffle, a total of 4580 KJ/(kg-mole)

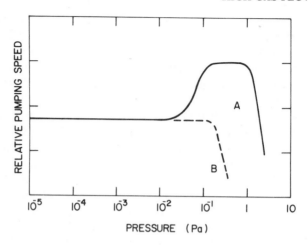

Fig. 12.7 Pressure dependence of cryogenic pumping speed: (*a*) free surface, (*b*) practical baffled pump.

would be removed. This value corresponds to 33% of the total heat that is removed during the pumping process. In practice the gas is not cooled to 80 K on impact with the warm stage because the accommodation coefficient is not unity. Adequate cooling does take place, however. If, for example, the warm stage had an efficiency of 0.5 for cooling the argon, then 85% of the total heat of the gas would remain to be removed by the 20 K stage. The cold stage removes at least 67% of the heat from argon because its vapor pressure so low and its heat of condensation is so large. Table 12.1 gives the approximate total enthalpy at ambient temperature at the nominal temperatures of the first and second stages of a cryogenic pump for argon and nitrogen, and for some gases frequently used in reactive ion etching. The maximum throughput for each gas is determined by the maximum power that can be removed by the stage on which the gas or vapor condenses. Cryogenic pumps do not show the same maximum throughput limits for all gases. CF_4 has a maximum throughput of half the value quoted by the manufacturer for argon. Several other gases used for reactive-ion etching deposit on the 80 K stage and can completely close the baffles and quickly render the pump useless for high vacuum pumping unless it is first regenerated.

At high flow the heat absorbed by the expander will be proportional to the total gas throughput until the pressure is large enough for heat conduction from the walls to be appreciable. Heat conduction from

Table 12.1   Enthalpy of Gases Frequently Pumped
at High Flow Rates[a]

Gas or Vapor	Total Enthalpy in Vacuum kJ/(kg-mole)		
	300 K	80 K	20 K
Ar	13,950	9,370	**88**
N_2	15,580	9,190	**134**
CCl_4	49,750	**2,950**	738
CF_4	26,000	16,000	**670**
CF_3H	30,750	**2,678**	670
CF_3Cl	31,300	**20,230**	678

[a]Approximate total enthalpy at ambient temperature and at
the nominal temperatures of the first and second stages
of a cryogenic pump for several gases and vapors used in
high gas flow applications. The enthalpy is shown in
bold for the surface on which the gas or vapor solidifies.

the walls begins at a Knudsen number of 1 and reaches its maximum
or high pressure value at Kn = 0.01. This heat flow is added to the
heat flow due to heat of condensation but not in a linear manner.
Most of the heat delivered by gas conduction flows to the 80 K stage.
This added load on the warm stage reduces the refrigeration capacity
of the cold stage. For high gas flow applications a cryogenic pump
should be designed to minimize the heat load from the walls so that
the expander may be most efficiently used to remove the heat of
condensation of the incoming gas. For a pump of typical dimensions
the Knudsen number for air at 0.2 Pa is about 0.3. At this value the
heat capacity is a substantial fraction of the high pressure value which
allows several watts to flow from the chamber walls to the warm stage.
To reduce this heat load the warm stage is completely insulated except
for the baffled entrance to the cold stage or surrounded with an added
liquid nitrogen shroud. See Fig. 12.8. In either case the heat flow to

the warm stage is reduced. The best insulation material is one whose surface has a low emissivity. A liquid nitrogen shroud provides an alternative sink for the conductive heat flow and substantially eliminates this load on the warm stage.

In some high gas flow situations continued pumping of hydrogen is important but does not always take place. Pumps are designed so that the inner surface of the cold stage is baffled from the warm stage and covered with a layer of activated charcoal. This surface will pump hydrogen, neon, and helium effectively if it is adequately cooled. At 10 K there is adequate sorption capacity for hydrogen on activated charcoal or molecular sieve. Argon will also be pumped on this surface. To avoid or reduce the argon pumping on the inner surface it is baffled from the remainder of the system. If the baffle is optically dense to keep argon reaching the inner surface, the hydrogen pumping speed will be reduced, and if the baffle is open the argon will readily pass through, condense on the inner surface, and cover the sorbent with solid argon. All pump designs are a compromise between these two concerns. In a high flow application, however, the sorbent surface will become coated with argon rather quickly, regardless of the nature of the baffling. Once the sorbent becomes coated, cryotrapping is the only mechanism by which hydrogen can be pumped. Hengevoss[17] has shown that the cryotrapping of hydrogen in argon is strongly temperature dependent and becomes nill for solid argon temperatures greater than 20 K. The value of argon throughput which will keep the second stage temperature below 20 K may be below the stated maximum value for a particular pump.

The constraints placed on a cryopump system for high gas flow are considerably different than those placed on pumps used for high vacuum. The system requires a throttle valve to keep the pressure in the

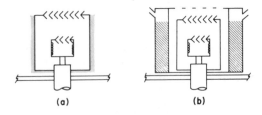

(a)   (b)

Fig. 12.8 Sectional view of (a) gas refrigerator cryopump with insulation around the warm stage, and (b) gas refrigerator cryopump with a liquid nitrogen shroud surrounding the warm stage.

cryopump below a value of 0.2 and 0.4 Pa and the pump must be able to handle the conductive heat load from the walls. One way of reducing this heat contribution is to use a two-stage gas refrigerator of moderate capacity surrounded by a liquid nitrogen baffle, whereas another design involves only the use of a large capacity gas refrigerator and insulation. Regeneration of a high gas flow system will obviously be more frequent in a pump used for high gas flow than in a pump used for high vacuum, but automatic controllers are available for performing this function on idle time. If a cryopump is used at the design-limit argon throughput, the cold stage will be heated to a temperature at which it will not pump hydrogen or helium; either supplemental pumping or throughput reduction will be necessary to pump these gases. Furthermore, the pump's maximum throughput is dependent on the vapor pressure, specific heat, and heat of sublimation of the gas or vapor being pumped. Cryogenic pumps also suffer from the phenomenon of overloading. Irreversible warming of a pump can be triggered by a gas burst entering a pump that is running near its throughput limit. It will then cease to pump and require regeneration. Cryogenic pumping of toxic or explosive gases presents serious safety concerns. If the pump were to suddenly warm, a large quantity of gas could be emptied into the exhaust system. If the pump is condensing an explosive gas, operation of an ionization gauge in the pump body during release of the gas presents a serious problem. The ion gauge tube should be located outside the pump body and interlocked so that its filament cannot be operated when the gas is being released. A prudent operator would not choose cryogenic pumping for certain gases. These concerns should be understood before cryopumps are chosen for high gas flow applications.

12.2 MEDIUM AND LOW VACUUM SYSTEMS

Flowing gas environments at reduced pressures are used in low pressure chemical vapor deposition, reduced pressure epitaxy, and several plasma processes. The common link in all these processes is the pumping of hazardous gases. Rotary and Roots pumps are ideally suited to maintain the pressure dynamically in the 10 to 10^4 Pa region. For LPCVD and plasma processes the flow of gas through the reactor is required primarily to replenish the reactant continuously which is consumed during the process of growth or etching. Polycrystalline silicon deposition, for example, typically requires a flow of 300 Pa-L/s to grow the layer at a rate of 20 nm/min on both sides of a batch of

100 wafers with 76-mm diameters [18]. The rate at which gas is consumed is dependent on the number of wafers, the area of the chamber, the layer growth rate, and the dilution and depletion of the reactant gas. In the numerical example just given about half the flow is hydrogen carrier gas. The remainder of the flow is reactant gas of which about 30% is consumed in the process.

In some commercial reduced pressure epitaxy processes a large gas flow dilutes the reactant to produce growth rather than etching. During low pressure silicon epitaxy a reactant gas flow of 300 Pa-L/s is typically diluted in approximately 10^5 Pa-L/s of hydrogen and flowed through the reactor at pressures of 10^3 to 10^4 Pa. Reduced-pressure epitaxy is a relatively new technique and it is not clear that a carrier gas flow of this magnitude will remain an essential part of the process.

Most of the gases or vapors used in these processes are toxic, explosive, or corrosive. In silicon technology LPCVD uses hydrides of silicon, phosphorus, arsenic, and boron, as well as O_2, HCl, and NH_3. Reduced pressure epitaxy uses hydrogen, silicon chlorides, and some of the aforementioned compounds. Plasma growth and etching processes also use fluoro- and chlorocarbons. These compounds, alone or in combination, are dangerous in low-pressure systems. They present a safety risk to operating personnel as well as to the operation and maintenance of the equipment. They can ignite, explode, corrode, escape to the atmosphere that the operating personnel breathe, react with residual gases and pump fluids to form deposits, and cause pump failure.

The deleterious effects of these gases on the equipment and personnel may be minimized by preventing the concentration of noxious gases and particulates, proper selection and maintenance of pumps and fluids, removal of residual gases, and safe design and operation of the equipment. Figure 12.9 [19] shows an idealized system for pumping large quantities of reactive gases. Gas flushing and venting are extensively used. The build up of poisonous or explosive gas concentrations can be prevented by flushing with a nonreactive gas such as nitrogen at the filter, at the mechanical pump ballast, over the oil reservoir, and at the exhaust line. It is important to purge the gas spaces in the filter, the piping, and over the oil reservoir when using explosive gases. If not done, an explosion could result on exposure to atmosphere. A side benefit of gas ballasting is that it causes the pump to work harder, heat, and more thoroughly expel dangerous high vapor pressure substances from the oil. A gas purge downstream from the Roots pump is necessary to maintain the backing pressure high enough to prevent backstreaming. The outlet of the Roots pump should always be main-

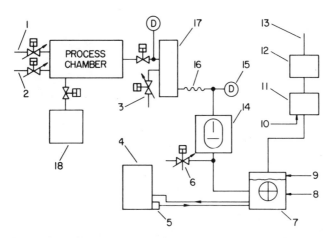

Fig. 12.9 Medium vacuum system for pumping reactive gases: (1) process gas inlet, (2) flush gas inlet, (3) flush gas inlet, (4) external oil reservoir, (5) oil filter, (6) roughing-line flush gas, (7) rotary piston pump, (8) gas ballast, (9) oil reservoir flush, (10) mist-filter flush, (11) mist filter, (12) effluent scrubber, (13) exhaust line, (14) Roots pump, (15) diaphragm gauge, (16) vibration isolation bellows, (17) particulate filter, (18) auxiliary high vacuum pump. Adapted with permission from P. Whelan, Verity Instrument Co., Richardson, TX 75080.

tained at a pressure greater than 15 to 20 Pa to prevent backstreaming. Nonreactive gas flushing at all these points, including the building exhaust, is needed whether or not the process gas is flowing in the system. If the flush gas flow is stopped when the system is not in use, air will back diffuse into the exhaust line, over the oil reservoir, and in through any leaks, where it will be ready to react instantly with the process gas. Systems using explosive gases should be prepurged with nitrogen for 15 minutes before initiating gas flow. A flame arrestor may also be added at the mechanical pump exhaust.

Particulate formation is a problem in some plasma and thermal deposition processes. Particulates can form as a result of a desired reaction between two process gases or as an undesired reaction between a process gas and residual gas. The reaction of silane with residual air and water vapor to form colloidal SiO_2 is one well-known undesired reaction. In such cases a woven fiberglass filter can be installed at the exit of the reaction chamber to collect some of the particulates. The filter should contain no paper or organic binder that could react with the process gas and generate more particulates [13]. Particulate formation may continue throughout the system as the reactant gases combine with residual gases from the chamber, leaks, or back diffusion. Cryogenic traps could be used to trap residual and unconsumed reactant gases, but their concentration and disposal presents a severe safety problem. In some processes the plasma decomposes the reactant gas and forms a fragment with a high vapor pressure. This gas may condense on the liquid nitrogen trap at high pressures and reemit at low pressures, thus limiting the system base pressure.

A water-mist scrubber is effective for removing effluent gases and particulates because many are water-soluble [13]. The liquid discharge from the effluent scrubber, which may contain dangerous chemicals such as hydrochloric or hydrofluoric acid, should be disposed of in a safe manner. The water-mist scrubber also has the advantage of generating a relatively low back pressure because it does not clog. Scrubbers which bubble the effluent through a liquid are not recommended because they can cause a high back pressure that can damage the mechanical pump. Chemical filtration stages must be added to remove substances not soluble in water.

The selection and maintenance of the pumps is as important as gas flushing and particulate filtration. Figure 12.2 outlines the regions in speed-flow space for which the Roots and rotary pumps are suitable. Rotary mechanical pumps are free of backstreaming for process pressures as low as 15 to 20 Pa. For corrosive gas applications the rotary *piston* pump is strongly recommended. The piston pump has larger clearances than the vane pump and its rotor is geometrically constrained from contacting the stator. It has no vanes to stick or rub against the stator housing and will pump as long as its rotor can be turned. Because of these reasons the piston pump is more reliable and tolerant of particulates than the vane pump. The ultimate pressure of the mechanical pump is of little concern, because it will be operated in the viscous flow region at pressures above its ultimate. In large systems a Roots pump will pump high gas flows or operate at somewhat lower pressures than are possible with the mechanical pump. Because many reduced pressure processes are thermal rather than plasma, and use hydrogen or a hydride, the low hydrogen compression ratio of a Roots pump does not present the problem in these processes that it does on sputtering.

Maintenance on hazardous gas pumping systems needs to be done with care. Safe handling of a pump that has been exposed to a toxic gas in a film deposition, etching, or ion implantation application is important. Toxic chemicals such as arsenic, boron, and phosphorus accumulate in the pump fluid. Not only should the fluid be safely discarded or stored but maintenance personnel must be made aware of the history of the fluid and the pump in order to implement safe handling procedures. The fluid should be removed and the pump flushed before sending it to a repair facility. The nature of the contamination should be clearly indicated on the pump and waste-fluid container. Whelan [19] has pointed out the clear need for designs with demountable pipe connections and access ports that would permit ease of maintenance.

Particulates can be generated all through the system, including the exhaust line. In plasma etching of silicon with CF_4, SiF_4 is generated. If it contacts air at the exhaust connection, it will react and form SiO_2 that can completely plug the exhaust line.

Particulate filters need routine replacement. They should be flushed with nitrogen for at least 15 min before changing to remove any gas which may burn when contacting air or harm nearby personnel. A dust mask and gloves should always be worn when changing filters.

Mechanical-pump fluids can react with the process gas and become unstable. They can thicken, acidify, break down, and fill with sludge deposits and particulates. Regardless of its initial viscosity, the viscosity of an oil will increase as it becomes contaminated. To date no solution has been devised that will totally prevent pump cleaning and fluid changing. To extend the oil change interval an external oil reservoir and oil filter are recommended. This will increase the volume of oil by a factor of 3. Diatomaceous earth or molecular sieve filters have the added advantage of removing chemicals. A choice can be made between hydrocarbon oil, and a phosphate ester or inert fluid. Hydrocarbon oils are dangerous in most applications, especially those requiring oxygen. They are safe for hydrogen but not for many of the gases used with a hydrogen carrier flow. Hydrocarbon oils quickly react with chlorine or fluorine. Silane will not react with a hydrocarbon oil, but it will react with any residual gas disolved in the oil. Phosphate esters, like hydrocarbons will react with many process gases, especially chlorine. They also react with water vapor to form acids. The only fluids that are completely safe with reactive gases are fluorochloro- and chlorocarbons which are highly recommended for these applications. The proper use of these fluids in mechanical pumps is discussed in Section 7.1. Inert chlorofluorocarbon and perfluoropolyether fluids are stable but must still be filtered and changed when they become contaminated. Because a rotary piston pump requires a fluid of greater viscosity than a rotary vane pump, it can take advantage of high viscosity, moderately priced fluorochlorocarbon fluids. For these applications the effect of the fluid type on the ultimate pressure (2 - 4 Pa) is unimportant, because the pump will be operated at pressures well above its ultimate (10 Pa). The Roots pump requires fluid only to lubricate the bearings and drive gears and not to make a seal between the rotor and stator. Even though the fluid does not circulate in the pumping cavity, process gas can enter the lubrication chambers and react with the fluid. Roots pumps, like rotary pumps, require a nonreactive lubricating fluid when pumping reactive gases.

One important cause of system- and pump-fluid particulate contamination is the reaction between the process and residual gases remaining in the reactor after loading the wafers. Figure 12.10 sketches a typical process. The initial pumping and backfill steps do not remove all the residual air from the process chamber. Pumping to 1 Pa before a 50 Pa process leaves a 2% contaminant level. The contaminants, which consist mainly of water vapor, are flushed through the system, where they can react to cause particulates and deposits. Installation of an auxiliary high vacuum system for initial evacuation of the reaction chamber will prevent much of the residual-gas contamination from reaching the main pumping system [13], especially when accompanied by a mild bake. This is an important and often overlooked method of reducing the otherwise high maintenance level usually associated with these systems. Alternatively, an atmospheric pressure helium flush and bake could be used. The addition of an auxiliary pumping or flushing package and the use of an inert oil will increase the system cost—an increase that will be repaid readily by longer maintenance intervals and

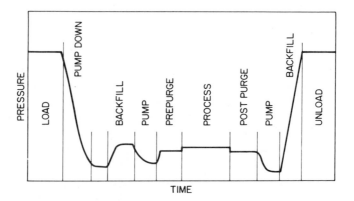

Fig. 12.10 Process steps in a typical medium vacuum reactive gas system. Reprinted with permission from M. T. Wauk, Applied Materials Inc., 3050 Bowers Avenue, Santa Clara, CA 95051.

increased product throughput. A small diffusion, turbomolecular, or cryogenic pump will serve well as an auxiliary pump.

Careful equipment design and personnel training are vital to the safe operation of reactive gas systems. Gas sources and associated piping should be constructed and interlocked according to established practices. In a multipurpose system a hazardous situation could result if gases from different processes made contact with one another or air. Careful interlocking and use of backcheck valves is necessary. Reactions between process gases and residual air in the system can be minimized by interlocks that prevent flow of the process gas without gas flushing. Interlocks are absolutely necessary if the process gas will explode or burn when contacting air. Many potential problems that could result from the use of these gases can be prevented by making the operating personnel aware of their dangers and of proper safety and emergency procedures.

REFERENCES

1. E. Stern and H. L. Caswell, *J. Vac. Sci. Technol.*, **4**, 128 (1967).

2. J. J. Cuomo, P. A. Leary, D. Yu, W. Reuter, and M. Frisch, *J. Vac. Sci. Technol.*, **16**, 299, (1979).

3. L. T. Lamont, *J. Vac. Sci. Technol.*, **10**, 251 (1973).

4. J. Visser, *Trans. Conference and School on Elements and Techniques and Applications of Sputtering*, Brighton, November 7–9, 1971, p. 105.

5. L. I. Maissel, *Physics of Thin Films*, **3**, G. Hass and R. E. Thun, Eds., Academic, New York, 1966, p 106.

6. G. A. Shirn and W. L. Patterson, *J. Vac. Sci. Technol.*, **7**, 453 (1970).

7. H. C. Theuerer and J. J. Hauser, *Appl. Phys.*, **35**, 554 (1964).

8. F. M. d'Heurle, *Metall. Trans.*, **1**, 625 (1970).

9. A. G. Blachman, *Metall. Trans.*, **2**, 699 (1971).

10. The author is indebted to Dr. G. Rettinghaus of Balzers High Vacuum for this discussion.

11. G. Rettinghaus and W. Huber, *Vacuum*, **24**, 249 (1974).

12. J. F. O'Hanlon, *J. Vac. Sci. Technol.*, **16**, 724 (1979).

13. J. Vossen, Dry Etching Seminar, New England Combined Chapter and National Thin Film Division of AVS., Oct. 10-11, 1978, Danvers, MA.

14. G. Davey, *Vacuum*, **26**, 17 (1976).

15. J. P. Dawson and J. D. Haygood, *Cryogenics*, **5**, 57 (1965).

16. M. E. Bland, *Cryogenics*, **15**, 639 (1975).

17. J. Hengevoss, *J. Vac. Sci. Technol.*, **6**, 58 (1969).

18. M. T. Wauk, Annual Symposium, Delaware Valley Chapter, AVS, May 18, 1978, Princeton, NJ.

19. P. Whelan, Annual Symposium, Delaware Valley Chapter, AVS, May 18, 1978, Princeton, NJ.

CHAPTER 13

Economic Analysis

This chapter discusses some of the economic realities of owning and operating vacuum systems from a user's perspective. Traditionally, this facet of plant operation has been a concern only to those working in a production environment who were interested in minimizing manufacturing costs. Until the commercial development of ion, turbomolecular, and cryogenic pumps comparative economic analysis was not possible because the only available pump was the diffusion pump. However, the development of viable alternatives to the diffusion pump coupled with the dramatic increase in utility costs during recent years, has made a wider range of users more conscious of the total cost of operating a vacuum plant. A production engineer is interested in minimizing the cost of a process step. The total production cost includes expenditures for capital, utilities and services, maintenance, and labor, as well as factors such as the space occupied by the equipment, its product throughput, and yield. The research or development engineer is becoming more aware of utility and liquid cryogen costs as laboratory management stresses efficiency as an alternative to spiraling operating costs.

The current state of the art of vacuum technology provides enough diversity so that more than one pumping technology generally satisfies the technical requirements of a particular application. The industrial user who is considering the installation of a large number of complex vacuum systems has the help of cost analysis and engineering departments that are unavailable to the small user. This chapter presents several examples of capital expenditures, operating costs, and investment choices but does not attempt to make absolute judgments. These examples will provide some understanding of the total cost of owning and operating a vacuum plant and the factors worth considering in the choice of a system.

For many users neither the methods nor the examples will be readily applicable because other factors will influence the decision. If cash is limited, the choice will have been made in favor of the system with the lower capital cost, regardless of other costs incurred over the life of the project. If a plant already has liquid nitrogen storage and distribution that is not operating at full capacity, the unit cost of adding a few systems requiring liquid nitrogen would be less than the unit cost of adding a large number of systems, that would require a new or expanded liquid nitrogen storage facility. If the user is forced to cut utility expenses, the equipment will be shut-down on idle time or replaced with more efficient equipment. This chapter reviews briefly the economics of owning and operating high vacuum and high gas flow systems. The economics of ultrahigh vacuum is not discussed because performance is the only concern.

13.1   HIGH VACUUM SYSTEMS

For clarity and simplicity this discussion is limited to one class of system; that is, high vacuum systems like those used for production of high-quality thin films. Typically, these systems are capable of rapid pump-down to the 10^{-4} or 10^{-5} Pa region and are fabricated with elastomer sealed joints. All the systems discussed here are capable of clean and dry ultimate pressures of approximately 5×10^{-6} Pa and are bakeable to 40 or $50°C$, a temperature easily attained by circulating hot water through coils attached to the surface of the work chamber. In systems of this type the dominant gas load is air, the water vapor absorbed on the work chamber, fixturing, and product, and the gas load evolved during the operation, for example, of a resistance, induction, or electron-beam heated source. The economics of diffusion, turbomolecular, ion, and cryogenic systems are considered here.

13.1.1   Capital Investment

Table 13.1 lists specifications of the systems for which capital cost data are presented. All pump stacks are complete, operational, and automatically controlled. The automatic controllers are capable of

Table 13.1 Components Included in the Capital Costs
of Four Types of Vacuum Systems

Components Common to All Systems

All piping and flanging, including baseplate, less work chamber and hoist. Electropneumatically controlled valves, automatic vacuum system controller, pressure gauging, and cabinet and electrical installation.

Components Unique to Each System

Diffusion pumped	*Ion pumped*
Diffusion pump	Sputter-Ion pump
Water baffle	Titanium sublimation
Liquid nitrogen trap	pump
Mechanical pump	Sorption pump
Turbomolecular pumped	*Cryogenic pumped*
Turbomolecular pump	Cryogenic pump
Mechanical pump	Helium compressor
Liquid nitrogen trap to	Mechanical pump
pump water vapor	Automatic regeneration

---

starting, stopping and fault protecting the pumps but are incapable of process control. All gauges, electrical wiring, plumbing, and cabinetry costs are incorporated, but the cost of the glass or steel chamber, tooling, process control, and hoist is not. Some of the components are unique to each system. The diffusion pumped systems are well trapped. Liquid nitrogen traps and water baffles are included, although some smaller systems use only a cold cap and not an additional partial water baffle. Ion pumped systems have titanium sublimation pumps with water-cooled shrouds and sorption pump roughing, and turbomolecular pumped systems have liquid nitrogen traps that will increase their water pumping speed. Both types of cryogenic pump, the gas refrigerator, and the liquid-nitrogen augmented gas refrigerator pump are included in these cost tabulations.

Figure 13.1 shows the results of this review of the single-unit costs of complete pump stacks as a function of their pumping speed at the entrance to the work chamber. It is more realistic to choose net

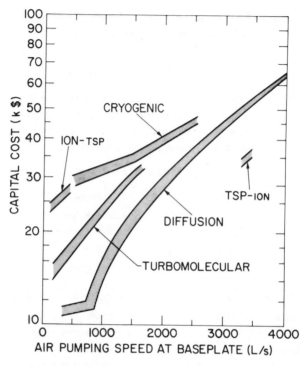

Fig. 13.1 Capital costs of automatically controlled vacuum systems as a function of their net air pumping speed at the chamber entrance. The components included in each system are described in Table 13.1.

pumping speed rather than throat pumping speed or flange size as the independent variable because of the differences in Ho coefficients. A cryogenic pump has its throat speed reduced only by the impedance of a gate valve, whereas the diffusion pump has additional traps and baffles. The range of costs for a particular pump type and size reflects the differences in pumping system design, quality, and pricing between vendors. Data points were not shown to preserve anonymnity within the group. In addition to the four categories discussed, a large titanium sublimation system is shown. The latter system also contains a liquid nitrogen reservoir and a smaller ion pump. The prices given are in 1979 dollars for single-lot quantities. Some price reductions are available on multiple-lot orders. There will be less price reduction for quantity diffusion pump orders than for other technologies because of the relatively advanced state of the development of diffusion pumps and competition in the market place. As the market for cryogenic and turbomolecular pumped systems expands, both will be manufactured with increasingly larger pumping speeds, and their price differences, in relation to each other and to the diffusion pump will change from the values shown in Fig. 13.1. Complete, automatically controlled ion pumped systems have more limited applications and are now available as stock items only in a speed range that is less than 500 L/s.

Clearly, diffusion pumps require less capital than most other systems of similar net pumping speed. The abrupt change in the slope of the diffusion pump costs for speeds less than 1000 L/s reflects the fact that the capital cost of the 4- and 6-in. pumps differs only by the difference in costs of the pump body, valve, trap and mechanical pump. For this reason many manufacturers are discontinuing 4-in. systems as standard catalog items. However, 4-in. pumps are still widely produced because they are suited to many other conductance-limited pumping applications. Systems with 16-in. and larger diffusion pumps are most often especially designed to meet the individual customer's needs, although many of the components are catalog items. Over the speed range shown the ion, turbomolecular, and cryogenic systems are between $7000 and $22,000 more expensive than similar diffusion pumped systems. The actual cost of custom systems may be greater than stock systems because of the one-time engineering costs incurred in their individualized design. A seemingly minor change in specifications from the standard design, such as the substitution for the standard forepump of one of increased capacity or different manufacture, would require reworking the foreline and electrical drawings before construction could begin. In extremely large systems such as space chambers, cryogenic pumping is the least expensive technique.

In this section we have taken care to present comparative data only for typical *complete* automatic pump stacks and not for the main component of each system. It is tempting but misleading to plot the cost of the high vacuum pump versus throat speed for the different pump types. As noted earlier, it would not be a true measure of the base plate speed nor would it be a true measure of the system cost differences. One pump type cannot be substituted for another without extensive changes in control circuitry, nor can the size of the high vacuum pump be increased without increasing the capacity of the ductwork, valves, forepump, and electrical distribution systems. Capital costs, of course, are but one factor in the total cost of operating a high vacuum plant. Nevertheless, this expenditure is of great interest to those users who are operating on a limited capital budget. Now let us examine typical utility costs.

13.1.2   Operating Costs

Utility costs are of more concern than capital costs to most vacuum
system operators. Representative operating costs for two sizes of
vacuum system, 1000 L/s and 2000 L/s, are presented here. The type
of pump, the pattern of filling cryogenic traps, and the utility rates all
affect the total operating costs. These examples illustrate how the
utility costs are distributed in a given system and compare the standing
costs of several systems.

Table 13.2   Operating Costs of High Vacuum Pumps[a]

Pump Type	Water ($0.17/m^3)	Electrical Power ($0.04/kWh)	($0.08/kWh)	Liquid Nitrogen ($0.12/L)	($0.50/L)
Diffusion					
Pump	4.80	14.75	29.50		
Trap				15.10	63.00
Mech. Pump		4.00	8.00		
Total	**4.80**	**18.75**	**37.50**	**15.10**	**63.00**
Cryogenic[b]					
Refrigerator	4.80	7.40	14.80		
Mech. Pump		0.10	0.20		
Total	**4.80**	**7.50**	**15.00**		
Cryogenic[c]					
Refrigerator	4.80	7.40	14.80		
Trap				14.00	58.50
Mech. Pump		0.10	0.20		
Total	**4.80**	**7.50**	**15.00**	**14.00**	**58.50**
Turbomolecular					
Turbo	4.80	4.00	8.00		
Trap				9.60	40.00
Mech. Pump		4.00	8.00		
Total	**4.80**	**8.00**	**16.00**	**9.60**	**40.00**

[a] Weekly operating costs of four types of high vacuum pumps having net air pumping speeds
of 1,000 L/s:

[b] gas refrigerator only,

[c] liquid nitrogen assisted cooling around the 80 K stage.

The first group of pumping systems for which operating cost data are presented consists of four systems, each with a net pumping speed of 1000 L/s at the entrance to the work chamber. In making these cost calculations, it was assumed that the systems were used to provide the vacuum environment for a process one shift each day, five days each week, with one pump-down from atmosphere each day. Furthermore, it was assumed that the diffusion pump and the turbomolecular pump were in operation seven days a week. The gas-refrigerator powered cryogenic pumps were assumed to be shut down one day a week for regeneration. This is a bit frequent. When liquid nitrogen costs were included, it was for eight hours a day, five days a week for the turbomolecular pump, 24h a day, seven days a week, for the diffusion pump, and 24h a day, six days a week, for the cryogenic pump. Table 13.2 details the costs of operating these pumping systems on the basis of the three highest priced utilities: electrical power, liquid nitrogen, and cooling water. Electrical rates of $0.04/kWh and $0.08/kWh were used to illustrate regional differences. Liquid nitrogen rates of $0.12/L and $0.50/L were assumed. These rates include boil-off losses due to transportation, storage, and transfer, as well as maintenance costs, and bracket the costs of the large and small user, respectively. The assumed loss rates for the two classes of user were quite different. The large user must maintain elaborate storage and distribution facilities. The small user loses only the boil-off from a dewar and the transfer loss from the dewar to the trap but pays a higher unit cost. Estimates by Gardner [1] show that the very small user may pay as much as three to four times the value assumed here for the small user. Electrical power consumption was taken from manufacturers' literature or from operating systems. The liquid-nitrogen consumption rates assumed were taken from typical operating systems. The amount of liquid nitrogen consumed in a production system with dirty traps of high emissivity and a large process heat load is much more than that in a system with a clean, dry bell jar and a cold water baffle between the trap and pump. The actual cooling water flow is usually two to three times the amount stated by the manufacturer because of the nature of the flow switches used for fault protection.

The helium gas refrigerator cryogenic pump has an advantage over other pumps in that many designs do not require the use of a nitrogen trap, but those pumps that have nitrogen traps connected to or surrounding the warm stage of the refrigerator do cool the gas more effectively and perform better in high gas-flow applications than pumps that do not use liquid nitrogen. These liquid nitrogen traps

Table 13.3   Annual Operating Costs of High Vacuum Pumps[a]

Pump Type	Electric: $0.04/kWh Liquid Nitrogen: ($0.12/L)	($0.50/L)	Electric: $0.08/kWh Liquid Nitrogen: ($0.12/L)	($0.50/L)
Diffusion	1932	4327	2870	5265
Cryogenic[b]	615	615	990	990
Cryogenic[c]	1315	3540	1690	3915
Turbomolecular	1120	2640	1520	3040

Example: Diffusion pump. $0.04/kWh electric; $0.12/L LN_2.
From Table 13.2: $(4.80 + 18.75 + 15.10)/W × 50
weeks = $1932/y.

[a] Annual operating costs for four types of high vacuum pump having net air pumping speeds of 1000 L/s:
[b] gas refrigerator only,
[c] liquid nitrogen assisted cooling around the 80 K stage.

must be continuously cooled. The data presented in Table 13.2 are summarized in Table 13.3 on an annual (50-week) basis. Examination of these data shows that the variation in electrical power and liquid nitrogen cost can have a two- to threefold effect on the annual operating cost. The diffusion pump is considerably more expensive to operate than either of the other pump types, based on the assumptions in these calculations. In some applications a Meissner trap is located in the processing chamber. Meissner traps must be vented each time the chamber is vented to atmosphere to avoid water-vapor condensation. In a large, rapid-cycle production system this results in liquid nitrogen consumption rates that are ten times greater than those in a trap located on the high vacuum side of the gate valve. Consumption rates of 75 to 100 liters per 8-h shift are not uncommon.

The second example compares the same costs for three systems of approximately 2000 L/s air pumping speeds. The 10-in. diffusion pump is liquid-nitrogen trapped and has a pumping speed at the entrance to the chamber of 2400 L/s. The 2500 L/s cryogenic pump is operated by a 5-kW helium gas refrigerator. The third is a 1900 L/s turbomolecular pump. All pumps are backed by a 60-m^3/h rotary vane pump, but the rotary pump attached to the cryogenic pump is operated only during the roughing cycle. Pumps of the sizes described here are found on chambers of ⅓ to ½ m³ volume. Table 13.4 tabulates the annual costs of electrical power, cooling water, and liquid nitrogen for the three systems on the assumption of a liquid nitrogen cost of $.12/L, cooling water cost of $.17/m³, and electrical power costs of $.04 and 0.08/kWh, respectively. Again, as in Table 13.3 for the smaller systems, the diffusion pumped system is the most expensive consumer of utilities. The gas refrigerator cryogenic pump, however, is the cheapest of the three because it uses no liquid nitrogen. Obviously, as the cost of liquid nitrogen increases the cost of operating the helium compressor becomes increasingly more attractive.

The cost studies for the three types of 1000 L/s and 2000 L/s pumping system, which are shown in Tables 13.2, 13.3, and 13.4, are not intended to be precise or judgmental. Instead, they give the ranges of utility costs to be expected for typical operating conditions. It is a simple matter for the user to repeat these calculations for a particular system of interest according to local utility costs and equipment consumption rates. In the preceding discussion ion pumps have not been mentioned. They are not generally used for quick-cycle, high-speed production systems but it should be noted that they are the least expensive high vacuum pump to operate. A 100 L/s ion pump power supply delivers only 50 mW at a pressure of 10⁻⁶ Pa.

13.1.3 Return on Investment

Return-on-investment calculations are performed in different ways in each industry but all strive to determine where to invest capital to obtain the best return. Before becoming engrossed in the details of a return-on-investment calculation we must be aware of the range of its application and usefulness. Return-on-investment calculations are expensive and are performed only on large projects. A great deal of speculation is involved in forecasting the annual inflation rate of utilities, labor, repair, and end-of-life salvage values of the equipment. The more accurate the estimate, the more time and money required for its preparation [2]. This is a special concern in newer technologies that have no long history on which to base accurate projections. The time to break even is not only dependent on these relative cost projections but also on the desired percentage return on investment. To some extent we can determine the outcome of the calculation by a judicious choice of the input variables. In addition, the proposed cost savings must be related on a per-unit-product basis, the total savings divided by the product throughput over the lifetime of the equipment.

Table 13.4   Annual Operating Costs for Three Types of Pumping Systems[a]

System	Electrical Power Cost	
	$0.04/kWh	$0.08/kWh
Diffusion Pumped (2400 L/s)	$3990	$5670
Cryogenic Pump (2500 L/s)	1750	3090
Turbomolecular Pumped (1900 L/s)	1850	3320
Turbomolecular Pumped (1900 L/s) plus LN_2	3310	4820

[a] Air pumping speeds are ~ 2000 L/s at the work chamber baseplate. Liquid nitrogen costs have been assumed to be $0.12/L and cooling water costs $0.17/m³.

If vacuum processing is a dominant part of the product's manufacturing cost, some savings may be affected by using the least costly tooling.

The return on investment calculation presented here does not follow methods used in industry. In practice taxes (about 50%) are never included in such a calculation; instead the desired return on investment is doubled for simplicity. This simple method is sufficient to compare projects that are competing for capital. There are always more demands for capital than there is capital available, therefore the simple method is used to rank projects. Cash is then invested in the highest ranking projects. We have chosen a detailed method just as an exercise to show the magnitude of annual expenditures necessary to maintain a system. Many vacuum system users may not be aware of the hidden expenses incurred in the operation and maintenance of this equipment.

In Table 13.5 the cash flow for the 1000 L/s diffusion pump is examined for a 10-year period. The operating costs for electrical power rates of $0.04/kWh and $0.50/L for liquid nitrogen are taken from Table 13.3. Each expenditure is expressed in terms of its annual value. The capital cost for this diffusion pumped system includes the cost of liquid nitrogen dewars and piping; certain one-time installation

Table 13.5 Diffusion Pump Present Worth[a]

Year	1	2	3	4	5	6	7	8	9	10
Capital equipment	18,000									
Capital installment	2,000									
Total capital	**20,000**									
Utility expense	4,327	4,760	5,236	5,759	6,0335	6,969	7,666	8,432	9,275	10,203
Maintenance expense	940	1,034	1,137	1,250	1,376	1,651	1,981	2,378	2,856	3,424
Total expense	**5,216**	**5,793**	**6,373**	**7,009**	**7,741**	**8,619**	**9,646**	**10,810**	**12,128**	**13,626**
Depreciation	3,272	2,945	2,619	2,291	1,964	1,631	1,309	981	655	328
Depr. tax credit	1,686	1,472	1,309	1,145	982	818	654	490	328	164
Investment credit	2,000									
Expense credits	2,633	2,896	3,186	3,504	3,870	4,309	4,823	5,405	6,064	6,813
Total tax credit	**6,269**	**4,368**	**4,495**	**4,649**	**4,852**	**5,127**	**5,477**	**5,895**	**6,392**	**6,977**
Total cash flow	**18,998**	**1,425**	**1,878**	**2,360**	**2,889**	**3,492**	**4,169**	**4,915**	**5,736**	**6,649**
PW (17%)	18,998	1,218	1,372	1,474	1,542	1,592	1,625	1,638	1,633	1,618
Cum. PW	**18,998**	**20,216**	**21,588**	**23,062**	**24,604**	**26,197**	**27,822**	**29,460**	**31,093**	**32,711**

[a] Present worth (dollars) of a diffusion pump system with a net air pumping speed of 1000 L/s over a period of 10 years. Utility rates: Electrical power, $0.04/kWh; liquid nitrogen, $0.50/L; cooling water, $0.17/m^3.

expenses are also capitalized. Depreciation practices vary widely with the industry and type of equipment involved. For this example the sum-of-the-years method (10 years) with no salvage value was assumed. Periods of seven or eight years with finite salvage values are often used. It was assumed that the equipment was purchased in the first month of the fiscal year so that the first deduction for depreciation was made in year 1. In practice the initial fractional year's ownership will be depreciated in year 1. The values for maintenance and spare parts were calculated according to two assumptions: (1) the cost of maintenance and spare parts was assumed to increase at an annual rate of 10% for the first five years and at 20% for the next five years and (2) the total maintenance and spares expenditure for the 10-year period was made equal to the capital cost of the equipment. The actual rate of increase within the 10-year period may not be exactly as assumed but it will increase more rapidly toward the end of the period than at the beginning. Even though we may not be able to predict the details of the rate of increase of these expenses accurately, history has shown it to be accurate to set the total maintenance and spares expenditure over the life of the project equal to the capital cost of the equipment.

The tax credits are also tabulated. A 50% tax rate was assumed. The investment tax credit was set at 10% but this value may fluctuate. From the sum of the capital costs and expenses, less the tax credits, the total cash flow for each year was calculated. The annual cash flow was then reflected to year 1 by the present worth factor. In this example an interest rate or desired return on investment of 17% was assumed. The last line, the cumulative present worth, gives the cost of operating the system in year-1 dollars for 10 years. A plot is shown in Fig. 13.2.

The results of similar calculations for the turbomolecular pump and liquid nitrogen dewar and the gas cryogenic pump are given in Fig. 13.2. These results are based on the data in Table 13.3. The same rule was used for calculating maintenance costs for these pumps and the diffusion pump. The total cost of maintenance and spares over the life of the project was made equal to the capital cost. Unlike the diffusion pump, these pumping systems have no extensive maintenance history on which accurate projections can be based. They may be more reliable but they are new. Helium gas refrigerators, for example, have an extensive maintenance history but the same cannot be said for cryogenic pumping systems. As in anything new, failures take a great deal of time to repair because maintenance personnel are unfamiliar with the equipment. The expense of training personnel to maintain and repair new equipment should also be included. Based on these considerations, it is reasonable to set the maintenance costs of new pumps proportionally *at least* as great as those of the diffusion pump. An expensive pump will cost more to maintain than an inexpensive pump regardless of the technology. Under these assumptions, the cryogenic pump, which had the highest capital expense of the three pumps considered, became cheaper than the turbomolecular pump after three years of operation. The diffusion pump was the cheapest for the first five years, after which the cryogenic pump became the least expensive. The relative present values are strongly dependent on utility costs. For electrical power costs of $0.04/kWh and liquid nitrogen costs of $0.12/L the diffusion pump always has the lowest present value, followed, respectively, by the turbomolecular and cryogenic pump.

In comparing the present values of these three systems, it was assumed that all had net stack speeds of 1000 L/s. That was intended

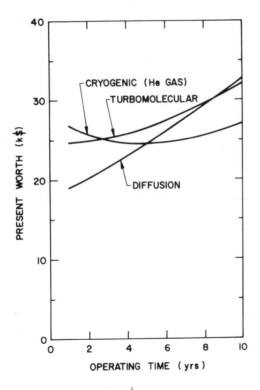

Fig. 13.2 Net present value of three pumping systems versus operating time. Each system has a net pumping speed for air of 1000 L/s. Utility costs: electrical power, $0.04/kWh; liquid nitrogen, $0.50/L; cooling water, $0.17/m^3.

to imply that the pumping time, hence the processing time, would be the same for all systems. Furthermore, it was assumed that the labor costs and floor-space requirements were identical. Given these constraints, the product throughput would be identical for all three systems. The present value is thus some measure of determining the most economical way of manufacturing a product.

13.2   HIGH GAS FLOW SYSTEMS

The procedure for calculating the return on investment of a system intended to handle a high gas throughput, such as a sputtering system, is the same as that outlined for high vacuum systems. Only the numbers are different. The capital costs of these systems will be greater than for high vacuum systems of the same net stack speed. Both diffusion pumps and turbomolecular pumps will use oversize backing pumps. Under high gas flow conditions the ratio of turbomolecular pumping speed to backing pump speed may be as low as 20:1. For large turbomolecular pumps the economics dictate that this backing pump be a Roots pump followed by a rotary vane pump. The cryogenic pumps which appear currently to be most suitable for sputtering are those that have a liquid nitrogen reservoir surrounding the warm stage of the helium refrigerator. Not only is this the more expensive of the two cryogenic pumps but the capital costs of liquid nitrogen storage and distribution must also be considered, just as they are for the diffusion pump. These added costs must be known in order to calculate the cost-per-unit gas throughput. If the gas throughput and backstreaming are identical for all pumps under consideration, the sputtering process will be independent of the pump. If the capital costs of systems with equal gas throughput are compared, it will be seen that the turbomolecular pump, followed by the diffusion pump and the cryogenic pump, is the least costly. These systems, however, will not have the same pumping speed because the maximum throat pressures are different for each pump. This means that the product cycle time will differ slightly because the pumping times will be somewhat different for each system. (The pumping time is primarily dependent on outgassing rate.) If the product throughput times are known for each system under consideration, the calculation can be completed. Increasing the pumping speed is only one way of decreasing the product throughput time. Load locks and in-line systems are alternatives. Any of these methods for decreasing the cycle time will result in additional capital investment. In-line systems also use more floor space, have greater installation costs, and, depending on the number of units in one location, may require more operators per machine. When these details are known, the cost per unit product for the various alternatives can be compared with a greater degree of accuracy. The latter has been considered by Poley [3].

13.3 ENERGY CONSERVATION

The preceding 20 years was a period of rapid development in the vacuum industry. It was also a period in which research, development, and manufacturing were adequately funded. Although large investments often brought about rapid technological advancement, they did not foster efficient equipment operation. Almost all vacuum technologists were taught to put a diffusion pump in operation and keep it running continuously under the illusion that the ultimate vacuum would degrade if the pump were shut down periodically. Not only is this concept invalid but procedures are available that can improve the system performance by cleaning the chamber during idle periods. Manufacturing facilities that operate two or more shifts per day will not save by idle-period shutdown, but the small operator who uses a system one shift per day on week days only can realize considerable savings. Power to the diffusion pump can be removed during idle periods by established procedures that do not compromise the quality of the vacuum system. Energy conservation therefore does not mean the same thing to all users. For the large production operator it means efficient operation of existing equipment and cost-efficient purchases of new. At the other economic extreme—the small operator on a limited capital budget—it implies idle-period shutdown of systems requiring a minimum of capital investment. In this section we compare the intrinsic efficiency of several pumps and discuss relative reductions in the operating costs made possible by idle-period shutdown.

The relative gas handling efficiencies of the major high vacuum pumps are considerably different. It is useful to define a speed efficiency (pumping speed per unit input power) and a throughput efficiency (gas throughput per unit input power) by which they may be compared. Table 13.6 lists values of these parameters for small pumps. Liquid nitrogen, cooling water, and rough pumping costs have not been included here, as they were in Section 13.1. The sputter-ion pump has the highest speed efficiency of this group, whereas the turbomolecular pump has the highest throughput efficiency. Ion pumps are not suitable for pumping large gas throughputs because they do not operate well when run continuously at pressures greater than 10^{-3} Pa.

The information contained in Table 13.6 is as sophisticated as the small user will require to make an intelligent selection of a new pumping system that will operate with minimum expense. The sputter-ion pump is the most efficient to operate on an ultrahigh vacuum system that is not frequently cycled to atmosphere. The next most efficient pump the helium gas refrigerator pump, is thus the best to use on a routine quick-cycled system even if liquid nitrogen is used. The turbo-molecular pump has the lowest operating cost of any pump for the most demanding high gas flow requirements. For some high gas flow requirements the throughput of the gas refrigerator cryogenic pump can be increased beyond the value stated in this example, provided that pumping of hydrogen, helium, and neon are not required, in which case the temperature of the second (cold) stage will reach as high as 35 K and the throughput efficiency of the pump will exceed that of a turbomolecular pump. These examples show that in many instances cryogenic, sputter-ion, and turbomolecular pumps can provide cost-efficient alternatives to the traditional diffusion pump system. This is not the only way, however, in which the user can reduce operational costs. Additional cost savings can be realized by turning off all systems during nonworking hours.

Diffusion pumps consume large amounts of energy, but surprisingly enough most operators seem reluctant to place them in a standby mode when not in use for fear that daily starting and stopping will quickly contaminate the system and reduce its performance. Power can be removed from all vacuum systems in ways that will not impair performance or cause contamination. The basic technique was described by Santeler [4] and more recently by Hoffman [5]. Power is removed from the diffusion pump, the liquid nitrogen trap is allowed to warm, and dry nitrogen gas is flowed through the chamber and diffusion pump. This gas flow is removed by the continually operating forepump. A flow of 100 Pa-L/s per 1000 L/s of diffusion pumping speed, measured at the pump throat, will produce a forepressure of 20 to 40 Pa. For a 6-in. diffusion pump backed by a 30-m³/h (8-L/s) forepump the recommended 200-Pa-L/s gas flow would produce a pressure of 25 Pa at the inlet to the forepump and stop oil backstreaming in the foreline. As the cold trap warms the evolved gases are flushed away by the purge gas. This purge gas is also adequate to flush away potential transient backstreaming which results from the cooling and subsequent warming of the vapor jets. When the system is needed again, the diffusion pump can be started, the trap cooled, and

Table 13.6   Relative Pumping Speed and Gas Throughput Efficiencies for Small
Diffusion, Turbomolecular, Helium Gas Cryogenic, and Sputter-Ion Pumps

Pump Type	Power Consumption[a] (W)	Pump Speed[b] (L/s)	Maximum Throughput (Pa-L/s)	Speed Efficiency (L/s)/W	Throughput Efficiency (Pa-L/s)/W
Diffusion	3000	1000	200	0.33	0.07
Turbo-molecular	1500	500	200	0.33	0.13
Cryogenic (He gas)	1000	1000	100	1.00	0.10
Sputter-Ion	100	500	-	5.00	-

[a] High vacuum and forepump power only; power to the gauging and controls not
included.
[b] Speed measured at the baseplate of the system.

the purge gas removed. With this simple technique the system can be
started and stopped with considerable savings in electrical power,
liquid nitrogen, and cooling water without added contamination. Some
additional savings in cooling water can be realized by the installing of
water-flow restrictors on the inlet of cooling water lines.

It has been proposed that diffusion pump energy can be saved by
reducing the power to the heater on the basis that the heater power
was chosen for the maximum throughput situation. Reduction of the
power will have only a small effect on backstreaming. It may increase
or decrease, depending on the pump, but it is not clear that power
reduction is compatible with the use of low vapor-pressure fluids such
as Santovac 5 or DC-705. These fluids already have a lower forepressure tolerance than higher vapor-pressure fluids and a further reduction in forepressure tolerance may not be acceptable. It is also not
clear that the fractionation stages of all pumps are effective at reduced
input power. Rather than reduce the power to a fraction of its design
value, it makes more sense to remove power completely when not in
use.

The turbomolecular pump can be shut down in an identical manner.
The flow rate of dry nitrogen is adjusted to keep the foreline in viscous flow. At the appropriate time the turbomolecular pump is shut

off and the flow of liquid nitrogen to the trap is stopped, whereas the mechanical pump continues to operate. The gas purge flushes away the desorbed vapors from the trap and prevents oil from backstreaming through the turbomolecular pump when its rotational velocity falls below 50% of its rated value. Starting is accomplished in the reverse manner; the turbomolecular pump is started and when it has reached full rotational velocity the dry nitrogen gas flow is removed. The liquid nitrogen trap may be cooled when added water-vapor pumping speed is needed.

Shutdown of the cryogenic pump is simpler than for the diffusion or turbomolecular, because no forepump is used. Power is removed from the compressor and dry nitrogen gas is admitted to the system after the cold stage has warmed to a temperature of 80 K. The gas purge will raise the pressure in the pump until the spring-loaded safety valve or other auxiliary valve releases the nitrogen gas to atmosphere. This gas flow is continued until regeneration is completed—typically an overnight process.

Energy conservation measures will have a profound effect on the operating costs of each of these systems. Figure 13.3 sketches the potential savings that may be realized from reduced operation of each of the four 1000 L/s systems described in Section 13.1. See Table 13.2. The cost of the purge gas was not added into Fig. 13.3. Often it is boil-off from the liquid tank and therefore free. Were it to be purchased it would cost about $150/y for weekend purging of a 6-in. diffusion pumped system. The largest of the operating costs shown for each system is the continuous operating condition in Table 13.2. The operating cost for each system was also calculated for system shutdown from 6 PM each Friday until 6 AM each Monday (weekend shutdown) and when the pumps were turned off from 6 PM to 6 AM each weekday (evening shutdown). The mechanical forepumps on the diffusion and turbomolecular pumped systems remained on during these periods to remove the purge gas.

The electrical power and cooling water savings scale linearly with the time each piece of equipment is turned off. This is not true of liquid nitrogen because an additional amount is consumed each time the trap is cooled. This is most noticeable for daily shutdown because each trap must be initially cooled five times a week. Thus the diffusion pump that uses liquid nitrogen at $0.50/L costs $63/week to run continuously, $44/week if shutdown on weekends, and $40/week if operated only 12 a day, five days a week. Therefore the percentage of savings is the greatest for the user who has access to low-cost liquid nitrogen but who must pay premium electric power rates. This is

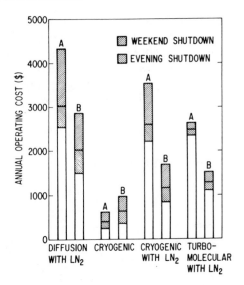

Fig. 13.3 Cost savings made possible oy weekend and evening vacuum system shutdown. (a) Electrical power; $0.04/kWh, liquid nitrogen; $0.50/L. (b) Electrical power; $0.08/kWh, liquid nitrogen; $0.12/L. Liquid is supplied to the diffusion pump and the liquid-nitrogen-assisted cryogenic pump while they are operating, and to the turbomolecular pump for 8 h each weekday.

illustrated in Fig. 13.3 by plotting the operational costs for Case *a* (electric power; $0.04/kWh; liquid nitrogen; $0.50/L) and Case *b* (electric power; $0.08/kWh, liquid nitrogen; $0.12/L). From this figure we observe that the fractional savings realized by idle-period shutdown are the greatest on the all electric cryogenic pump because the electric and cooling water rates scale linearly with operating time. The fractional savings from the turbomolecular pump are the lowest because the liquid nitrogen usage (12 h/day) and the mechanical pump usage are constant for all three operational modes; only the turbomolecular pump electrical power and cooling water are reduced with idle-time shutdown. The time between turbomolecular pump oil changes, however, can be extended by a factor of 1.5 for weekend shutdown and 2.8 for evening shutdown.

Because they are the most costly pumps to operate, the greatest dollar savings are possible on diffusion pumps used only one shift per work day. Obviously little can be saved on production equipment that operates two or three shifts a day. Ion pumps were not included in this discussion. A 500-L/s ion pump would cost about $50/y less to operate if it were shutoff each weekend than if it were operated continuously—a savings hardly worth considering. Perhaps the most important conclusion to be drawn from this discussion of energy conservation is that the greatest financial savings on existing equipment can be realized by the simple expedient of removing power from diffusion pumps and purging them with dry nitrogen when they are not needed.

REFERENCES

1. L. Gardner, *Ind. Res./Dev.*, December 1978, p. 93.
2. W. R. Park, *Cost Engineering Analysis*, Wiley, New York, 1973, p. 133.
3. N. Poley, *J. Vac. Sci. Technol.*, **14**, 630 (1977).
4. D. J. Santeler, *J. Vac. Sci. Technol.*, **8**, 299 (1971).
5. D. Hoffman, *J. Vac. Sci. Technol.*, **16**, 71 (1979).

Symbols

Symbol	Quantity	Units
A	Area	m^2
B	Magnetic field strength	T (Tesla)
C	Conductance (gas)	L/s
D	Diffusion constant	m^2/s
E_o	Heat transfer	$J\text{-}s^{-1}\text{-}m^{-2}$
F	Force	N (Newton)
G	Electron multiplier gain	
H	Heat flow	J/s
K	Compression ratio (gas)	
K_p	Permeability constant	m^2/s
Kn	Knudsen's number	
K_R	Radiant heat conductivity	$J\text{-}s^{-1}\text{-}m^{-1}\text{-}K^{-1}$
K_T	Thermal conductivity	$J\text{-}s^{-1}\text{-}m^{-1}\text{-}K^{-1}$
M	Molecular weight	
N	Number of molecules	
P	Pressure	Pa (Pascal)
Q	Gas flow	$Pa\text{-}m^3/s$
R	Reynold's number	
S	Pumping speed	L/s
S'	Gauge sensitivity	Pa^{-1}
T	Absolute temperature	K (Kelvin)
U	Average gas stream velocity	m/s
U	Mach number	
V	Volume	m^3
V_a	Acceleration potential	V

349

V_b	Linear blade velocity	m/s
W	Ho coefficient	
a	Transmission probability	
b	Turbomolecular pump blade chord length	m
c	Condensation coefficient	
c_p	Specific heat at constant pressure	J-(kg-mole)$^{-1}$-K^{-1}
c_v	Specific heat at constant volume	J-(kg-mole)$^{-1}$-K^{-1}
d	Diameter dimension	m
d_o	Molecular diameter	m
d^1	Average molecular spacing	m
i_e	Emission current	A
i_p	Plate current	A
l	length dimension	m
m	mass of molecule	kg
n	gas density	m^{-3}
q	outgassing rate	W/m^2
q_k	Permeation rate	W/m^2
r	radius	m
s	turbomolecular pump blade spacing	m
s_r	Turbomolecular pump blade speed ratio	
u	Local gas stream velocity	m/s
v	Average particle velocity	m/s
Γ	Particle flux	m^{-2}-s^{-1}
Λ	Free molecular heat conductivity	J-s^{-1}-m^{-2}-K^{-2}-Pa^{-1}
α	Accommodation coefficient	
β	Molecular slip constant	
γ	Ratio c_p/c_v	
ε	Emissivity	
λ	Mean free path	m
η	Dynamic viscosity	Pa-s
ρ	Mass density	kg/m^3
ω	Angular frequency	rad/s
ϕ	Turbomolecular pump blade angle	deg.

Appendixes

APPENDIX A

Units and Constants

Appendix A.1 Physical Constants

k	Boltzmann's constant	1.3804×10^{-23} J/K
m_e	Rest mass of electron	9.108×10^{-31} kg
m_p	Rest mass of proton	1.672×10^{-27} kg
N_o	Avogadro's number	6.02252×10^{26}/(kg-mole)
R	Gas constant	8314.3 J-(kg-mole)$^{-1}$-K^{-1}
V_o	Normal specific volume of an ideal gas	22.4136 m^3/(kg-mole)
σ	Stefan-Boltzmann constant	5.67×10^{-8} J-s^{-1}-m^{-1}-K^{-4}

Appendix A.2 SI Base Units

Length	meter	m
Mass	kilogram	kg
Time	second	s
Electric current	Ampere	A
Thermodynamic temperature	Kelvin	K
Amount of substance	mole	mole

Appendix A.3   Conversion Factors

Conventional unit →	multiply by →	to get SI unit
	Mass	
lb	0.45359	kg
	Length	
micrometer	0.000001	m
mil	0.00254	cm
inch	0.0254	m
foot	0.3048	m
Angstrom	1.0×10^{-10}	m
	Area	
ft^2	0.0929	m^2
$in.^2$	6.452	m^2
ft^2	929.03	m^2
	Volume	
cm^3	0.001	L
$in.^3$	0.0164	L
gal. (US)	3.7879	L
ft^3	28.3	L
L	1000.	m^3
	Pressure	
micrometer (Hg)	0.13332	Pa
N/m^2	1.0	Pa
millibar	100.	Pa
torr	133.32	Pa
in. (Hg)	3386.33	Pa
lb/in^2	6895.3	Pa
atmosphere	101,323.2	Pa
	Conductance or pumping speed	
L/h	0.000277	L/s
L/s	0.001	m^3/s
L/m	0.0166	L/s
m^3/h	0.2778	L/s
ft^3/m	0.4719	L/s
ft^3/m	1.6987	m^3/h
to get conventional unit ←	divide by ←	SI unit

Conventional unit	→ multiply by →	to get Si unit
	Gas flow	
micron - L/s	0.13332	Pa-L/s
Pa-L/s	3.6	Pa-m^3/h
Atm-cc/s	101.323	Pa-L/s
torr-L/s	133.32	Pa-L/s
torr-L/s	0.133	J/s
Watt	1,000	Pa-L/s
(kg-mole)/s	2.48×10^9	Pa-L/s
molecules/s	4×10^{-18}	Pa-L/s
	Outgassing rate	
(Pa-L)/(m^2-s)	0.001	W/m^2
(Pa-m^3)/(m^2-s)	1.0	W/m^2
μL/(cm^2-s)	1.33	W/m^2
(torr-L)/(cm^2-s)	1,333.2	W/m^2
	Dynamic viscosity	
poise	10	Pa-s
Newton-s/m^2	1	Pa-s
	Kinematic viscosity	
centistoke	1	mm^2/s
	Diffusion constant	
cm^2/s	0.0001	mm^2/s
	Heat conductivity	
Watt-cm^{-1}-K^{-1}	100	J-s^{-1}-m^{-1}-K^{-1}
	Specific Heat	
cal-(gm-mole)$^{-1}$-K^{-1}	4184.	J-(kg-mole)$^{-1}$-K^{-1}
J-kg^{-1}-K^{-1}	M	J-(kg-mole)$^{-1}$-K^{-1}
BTU-lb^{-1}-°F^{-1}	$4186M$	J-(kg-mole)$^{-1}$-K^{-1}
	Heat capacity	
cal-(gm mole)$^{-1}$	4184	J-(kg-mole)$^{-1}$
J/kg	M	J-(kg-mole)$^{-1}$
BTU/lb	$2325.9M$	J-(kg-mole)$^{-1}$
	Energy, work, or quantity of heat	
kW-h	3.6	MJ
kcal	4184	J
BTU	1055	J
ft-lb	1.356	J
to get Conventional unit ←	divide by ←	SI unit

APPENDIX B

Gas Properties

Appendix B.1   Mean Free Paths of Gases as a Function of Pressure[a]

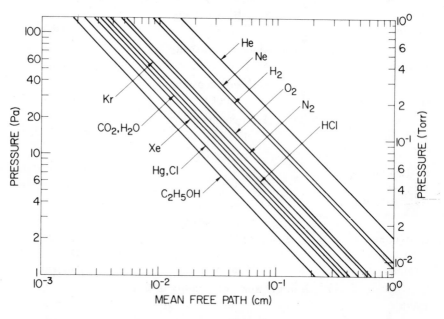

Source. Reprinted with permission from *Vacuum Technology*, p.505, A. Guthrie. Copyright 1963, John Wiley & Sons, New York, 1963.
[a] $T = 20°C$.

Gas	Symbol	MW[a]	Molecular Diameter[b] (nm)	Average Velocity[c] (m-s^{-1})	Thermal Cond.[a] (mJ-s^{-1}-K^{-1})	Dynamic Viscosity[a] (μPa-s)	Diffusion in Air[d] (μm^2-s^{-1})
Helium	He	4.003	0.218	1197.0	142.0	18.6	58.12
Neon	Ne	20.183	0.259	533.0	45.5	29.73	27.63
Argon	Ar	39.948	0.364	379.0	16.6	20.96	17.09
Krypton	Kr	83.8	0.416	262.0	6.81[e]	23.27	13.17
Xenon	Xe	131.3	0.485	209.0	4.50[f]	21.9	10.60
Hydrogen	H$_2$	2.016	0.274	1687.0	173.0	8.35	63.4[a]
Nitrogen	N$_2$	28.0134	0.375	453.0	24.0	16.58	18.02
Air		28.966	0.372	445.0	24.0	17.08	18.01
Oxygen	O$_2$	31.998	0.361	424.0	24.5	18.9	17.8[a]
Hydrogen chloride	HCl	36.46	0.446	397.0	12.76	14.25[g]	14.11
Water vapor	H$_2$O	18.0153	0.46	564.0	24.1[h]	12.55[h]	23.9[a,i]
Hydrogen sulfide	H$_2$S	34.08	0.47[j]	412.0	12.9	11.66	14.62[j]
Nitric oxide	NO	30.01	0.372[j]	437.0	23.8	17.8	19.3[j]
Nitrous oxide	N$_2$O	44.01	0.47[j]	361.0	15.2	13.5	13.84[j]
Ammonia	NH$_3$	17.03	0.443	581.0	21.9	9.18	17.44
Carbon monoxide	CO	28.01	0.312[a]	453.0	23.0	16.6	21.49
Carbon dioxide	CO$_2$	44.01	0.459	361.0	14.58	13.9	13.9[a]
Methane	CH$_4$	16.4	0.414	592.0	30.6	10.26	18.98
Ethylene	C$_2$H$_4$	28.05	0.495	452.0	17.7	9.07	13.37
Ethane	C$_2$H$_6$	30.07	0.53	437.0	16.8	8.48	12.14

[a] Reprinted with permission from *Handbook of Chemistry and Physics*, 58 ed., R. C. Weast, Ed. Copyright 1977, Chemical Rubber Co., CRC Press, West Palm Beach, FL.
[b] Reprinted with permission from *Kinetic Theory of Gases*, E. H. Kennard, p. 149. Copyright 1938, McGraw-Hill, New York.
[c] Calculated from Eq. 2.1.
[d] Calculated from Eq. 2.25.
[e] $T = 210$ K, Reprinted with permission from *Cryogenic and Industrial Gases*, May/June 1975, p 62. Copyright 1975, Thomas Publishing Co., Cleveland, Ohio.
[f] Ref. e, $T = 240$ K;
[g] $T = 18°C$;
[h] $T = 100°C$;
[i] $T = 8°C$;
[j] Calculated from viscosity data.

Appendix B.3   Cryogenic Properties of Gases

Property	Units	He	H_2	Ne	N_2	Ar	O_2	Xe	CF_4
nbp liq.[a]	K	4.125	20.27	27.22	77.35	87.29	90.16	164.83	145.16
mp (1 atm)[b]	K		14.01	24.49	63.29	83.95	54.75	161.25	123.16
den liq nbp[a]	kg/m^3	124.8	70.87	1208.	810.0	1410.	1140.	3058.0	1962.0
vol liq nbp	$(m^3/kg) \times 10^{-3}$	8.01	14.1	0.83	1.24	0.709	0.877	0.327	0.597
vol gas at 273 K[a]	m^3/kg	5.602	11.12	1.11	0.79	0.554	0.698	0.169	0.274
Ratio V^g_{273}/V^l_{nbp}		699.4	788.7	1337.	637.5	781.5	796.3	516.8	458.3
h of vap nbp[b]	kJ/(kg-mole)	95.8	911.0	1740.	5580.	6502.	6812.	12640.	12,000.
h of fus mp[b]	kJ/(kg-mole)	16.75	118.0	338.0	714.0	1120.0	438.0	1812.0	699.0
sp h, c_p^{vap}, 300 K[b]	kJ-(kg-mole)$^{-1}$-K^{-1}	20.94[c]	28.63	20.85	29.08	20.89	29.45	20.85	62.23

[a] Reprinted with permission from *Cryogenic and Industrial Gases*, May/June 1975. Copyright 1975, Thomas Pub. Co., Cleveland, Ohio.

[b] Reprinted with premission from *Handbook of Chemistry and Physics*, 58th ed., R. C. Weast, Ed. Copyright 1977, Chemical Rubber Co., CRC Press, West Palm Beach, FL;

[c] at -180°C.

Dimensionless numbers

Knudsen's number

$$Kn = \frac{\lambda}{d}$$

$$Kn = \frac{6.6}{P(\text{Pa})d(\text{mm})} \quad \text{air, } 22^\circ\text{C}$$

Reynold's number

$$\mathbf{R} = \frac{4m}{\pi k T \eta} \frac{Q}{d}$$

$$\mathbf{R} = 7.27 \times 10^{-4} \frac{Q}{d} \frac{(\text{Pa}-\text{L/s})}{(\text{m})}$$

Mach number

$$\mathbf{U} = \frac{4Q}{\pi d^2 P U_{sound}}$$

Quantities from kinetic theory

Average velocity

$$v = \left(\frac{8kT}{\pi m} \right)^{1/2}$$

$$v = 463 \text{ m/s} \quad \text{air, } 22^\circ\text{C}$$

Mean free path, one component gas

$$\lambda = \frac{1}{2^{1/2} \pi d_o^2 n}$$

$$\lambda(\text{mm}) = \frac{6.6}{P} \quad \text{air, } 22^\circ\text{C}$$

Ideal gas law

$$P = nkT$$

$$\frac{P_1 V_1}{T_1} = \frac{P_2 V_2}{T_2}$$

Viscosity at normal pressures

$$\eta = \frac{0.499(4mkT)^{1/2}}{\pi^{3/2}d_o^2}$$

Heat conductivity at normal pressures

$$K = \frac{1}{4}(9\gamma-5)\eta c_v$$

Diffusion constant for two gases

$$D_{12} = \frac{8\left(\frac{2kT}{\pi}\right)^{1/2}\left(\frac{1}{m_1}+\frac{1}{m_2}\right)^{1/2}}{3\pi n(d_{01}+d_{02})^2}$$

Diffusion constant, self diffusion

$$D_{11} = \frac{4}{3\pi nd_o^2}\left(\frac{kT}{\pi m}\right)^{1/2}$$

Speed of sound in a gas

$$U(m/s) = v\left(\frac{\pi\gamma}{8}\right)$$

$$\gamma \sim 1.4 \quad \text{diatomic gas}$$

$$\gamma \sim 1.667 \quad \text{monatomic gas}$$

$$U = 303.1 \quad m/s \quad \text{air, } 22°C$$

Flow regimes

Turbulent flow	R > 2200
Viscous flow	R < 1200, and Kn < 0.01
Molecular flow	Kn > 1

Conductance formulas and definitions

Conductance

$$C = \frac{Q}{P_2-P_1}$$

Pumping speed

$$S = \frac{Q}{P_2}$$

Parallel conductances

$$C_T = C_1 + C_2 + C_3 + \dots$$

Series conductances (isolated)

$$\frac{1}{C_T} = \frac{1}{C_1} + \frac{1}{C_2} + \frac{1}{C_3} + \dots$$

Viscous conductance, long circular tube

$$C(m^3/s) = \frac{\pi d^4}{128\eta\ell}\frac{(P_1 + P_2)}{2}$$

$$C(L/s) = 1.38 \times 10^6 \frac{d^4}{\ell}\frac{(P_1 + P_2)}{2}$$

Molecular conductance, long circular tube

$$C = \frac{\pi}{12}v\frac{d^3}{\ell}$$

$$C\ (m^3/s) = 121\frac{d^3}{\ell}$$

Molecular conductance, long rectangular duct

$$C = \frac{2}{3}v\frac{a^2 b^2 K'}{(a+b)\ell}$$

where K' is tabulated as

b/a	1.0	0.8	0.6	0.4	0.2	0.1
K'	1.1	1.12	1.13	1.17	1.29	1.44

Molecular conductance, small aperture

$$C = \frac{v}{4}A$$

$$C(L/s) = 1.16 \times 10^5 A\ (m^2) = 11.6A\ (cm^2)$$

Molecular conductance, circular tube of arbitrary length. Clausing's solution

$$C(m/s) = \frac{a'v}{4}A$$

$$C(L/s) = 1.16 \times 10^5 a'A(m^2)$$

where a' is given in Table 2.2

Molecular conductance, circular tube of arbitrary length, Dushman's solution

$$\frac{1}{C_{total}} = \frac{1}{C_{tube}} + \frac{1}{C_{ap}}$$

Molecular conductance, transmission probability method, arbitrary shape

$$C(\text{m/s}) = v\frac{A}{4}a$$

$$C(\text{m}^3/\text{s}) = 11.6aA(\text{cm}^2) \quad \text{air, } 22°C$$

Transmission probability a, of a combination of nonisolated conductances in molecular flow. Oatley's method

$$\frac{1-a}{a} = \frac{1-a_1}{a_1} + \frac{1-a_2}{a_2} + \dots$$

Transition conductance, Knudsen's method

$$C = \frac{Q}{(P_2-P_1)}$$

$$Q = Q_{viscous} + Z'Q_{molecular}$$

where

$$Z' = \frac{1 + 2.507\left(\frac{d}{2}\lambda\right)}{1 + 3.095\left(\frac{d}{2}\lambda\right)}$$

Appendix B.5   Vapor Pressure Curves of Common Gases

VAPOR PRESSURE CURVES OF COMMON GASES — SHEET A

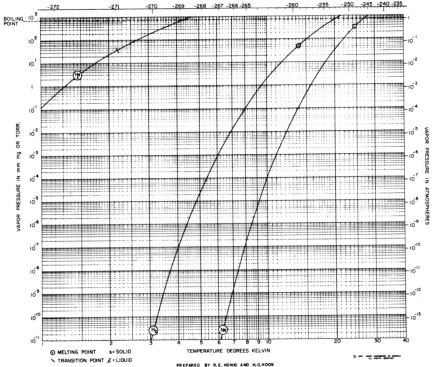

MELTING POINT   s = SOLID
TRANSITION POINT   ℓ = LIQUID

PREPARED BY R.E. HONIG AND H.O. HOOK

RADIO CORPORATION OF AMERICA   PRINCETON, N.J.

VAPOR PRESSURE CURVES OF COMMON GASES — SHEET B

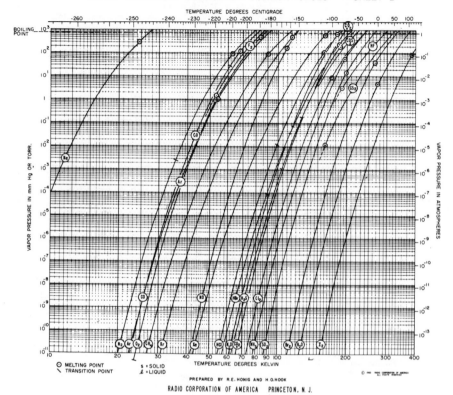

PREPARED BY R.E. HONIG AND H.O.HOOK

RADIO CORPORATION OF AMERICA   PRINCETON, N.J.

Source. Reprinted with permission from *RCA Review*, **21**, p,360, Sept. 1960, *Vapor Pressure Data for Some Common Gases*, by R. E. Honig and H. O. Hook. Copyright 1960, RCA Corp.

Appendix B.6 Appearance of Discharges in Gases and Vapors at Low Pressures

Gas	Negative Glow	Positive Column
Argon	Blue	Violet
Carbon tetrachloride	Light green	Whitish green
Carbon monoxide	Greenish white	White
Carbon dioxide	Blue	White
C_2H_5OH	-	Whitish
Cadmium	Red	Greenish blue
Hydrogen	Light blue	Pink
Mercury	Whitish yellow	Blue green
Potassium	Green	Green
Krypton	Violet	Yellow pink
Air	Blue	Reddish
Nitrogen	Blue	Red-yellow
Sodium	Whitish	yellow
Oxygen	Yelowish white	Lemon yellow with pink core
Thallium	Green	Green
Xenon	Pale blue	Blue violet

Source. Reprinted with permission from *Materials for High Vacuum Technology*, **3**, p. 393, W. Espe. Copyright 1968, Pergamon Press.

APPENDIX C

Material Properties

Appendix C.1   Outgassing Rates of Vacuum Baked Metals

Material	Treatment	q $(10^{-11}$ W/m$^2)$
Aluminum [a]	15 h at 250°C	53.0
Aluminum [b]	20 h at 100°C	5.3
6061 Aluminum [c]	glow disch. + 200°C bake	1.3
Copper [b]	20 h at 100°C	146.0
304 Stainless Steel [a]	30 h at 250°C	400.0
Stainless Steel [d]	2 h at 850/900°C vac. furnace	27.0
316L Stainless Steel [e]	2 h at 800°C vac. furnace	46.0
U15C Stainless Steel [f]	3 h vac. furn. 1000°C + 25-h *in situ* vac. bake at 360°C	2.1

Source. Adapted with permission from *Vacuum*, **25**, p. 347, R. J. Elsey. Copyright 1975, Pergamon Press.
[a] J. R. Young, *J. Vac. Sci. Technol.*, **6**, 398 (1969);
[b] G. Moraw, *Vacuum*, **24**, 125 (1974);
[c] H. J. Halama and J. C. Herrera, *J. Vac. Sci. Technol.*, **13**, 463 (1976);
[d] R. L. Samuel, *Vacuum*, **20**, 295 (1970);
[e] R. Nuvolone, *J. Vac. Sci. Technol.*, **14**, 1210 (1977);
[f] R. Calder and G. Lewin, *Brit. J. Appl. Phys.*, **18**, 1459 (1967).

Appendix C.2 Outgassing Rates of Unbaked Metals[1]

Material	q_1 (10^{-7} W/m^2)	α_1	q_{10} (10^{-7} W/m^2)	α_{10}
Aluminum (fresh)[a]	84.0	1.0	8.0	1.0
Aluminum (degassed 24-h)[a]	55.2	3.2	4.08	0.9
Aluminum (3-h in air)[a]	88.6	1.9	6.33	0.9
Aluminum (fresh)[a]	82.6	1.0	4.33	0.9
Aluminum (anodized 2 μm pores)[a]	3679.0	0.9	429.0	0.9
Aluminum (bright rolled)[b]	-	-	100.0	1.0
Duraluminum[b]	2266.0	0.75	467.0	0.75
Brass (wave guide)[b]	5332.0	2.0	133.0	1.2
Copper (fresh)[a]	533.0	1.0	55.3	1.0
Copper (mech. polished)[a]	46.7	1.0	4.75	1.0
OHFC copper (fresh)[a]	251.0	1.3	16.8	1.3
OHFC copper (mech. polished)[a]	25.0	1.1	2.17	1.1
Gold (wire fresh)[a]	2105.0	2.1	6.8	1.0
Mild steel[b]	7200.0	1.0	667.0	1.0
Mild steel (slightly rusty)[b]	8000.0	3.1	173.0	1.0
Mild steel (chromium plated polished)[b]	133.0	1.0	12.0	-
Mild steel (aluminum spray coated)[b]	800.0	0.75	133.0	0.75
Steel (chromium plated fresh)[a]	94.0	1.0	7.7	1.0
Steel (chromium plated polished)[a]	121.0	1.0	10.7	1.0
Steel (nickel plated fresh)[a]	56.5	0.9	6.6	0.9
Steel (nickel plated)[a]	368.0	1.1	3.11	1.1
Steel (chemically nickel plated fresh)[a]	111.0	1.0	9.4	1.0
Steel (chemically nickel plated polished)[a]	69.6	1.0	6.13	1.0
Steel (descaled)[a]	4093.0	0.6	3933.0	0.7
Molybdenum[a]	69.0	1.0	4.89	1.0
Stainless steel EN58B (AISI 321)[b]	-	-	19.0	1.6
Stainless steel 19/9/1-electropolished[c]	-	-	2.7	-
-vapor degreased[c]	-	-	1.3	-
-Diversey cleaned[c]	-	-	4.0	-
Stainless steel[b]	2333.0	1.1	280.0	0.75
Stainless steel[b]	1200.0	0.7	267.0	0.75
Stainless steel ICN 472 (fresh)[a]	180.0	0.9	19.6	0.9
Stainless steel ICN 472 (sanded)[a]	110.0	1.2	13.9	0.8
Stainless steel NS22S (mech. polished)[a]	22.8	0.5	6.1	0.7
Stainless steel NS22S (electropolished)[a]	57.0	1.0	5.7	1.0
Stainless steel[a]	192.0	1.3	18.0	1.9
Zinc[a]	2946.0	1.4	429.0	0.8
Titanium[a]	150.0	0.6	24.5	1.1
Titanium[a]	53.0	1.0	4.91	1.0

Source. Reprinted with permission from *Vacuum*, **25**, p 347, R. J. Elsey. Copyright 1975, Pergamon Press.

[1] $q_n = qt^{-\alpha_n}$, where n is in hours.

[a] A. Schram, *Le Vide*, No. 103, 55 (1963),

[b] B. B. Dayton, *Trans. 6th Nat. Vac. Symp. (1959)*, Pergamon Press, New York, 1960, p. 101,

[c] R. S. Barton and R. P. Govier, *Proc. 4th Int. Vac. Congr. (1968)*, Institute of Physics and the Physical Society, London, 1969, p. 775, and *Vacuum*, **20**, 1 (1970).

Appendix C.3   Outgassing Rates of Ceramics and Glasses[1]

Material	q_1 (10^{-7} W/m^2)	α_1	q_{10} (10^{-7} W/m^2)	α_{10}
Steatite[a]	1200.0	1.0	127.0	-
Pyrophyllite[b]	2667.0	1.0	267.0	-
Pyrex (fresh)[c]	98.0	1.1	7.3	-
Pyrex (1 month in air)[c]	15.5	0.9	2.1	-

Source. Reprinted with permission from *Vacuum*, **25**, p. 347, R. J. Elsey. Copyright 1975, Pergamon Press.
[1] $q_n = qt^{-\alpha_n}$, where n is in hours.
[a] R. Geller, *Le Vide*, No. 13, 71 (1958);
[b] R. Jaeckel and F. Schittko, quoted by Elsey;
[c] B. B. Dayton, *Trans. 6th Nat. Symp. Vac. Technol. (1959)*, Pergamon Press, New York, 1960, p. 101.

Appendix C.4   Outgassing Rates of Elastomers[1]

Material	q_1 (10^{-5} W/cm^2)	α_1	q_4 (10^{-5} W/m^2)	α_4
Butyl DR41[a]	200.0	0.68	53.0	0.64
Neoprene[a]	4000.0	0.4	2400.0	0.4
Perbunan[a]	467.0	0.3	293.0	0.5
Silicone[b]	930.0	-	267.0	-
Viton A (fresh)[c]	152.0	0.8	-	-
Viton A (bake 12 h at 200°C)[d]	-	-	0.027[e]	-
Polyimide (bake 12 h at 300°C)[d]	-	-	0.005[e]	-

Source. Adapted with permission from *Vacuum*, **25**, p. 347, R. J. Elsey. Copyright 1975, Pergamon Press.
[1] $q_n = qt^{-\alpha_n}$, where n is in hours.
[a] J. Blears, E. J. Greer and J. Nightengale, *Adv. Vac. Sci.Technol.*, **2**, E. Thomas, Ed., Pergamon Press, 1960, p. 473;
[b] D. J. Santeler, et al., *Vacuum Technology and Space Simulation*, NASA SP-105, National Aeronautics and Space Administration, Washington, DC, 1966, p. 219;
[c] A. Schram, *Le Vide*, No. 103, 55 (1963);
[d] P. Hait, *Vacuum*, **17**, 547 (1967);
[e] Pumping time is 12 h.

Appendix C.5 Outgassing Rates of Polymers[1]

Material	q_1 (10^{-5} W/cm^2)	α_1	q_{10} (10^{-5} W/m^2)	α_{10}
Araldite (molded)[a]	155.0	0.8	47.0	0.8
Araldite D[b]	253.0	0.3	167.0	0.5
Araldite F[b]	200.0	0.5	97.0	0.5
Kel-F[c]	5.0	0.57	2.3	0.53
Methyl Methacrylate[d]	560.0	0.9	187.0	0.57
Mylar (24-h at 95% RH)[e]	307.0	0.75	53.0	-
Nylon[f]	1600.0	0.5	800.0	0.5
Plexiglas[g]	961.0	0.44	36.0	0.44
Plexiglas[b]	413.0	0.4	240.0	0.4
Polyester-glass Laminate[c]	333.0	0.84	107.0	0.81
Polystyrene[c]	2667.0	1.6	267.0	1.6
PTFE[h]	40.0	0.45	26.0	0.56
PVC (24-h at 95% RH)[e]	113.0	1.0	2.7	-
Teflon[g]	8.7	0.5	3.3	0.2

Source. Reprinted with permission from *Vacuum*, **25**, p. 347, R. J. Elsey, Copyright 1975, Pergamon Press.
[1] $q_n = qt^{-\alpha_n}$, where n is in hours.
[a] A. Schram, *Le Vide*, No. 103, 55 (1963);
[b] R. Geller, *Le Vide*, No.13, 71 (1958);
[c] B. B. Dayton, CVC Technical Report;
[d] J. Blears, E. J. Greer and J. Nightengale, *Adv. Vac. Sci. Technol.*, **2**, E. Thomas, Ed., Pergamon Press, 1960, p. 473;
[e] D. J. Santeler, *Trans. 5th Symp. Vac. Tech. (1958)*, Pergamon Press, New York, 1959, p. 1;
[f] B. D. Power and D. J. Crawley, *Adv. Vac. Sci. Technol.*, **1**, E. Thomas, Ed., Pergamon Press, New York, 1960, p. 207;
[g] G. Thieme, *Vacuum*, **13**, 55 (1963);
[h] B. B. Dayton, *Trans. 6th Nat. Vac. Symp. Vac. Technol. (1959)*, Pergamon Press, New York, 1960, p.101.

Appendix C.6. Permeability of Polymeric Materials[a]

Material	Permeability (10^{-12} m^2/s)					
	Nitrogen	Oxygen	Hydrogen	Helium	Water Vapor	Carbon Dioxide
PTFE[b]	2.5	8.2	20.0	570.0	-	-
Perspex[b]	-	-	2.7	5.7	-	-
Nylon 31[b]	-	-	0.13	0.3	-	-
Neoprene CS2368B[b]	0.21	1.5	8.2	7.9	-	-
Viton-A[c]	0.05	1.1	2.2	8.9	-	5.9
Kapton[c]	0.03	0.1	1.1	1.9	-	0.2
Buna-S[d]	4.8 (30)	-	-	-	-	940.0 (30)
Perbunan[d]	0.8	-	-	-	-	23.0 (30)
Delrin[d]	-	48.0	-	-	17.0	93.0
Kel-F[d]	0.99 (30)	0.46 (30)	-	-	0.22 (25)	-

[a] Measurements made at 23°C unless noted in parenthesis after the value.
[b] Reprinted with permission from *Vacuum*, **25**, p. 469, G. F. Weston. Copyright 1975, Pergamon Press. Data derived by Weston from measurements made by Barton reported by J. R. Bailey in *Handbook of Vacuum Physics*, **3**, Part 4, Pergamon Press, Oxford, 1964;
[c] Reprinted with permission from *J. Vac. Sci. Technol.*, **10**, p. 543, W. G. Perkins. Copyright 1973, The American Vacuum Society;
[d] Reprinted with permission from *Vacuum Science and Space Simulation*, D. J. Santeler et al, NASA SP-105, National Aeronautics and Space Administration, Washington, DC, 1966, p. 216.

Appendix C.7 Vapor Pressure Curves of the Solid and Liquid Elements

VAPOR PRESSURE CURVES OF THE ELEMENTS

SHEET A

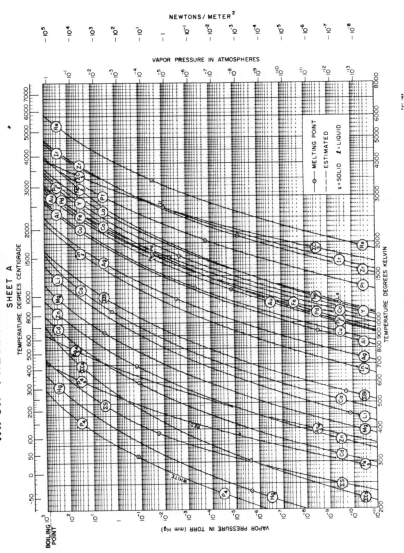

RC/I Laboratories

Prepared by Richard E. Honig and Dean A. Kramer

Princeton, N. J. 08540

371

VAPOR PRESSURE CURVES OF THE ELEMENTS
SHEET B

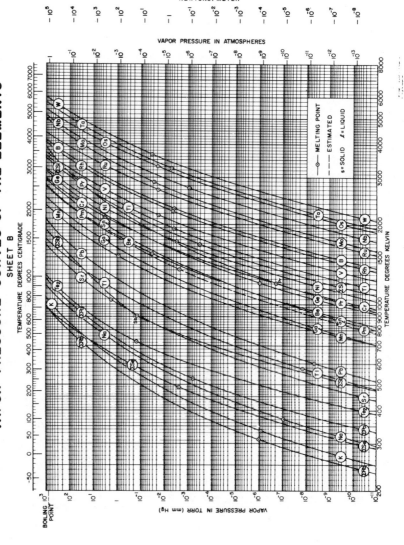

VAPOR PRESSURE CURVES OF THE ELEMENTS

SHEET C

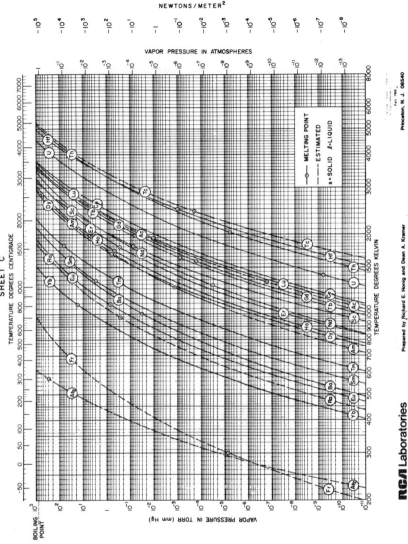

RCA Laboratories Prepared by Richard E. Honig and Dean A. Kramer

Source. Reprinted with permission from *RCA Review*, **30**, p.285, June 1969, *Vapor Pressure Data for the Solid and Liquid Elements*, by R. E. Honig and D. A. Kramer. Copyright 1969, RCA Corp.

AISI type	\<span\>Performance - recommend for\</span\> General Forming	High temperature use	Cryogenic use	Resist carbide precipitation	High yield strength	Non-magnetic at cryogenic temperatures	Free machining	Weldability	Tensile properties Typical, at R. T. Annealed 0.2% proof (yield) kgf mm^2	Ultimate TS kgf mm^2	Impact strength at -196°C Charpy-V J	Price index
302	x							E	22	60	70	100
303S Se							x	P	22	60		105
304	x		x					E	20	50	70	100
304L	x		x	x				E	18	48	80	115
304N	x		x		x			E	30	55-75	min 63	110
304LN	x		x	x	x	x		E	28	55-75	min 55	125
310	x	x			x	x		G	23	65		210
316	x		x				~	E	22	50		130
316L	x		x	x			~	E	20	45	120	150
316N	x		x		x	x		E	30	60		140
316LN	x		x	x	x	x		E	30	60-80	min 55	160
317	x	x					~	E	24	60		200
321	x	x		x				E	21	50	65	130
347	x	x		x				G	22	50	65	150

Source. Reprinted with permission from *Vacuum*, **26**, p. 287, C. Geyari. Copy
[a] Selection guide for vacuum and cryogenic equipment. X = recommended,
P = poor.

Austenitic Stainless Steels [a]

Typical composition, %						
C max	Cr	Ni	Mo	N	Others	Description
0.12	17-19	8-10				General purpose. Good resistance to atmosphere corrosion. Good mechanical properties.
0.15	17-19	8-10			S, Se≥0.15	Free machining type. Good corrosion resistance.
0.08	18-20	8-12				Low C variation of 302. Improved corrosion resistance after welding.
0.03	18-20	8-12				Extra low C prevents carbide precipitation.
0.06	18.5	9.5		0.25	Mn 2	Improved mechanical properties.
0.03	18.5	9.5		0.18	Mn 2	Improved mechanical properties. Extra low C prevents carbide precipitation.
0.25	24-26	19-22				High scale resistance. Superior corrosion resistance.
0.08	16-18	10-14	2-3			Very good corrosion resistance in most media.
0.03	16-18	10-14	2-3			Extra low C variation of 316.
0.07	17-18	10-13	2.5-3	0.2	Mn 0.5-2	Improved mechanical properties.
0.03	17.5	13	2.8	0.18	Mn 2	Improved mechanical properties. Extra low C prevents carbide precipitation.
0.08	18-20	11-15	3-4			Higher alloy content improves creep and corrosion resistance of 316.
0.08	17-19	9-12			Ti≥5xC	Stabilized--Ti prevents carbide precipitation. Improved corrosion resistance after welding.
0.08	17-19	9-13			Nb, Ta≥10xC	Stabilized--Nb, Ta prevent carbide precipitation. Improved corrosion resistance after welding.

right 1976, Pergamon Press.
~ = probable–should be tested, E = excellent, G = good with precautions,

Isotopic Abundances

Element	AMU	Relative Abundance	Element	AMU	Relative Abundance
H	1	99.985	Al	27	100.0
	2	0.015	Si	28	92.27
He	3	0.00013		29	4.68
	4	~100.0		30	3.05
Li	6	7.42	P	31	100.0
	7	92.58	S	32	95.06
Be	9	100.0		33	0.74
B	10	19.78		34	4.18
	11	80.22		36	0.016
C	12	98.892	Cl	35	75.4
	13	1.108		37	24.6
N	14	99.63	Ar	36	0.337
	15	0.37		38	0.063
O	16	99.759		40	99.600
	17	0.0374	K	39	93.08
	18	0.2039		40	0.0119
F	19	100.0		41	6.91
Ne	20	90.92	Ca	40	96.97
	21	0.257		42	0.64
	22	8.82		43	0.145
Na	23	100.0		44	2.06
Mg	24	78.60		46	0.0033
	25	10.11		48	0.185
	26	11.29	Sc	45	100.0

Element	AMU	Relative Abundance	Element	AMU	Relative Abundance
Ti	46	7.95	Br	79	50.52
	47	7.75		81	49.48
	48	73.45	Kr	78	0.354
	49	5.51		80	2.27
	50	5.34		82	11.56
V	50	0.24		83	11.55
	51	99.76		84	56.90
Cr	50	4.31		86	17.37
	52	83.76	Rb	85	72.15
	53	9.55		87	27.85
	54	2.38	Sr	84	0.56
Mn	55	100.0		86	9.86
Fe	54	5.82		87	7.02
	56	91.66		88	82.56
	57	2.19	Y	98	100.0
	58	0.33	Zr	90	51.46
Co	59	100.0		91	11.23
Ni	58	67.76		92	17.11
	60	26.16		94	17.4
	61	1.25		96	2.8
	62	3.66	Nb	93	100.0
	64	1.16	Mo	92	15.86
Cu	63	69.1		94	9.12
	65	30.9		95	15.70
Zn	64	48.89		96	16.50
	66	27.82		97	9.45
	67	4.14		98	23.75
	68	18.54		100	9.62
	70	0.617	Ru	96	5.47
Ga	69	60.2		98	1.84
	71	39.8		99	12.77
Ge	70	20.55		100	12.56
	72	27.37		101	17.10
	73	7.67		102	31.70
	74	36.74		104	18.56
	76	7.67	Rh	103	100.0
As	75	100.0	Pd	102	0.96
Se	74	0.87		104	10.97
	76	9.02		105	22.23
	77	7.58		106	27.33
	78	23.52		108	26.71
	80	49.82		110	11.81
	82	9.19			

Element	AMU	Relative Abundance	Element	AMU	Relative Abundance
Ag	107	51.82	Ba	130	0.101
	109	48.18		132	0.097
Cd	106	1.22		134	2.42
	108	0.87		135	6.59
	110	12.39		136	7.81
	111	12.75		137	11.32
	112	24.07		138	71.66
	113	12.26	La	138	0.089
	114	28.86		139	99.911
	116	7.85	Ce	136	0.193
In	113	4.23		138	0.250
	115	95.77		140	88.48
Sn	112	0.95		142	11.07
	114	0.65	Pr	141	100.0
	115	0.34	Nd	142	27.13
	116	14.24		143	12.20
	117	7.57		144	23.87
	118	24.01		145	8.30
	119	8.58		146	17.18
	120	32.97		148	5.72
	122	4.71		150	5.62
	124	5.98	Sm	144	3.16
Sb	121	57.25		147	15.07
	123	42.75		148	11.27
Te	120	0.089		149	13.84
	122	2.46		150	7.47
	123	0.87		152	26.63
	124	4.61		154	22.53
	125	6.99	Eu	151	47.77
	126	18.71		153	52.23
	128	31.79	Gd	152	0.20
	130	34.49		154	2.15
I	127	100.0		155	14.73
Xe	124	0.096		156	20.47
	126	0.090		157	15.68
	128	1.92		158	24.87
	129	26.44		160	21.90
	130	4.08	Tb	159	100.0
	131	21.18	Dy	156	0.052
	132	26.89		158	0.090
	134	10.44		160	2.294
	136	8.87		161	18.88
Cs	131	100.0		162	25.53
				163	24.97

Element	AMU	Relative Abundance	Element	AMU	Relative Abundance
	164	28.18		187	1.64
Ho	165	100.0		188	13.3
Er	162	0.136		189	16.1
	164	1.56		190	26.4
	166	33.41		192	41.0
	167	22.94	Ir	191	37.3
	168	27.07		193	62.7
	170	14.88	Pt	190	0.012
Tm	169	100.0		192	0.78
Yb	168	0.140		194	32.8
	170	3.03		195	33.7
	171	14.31		196	25.4
	172	21.82		198	7.21
	173	16.13			
	174	31.84	Au	197	100.0
	176	12.73	Hg	196	0.15
Lu	175	97.40		198	10.02
	176	2.60		199	16.84
Hf	174	0.18		200	23.13
	176	5.15		201	13.22
	177	18.39		202	29.80
	178	27.08		204	6.85
	179	13.78			
	180	35.44	Tl	203	29.50
Ta	180	0.012		205	70.50
	181	99.988	Pb	204	1.48
W	180	0.135		206	23.6
	182	26.4		207	22.6
	183	14.4		208	52.3
	184	30.6	Bi	209	100.0
	186	28.4	Th	232	100.0
Re	185	37.07	U	234	0.0057
	187	62.93		235	0.72
Os	184	0.018		238	99.27
	186	1.59			

Source. Reprinted with permission from *Mass Spectroscopy for Science and Technology*, F. A. White, p. 339. Copyright 1968, John Wiley & Sons.

Cracking Patterns

Appendix E.1   Cracking Patterns of Pump Fluids

AMU	Welch 1407[a]	Fomblin Y-25[b]	DC-704[c]	DC-705[c]	Octoil-S[c]	Convalex-10[c]
18					1.86	4.53
27					14.4	4.24
28			17.19	67.18	8.37	
29				2.56	28.53	
30				3.52	1.1	
31		31.49			1.27	0.72
32			2.15	10.08		
35						2.16
36						5.07
37						2.76
38					0.65	
39				2.22	8.82	
40				4.40	2.21	
41	40			3.59	54.03	
42					17.12	
43	74		2.4		66.58	
44			1.98		3.17	
45					1.98	
47		20.67				
50		15.30			4.	17
51		27.66	2.61		0.84	
52		2.92				
53	3				3.41	
54					4.34	
55	70		1.31		54.56	
56	23				22.18	
57	100		2.13		88.2	
59					4.59	

AMU	Welch 1407[a]	Fomblin Y-25[b]	DC-704[c]	DC-705[c]	Octoil-S[c]	Convalex-10[c]
60					2.0	
61						5.68
62						3.38
63						1.34
64						8.3
65						7.1
66		2.23				3.6
67	40				4.52	3.9
68	14			4.	2311.	3
69	91	100		3.59		18.5
70	31	3.39			56.83	9.5
71	83			2.9	46.92	
72					3.14	
73		1.63			2.98	
76						3.4
77	2		2.2			25.6
78	0.7			4.62		4.8
81	22	3.1		1.88	3.8	3.6
82	10				3.91	2.4
83	30			2.9	19.78	11.3
84	12				13.78	4.8
85		2.23		2.05	2.25	
87					1.82	
91		2.03	3.39	3.76		3.0
92						3.6
93	2				1.18	
94	7					3.6
95	3			2.05	2.87	2.4
96	10				2.11	
97		16.37			10.5	3.0
98					19.87	5.4
100		14.63			2.24	
101		14.63			2.0	
108						12.5
112					45.46	
113					25.01	61.3
119		13.88	3.87	3.24		
131		2.36				2.4
135		5.19	20.59	10.26		

[a] Data taken on UTI-100B quadrupole with V_{EE} = -60 V, only major peaks shown. Reprinted with permission from Uthe Technology Inc., 325 N. Mathilda Avenue, Sunnyvale, CA 94086;

[b] Sector data. Adapted with permission from *Vacuum*, **22**, p. 315, L. Holland, L. Laurenson, and P. N. Baker. Copyright 1972, Pergamon Press;

[c] Sector data. Reprinted with permission from *J. Vac. Sci. Technol.*, **6**, p. 871, G. M. Wood, Jr., and R. J. Roenig. Copyright 1969, The American Vacuum Society.

Note. Only peaks up to 135 AMU are shown for the data taken from source *b*; the largest mass peak (100%) occurs at a higher mass number. The data are *not* renormalized for the range tabulated here.

Appendix E.2   Cracking Patterns of Gases

AMU	Hydrogen[a] H_2	Helium[b] He	Neon[b] Ne	Carbon Monoxide[a] CO	Nitrogen[a] N_2	Oxygen[a] O_2	Argon[a] Ar	Carbon Dioxide[a] CO_2
1	2.7							
2	100	0.12						
3	0.31							
4		100						
6				0.0008				0.0005
7					0.0006			
8				0.0001		0.0013		0.0005
12				3.5				6.3
13								0.063
14				1.4	9			
15					0.026			
16				1.4		14		
17						0.0052		
18						0.028	0.071	0.0088
19							0.016	
20			100				5.0	
21			0.33					
22			9.9					0.52
22.5								0.0047
23								0.0012
28				100	100			15
29				1.2	0.71			0.15
30				0.2	0.0014			0.029
32						100		
33						0.074		
34						0.38		
36						0.0023	0.36	
38							0.068	
40							100	
44								100
45								1.2
46								0.38
47								0.0034
48								0.0005

[a] Data taken on UTI-100C-02 quadrupole residual gas analyzer. Typical parameters, V_{EE} = 70 V, V_{IE} = 15 V, V_{FO} = -20 V I_E = 2.5 mA, resolution potentiometer = 5.00. Reprinted with permission from Uthe Technology Inc., 325 N. Mathilda Avenue, Sunnyvale, CA 94086.

[b] Sector data. Reprinted with permission from E. I. du Pont de Nemours & Co., Wilmington, DE 19898

Appendix E.3 Cracking Patterns of Common Vapors

AMU	Water Vapor[a] H_2O	Methane[b] CH_4	Acetylene[b] C_2H_2	Ethylene[b] C_2H_4	Ethane[b] C_2H_6	Cyclopropane[b] C_3H_6
1	0.1	3.8	3.8	6.4	3.2	1.4
2		0.64	1.2	1.1	0.93	32.
3		0.009	0.002	0.022	0.15	0.10
6		0.0003	0.0006	0.0002		
7		0.0013		0.0018		
12		2.1	4.5	2.3	0.47	0.85
13		7.4	7.6	4.0	1.1	1.6
14		15.	0.86	8.1	3.4	5.6
14.5					0.24	
15		83.			5.7	8.1
16	3.07	100.			0.53	2.0
17	27.01	1.3				0.07
18	100.					
19	0.19					2.7
19.5						1.3
20						2.3
20.5						0.68
24			7.1	3.2	0.52	0.35
25			23.	12.	3.5	2.1
26			100.	61.	24.	17.
27			2.5	59.	33.	46.
28				100.	100.	18.
29				2.8	21.	11.
30					24.	0.29
31					0.54	
36						1.4
37						11.
38						15.
39						69.
40						30.
41						100.
42						90.
43						18.

[a] Sector data. Reprinted with permission from E. I. du Pont de Nemours & Co., Wilmington, DE 19898;

[b] Quadrupole data, same conditions as given in Appendix E1a. Reprinted with permission from Uthe Technology Inc, 325 N. Mathilda Avenue, Sunnyvale CA 94086.

AMU	Methyl Alcohol[a]	Ethyl Alcohol[a]	Acetone[a]	Isopropyl Alcohol[a]	Trichloro-ethylene[a]	Gentron-142B[b]
2						3.4
12						2.9
13						3.2
14						9.2
15						16.0
18	1.9	5.5				
19		2.3		6.6		3.3
20						1.7
25						6.5
26		8.3	5.8			17.0
27		23.9	8.0	15.7		4.4
28	6.4	6.9				
29	67.4	23.4	4.3	10.1		
30	0.8	6.0				
31	100.	100.		5.6		17.0
32	66.7					1.0
35					39.9	5.2
36						1.8
37			2.1		12.8	1.6
38			2.3			0.9
39			3.8	5.7		
41			2.1	6.6		
42		2.9	7.0	4.0		
43		7.6	100.	16.6		0.5
45		34.4		100.		53.0
46		16.5				3.5
47					25.8	1.1
48						0.5
49						1.4
50						2.9
51						1.9
58			27.1			
59				3.4		
60					64.9	0.6
62					20.9	0.5
63						3.4
64						8.6
65						100.0
66						2.5
87						1.8
95					100.0	
97					63.9	
130					89.8	
132					84.8	
134					26.8	

[a] Sector data. Reprinted with permission from VG-Micromass Ltd., 3 Tudor Road, Altringham, Cheshire, TN34-1YQ, England.
[b] Quadrupole data, same conditions as given in Appendix E1a. Reprinted with permission from Uthe Technology Inc., 325 N. Mathilda Avenue, Sunnyvale, CA 94086.

AMU	Arsine AsH$_3$	Silane SiH$_4$	Phosphine PH$_3$	Disilane Si$_2$H$_6$	Diphosphine P$_2$H$_4$	Diborane B$_2$H$_6$
1						21.1
2						134.7
3						0.35
10						9.72
11						39.4
12						26.4
13						34.9
14.		0.4				2.23
14.5		0.5				
15		0.4				
15.5		0.1	0.23			
16			0.62			
16.5			0.13			
17			0.48			
20						0.22
21						1.85
22						11.8
23						48.5
24						94.0
25						57.7
26						100.0
27						95.2
28		28.				16.7
29		32.				
30		100.0				
31		80.	26.7			
32		7.3	100.0			
33		1.5	25.4			
34		0.2	76.7			
56				33.		
57				48.		
58				82.		
59				37.		
60				100.0		
61				40.		
62				42.	100.0	
63				5.7	58.8	
64				4.0	70.6	
65					26.5	
66					1.5	
75	38.5					
76	100.0					
77	28.8					
78	92.3					

[a] Quadrupole data, same conditions as given in Appendix E1a. Reprinted with permission from Uthe Technology Inc., 325 N. Mathilda Avenue, Sunnyvale, CA 94086.

Pump Fluid Properties

Appendix F.1   Vapor Pressure Curves of Mechanical Pump Fluids[a]

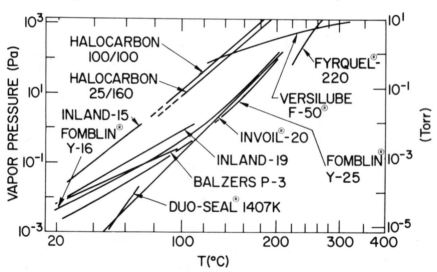

[a] Duo-Seal measured with a trapped McLeod gauge, Invoil-20 measured with an ionization gauge; all others are measured with a Pirani gauge. Duo-Seal 1401 [1], Invoil-20 [2], Halocarbon [3], Inland-15 and 19 [2], Fomblin [4], Fyrquel [5], Versilube [6], and Balzers P-3 [7]. Invoil-20 is a hydrocarbon diffusion pump oil that may be used in rotary vane pumps.

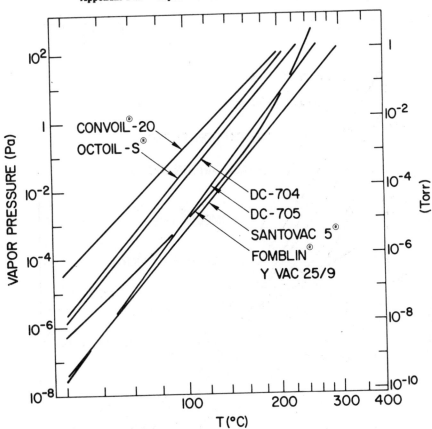

[a] Convoil-20 [8], Octoil-S [8], DC-704 and 705 [9], Santovac 5 [10], and Fomblin [4].

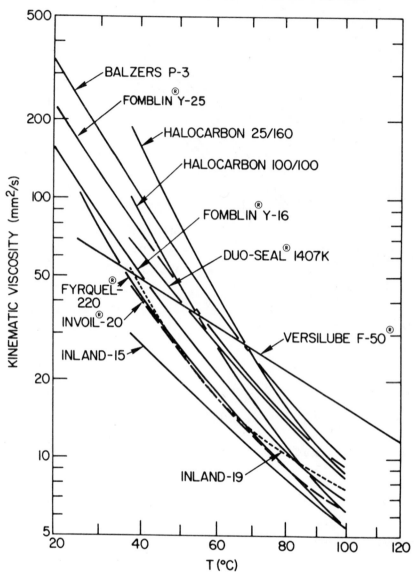

[a] Balzers P-3 [7], Halocarbon [3], Fomblin [4], Duo-Seal 1401K [1], Versilube F-50 [6], Fyrquel [5], Inland and Invoil [2]. Invoil-20 is a diffusion pump oil that may be used in rotary vane pumps.

REFERENCES

1. Reprinted with permission from Sargent-Welch Scientific Co., Vacuum Products Division, 7300 N. Linder Ave. Skokie, IL 60077

2. Reprinted with permission from Inland Vacuum Industries, Inc., 35 Howard Ave., Churchville, NY 14428

3. Reprinted with permission from Halocarbon Corporation, 82 Burlews Court, Hackensack, NJ 07601

4. Reprinted with permission from Montedison, USA, Inc., 1114 Ave. of the Americas, New York, NY 10036

5. Reprinted with permission from Stauffer Chemical Company, Specialty Chemical Division, Westport, CN 06880

6. Reprinted with permission from General Electric Company, Silicone Products Department, Waterford, NY 12188

7. Reprinted with permission from Balzers High Vacuum, Furstentum, Liechtenstein.

8. Reprinted with permission from CVC Products, Inc., 525 Lee Rd. Rochester, NY 14603

9. Reprinted with permission from Dow Corning Company, Inc., 2030 Dow Center, Midland, MI 48640

10. Reprinted with permission from Monsanto Company, 800 N. Lindbergh Blvd. St. Louis, MO 63166

Index

145682

BERKELEY
LIBRARY
UNIVERSITY OF
CALIFORNIA

PHYSICS
LIBRARY

}